Evolution
of
Sedimentary
Rocks

W. W. NORTON & COMPANY. INC. New York

Evolution
of
Sedimentary
Rocks

ROBERT M. GARRELS
Scripps Institution of Oceanography

FRED T. MACKENZIE
Northwestern University

Copyright©1971 by W. W. Norton & Company, Inc.
First Edition
SBN 393 09959 8
Library of Congress Catalog Card No. 76-129500
Printed in the United States of America
1 2 3 4 5 6 7 8 9 0

To Our Parents

Contents

Preface In these days of revolution in natural science, with the moon invaded by man, the planets metered by probes, and the great ocean basins stripped of their permanence, there should be, ideally, a continuous reflection of the revolution at every level of teaching. But the course content in the undergraduate curriculum of departments called Geology, Geological Sciences, and Earth Science, if judged by textbooks now available, has not changed drastically in response.

—Therefore we have addressed ourselves to the question of what should be the present-day content of a book designed to introduce an informed reader to the earth. What are the major concepts that should be included? One of our first decisions was that, as compared to the usual introduction, only first-order phenomena should be treated. Emphasis should be on global phenomena. The reader needs to know the gross structure of the earth, its origin, the outlines of its physical and biological history, its age. At the same time, we wanted to show the importance of an understanding of present-day processes in interpreting earth history. The revolution in geology has come largely from studies of today's heat flow, today's sea floor spreading, today's distribution of sediments in the oceans.

We set about to develop major concepts, trying to digest and summarize the recent work on earth origin and internal structure, on crustal movement and ocean-continent relations, and to add to and integrate a synthesis of present-day sedimentation and continental erosion. Apparently the last aim caused the horse to trot away a bit with the cart.

Evolution of Sedimentary Rocks was originally planned as an introductory text, designed especially for students who knew some introductory mathematics, chemistry, and physics, but we found when we had finished writing it that our critics considered it too uneven in the treatment of the usual introductory topics, too slanted toward material on sediments and sedimentary geochemistry involving our own interests, too full of new and controversial material, and written, all too often, assuming a grasp of geologic concepts and of vocabulary that could come only from having read an introductory text.

We were not in entire agreement with our critics, having successfully taught the material during three years to beginning students at Northwestern University who were all nonscientists; we had little difficulty in filling in required vocabulary and background during the lectures or in the laboratory. On the other hand, we had to admit to our publisher that we exercised great care in presenting the numerous tables and graphs in the book to an audience initially unsympathetic to such devices.

At any rate, the book evolved remarkably over a period of two years or so, as we attempted to react constructively to the reasonable pressures from our publisher and from an anonymous biting but constructive critic.

We ended by extending material on sedimentary rocks, much of which is either a synthesis from many sources or is new, as the major theme of the book and tried to put the sediments into a modern tectonic framework.

So the book grew with a few chapters setting the physical framework on to a discussion of present-day erosional and depositional relations, using them to interpret conditions of the past, all developed on a global basis. We stayed almost entirely on the physical side, finding that a parallel treatment of organic relations and history would require more space and knowledge than we possessed.

As we see it now, the background required to *read* the book successfully is a year of college-level chemistry and a short course in physical geology at either high school or college level. We think it can be *taught* successfully at a fairly elementary level, with the lecturer presenting the major concepts in a way that is palatable to his particular audience, using the abundant data in the book as he sees fit for documentation.

We see a variety of slots into which this book could be fitted. It could be used in conjunction with a more classic text in a beginning course or it could stand on its own in a course under the book's title to be taught at any more advanced level; it could be used together with texts on sedimentation, on sedimentation and stratigraphy, or on sedimentary rocks if the teacher wished to combine our "global" approach with details of rock occurrences. We recognize, and have tried to indicate objectively, the unevenness of the data we have attempted to integrate. On the other hand, the attempt has developed exciting problems that must be resolved. We, at least, have been pleased with the education we have obtained in the effort.

When a book is written and finally printed, the authors are always shocked to discover how little their own contribution is and how important is that of their colleagues, their students, and the people who presumably were only typing or drafting. We attempt below to give at least partial credit to the many people who have made an important contribution to this book. We are perfectly aware that there are many others whose ideas, suggestions, and comments have been incorporated and we wish there were some way to include them all.

We are especially indebted to our editor, Kenneth B. Demaree, who promoted an ideal relationship between authors and publisher. Only by our continuous interaction with him was it possible for this book to progress beyond the stage of an idea.

Our colleagues at Northwestern provided stimulus and discussion. W. C. Krumbein helped us with the computer programming of sedimentary mass models. L. L. Sloss was a continuous helpful critic of some of the ideas presented in this book. H. C. Helgeson tried to keep us "honest." Our students both in the introductory and advanced courses at Northwestern reacted critically to our ideas and helped us to refine them. We are particularly indebted to G. M. Lafon, D. C. Thorstenson, W. Scherer, J. Sonderegger, and S. Derks.

We have exchanged ideas with many colleagues and friends; it is impossible to give all formal acknowledgment. The partial list below only emphasizes to a greater degree those individuals who are really responsible for this book. A. B. Ronov of the Academy of Sciences of the U.S.S.R. exchanged thoughts with us on many topics and has already propounded many of the concepts we talk about. The germ of the concept of differential cycling of various sedimentary rock types was instilled by W. S. Broecker of Lamont-Doherty Geological Observatory. G. D. Nicholls of the University of Manchester discussed with us the details of Precambrian sedimentary rock compositions. In innumerable discussions, H. D. Holland of Princeton University was especially critical and helped us to focus our ideas and thoughts. We thank R. Siever of Harvard University for his suggestions and advice throughout the course of preparation of this book. Jack Stark provided the idea for the devil in Figure 12-14. Many others read and commented on rough (very rough at times!) drafts of the text before finalization; we thank O. P. Bricker of the Johns Hopkins University, B. Gregor of West Georgia College, L. S. Land of the University of Texas, G. K. Billings of Louisiana State University, R. Wollast of the University of Brussels, and so many others that we cannot name them all.

Our sincere thanks go to our innumerable typists and illustrators, including L. E. Arnold for typing and help with exposition, I. L. Cheverton for imaginative illustrations, C. Hunt for editing and prodding, and E. J. Faulkner, S. Riker, M. Emmott, M. E. Hastings, S. A. Ball, and L. Meinhardt for typing and many seemingly little but important tasks.

Our deepest thanks go to Mary Mackenzie for her encouragement throughout the course of this work.

In attempting to pull together data on such things as the composition of average igneous rock, or in trying to get the average composition of Precambrian sedimentary rocks, we have undoubtedly missed many references containing valuable material simply because the title did not lead us to the article, and we apologize for pertinent work not cited.

Much of the material in this book has been drawn from research supported by the National Science Foundation and by the Petroleum Research Fund administered by the American Chemical Society.

ROBERT M. GARRELS
FRED T. MACKENZIE

La Jolla, California
Evanston, Illinois
October, 1970

1

The Earth as a Whole

The day may come when we are visited by a spaceship from far out in our galaxy, and if the creatures aboard are intelligent and friendly, they may want to know about the history of the earth. Hopefully, by the time they arrive, two motion pictures will be available to show them. The first would show the earth as seen from an orbiting satellite, so that the visitors could see the changing patterns of continents and seas, the sinking of parts of the crust and the rising of others, and the crumbling of mountains and the filling of basins. The second would show the experiences of a time-traveler at the earth's surface, who could watch life evolve, could boat on the rivers and taste the seas, and could journey across deserts and swamps. By using flashbacks from the first movie, the traveler's experiences could be oriented in time and place.

In the past, most of the geologists' work has been devoted to researches suitable for inclusion in the film on surface travel, whereas the view from orbit has been neglected. But in the last few years, perhaps in part because there are in fact satellites sending us pictures of the earth, emphasis has swung toward solving problems necessary for making the film of the earth as a whole. Some of the more important recent developments are described briefly below.

The vast space program has "spun off" a remarkable amount of information concerning the earth itself, and oceanographic research is beginning to fill in the near-void concerning the 70 percent of the earth's surface covered by water. The development of increasingly sensitive and varied instruments for looking into the earth, such as gravimeters and seismometers, as well as the achievement of high pressures and temperatures in laboratory studies of rock materials, is finally giving us the necessary expansion of our direct sensing that we need to make useful guesses about the earth's interior.

Observations of the orbits of earth-circling artificial satellites have provided precise measurements of the earth's gravitational field and of its shape. These kinds of measurements have been applied to estimating the earth's response to stress differences such as those induced by tidal forces.

Measurements of earth tremors have been expanded both in sensitivity and in types of phenomena that are recorded; seismograph stations all over the world collaborate in studies of the transmissions of energy through the earth. Some scientists concern themselves with the delicate tremors that are induced by tiny forces such as those resulting from changes in atmospheric pressure; others study the oscillations of the whole earth as it rings like a huge bell when major earthquakes occur.

For the first time, promising hypotheses to explain the earth's magnetic field have been developed; studies of the magnetic properties of rocks indicate that the polarity of the magnetic poles of the earth has reversed many times in the past. Also, there is increasing evidence that the magnetic field not only reverses but may possibly migrate with respect to the earth's rotational axis.

Measurements of the heat flowing out of the earth, coupled with soundings of the sea floor and sampling of oceanic rocks, indicate that great crustal rifts form a connected pattern throughout the ocean basins. These rifts are active zones of earthquakes, volcanoes, and of the emergence of lavas from the depths; they seem to be widening at rates of 1–5 cm/year. If this is indeed so and the ocean floor is spreading, continental drift, in which the continental masses are moving with respect to each other, may be occurring today. The postulated present rate of sea-floor spreading is such that today's ocean basins may have formed during a relatively short period of earth history.

These investigations into the physics and chemistry of the earth reveal that the earth is apparently remarkably active today. Most geologists have grown accustomed to the idea of mountains being raised up and basins being filled with debris and sinking. However, strong evidence of major lateral crustal movement, movement of subcrustal materials, short-period reversals of the polarity of the earth's magnetic field, and apparent material transfer between the crust and the underlying mantle have contributed to the development of a new image of a dynamic earth. This "New Earth" is an exciting one; we are observing the start of an era in which there is the same kind of scientific stir that was generated in physics by the discovery of radioactivity at the turn of the century.

Some of the key words in the preceding brief summary are *gravitational field, magnetic field, earthquake waves, heat flow, crustal rifts, sea floor spreading,* and *subcrustal materials.* It is noteworthy that investigation of these phenomena depends heavily on sophisticated instrumentation and methods for storing and processing data. Missing from the list are terms we think are critical in writing the global movie script—*continental erosion, ocean basin filling, sedimentary rock mass, atmosphere-continent-ocean interactions,* and similar words descriptive of the processes and results of the continuous transfer of material from land to sea and back again. Most of the work on sedimentary rocks seems to have been done from the view of the time-traveler, and it is hopefully not too soon to look at surface processes and their relations to deeper ones from the orbital view.

With this goal in mind, we organized this book as follows. First, we have set the framework by a chapter on the origin of the earth and the development of its internal structure, with emphasis on those aspects, such as the probable loss of a primordial atmosphere and the development of the present atmosphere and oceans as a result of degassing of the interior, that have been important factors in the development of the present-day system of material transfer. Next, we have looked at the crust as the complex and dynamic base on which sediments are deposited and with which they interact mechanically and chemically. Finally, in setting the stage, we have reviewed briefly the geologic time scale and use it as a vehicle for discussion of radioactive and stable isotopes.

In accord with true Huttonian principles, the next few chapters are directed to the sedimentary processes of today. From an assessment of the relative importance of the various agents that are contributing material to the oceans, we have moved on to denudation by streams on a continental scale, with an analysis of rates and controls. Then chemical erosion is discussed in detail, followed by a complementary treatment of deposition in terms of environments and sediment types.

After "today" we have summarized what is known or can be deduced about the mass, origin, composition, and age distribution of sedimentary rocks and then have proposed some simple models to account for these distributions in terms of rates of deposition and destruction of sediments throughout earth history. Then we integrate this knowledge into an interpretation of the genesis and history of sea water.

In the concluding chapter we have tried to write a synopsis of the satellite movie, weaving the results of the study of sedimentary cycles into the rest of the current views on gross earth history. The chief function of these chapters is to show that 99 percent of the scenes cannot yet be shot. Our visitors from Arcturus will have to be satisfied for a while with a still-life or two, plus some speculation on what might have been.

Appendices A (Minerals) and B (Mineral Chemistry) provide the reader with some basic information necessary for understanding this book.

THE SOLAR SYSTEM

The solar system includes the sun and all materials under the continuous influence of its gravitational field. Some of the many kinds of bodies included in this definition are the 9 planets and their 32 satellites, asteroids, and comets. In the space between the bodies of the solar system there is a tenuous cloud of gas, chiefly hydrogen, with a density of about 1–6 atoms/cm^3. In addition, there are irregular streamers of charged particles, such as electrons, plus widely scattered bits of solids, most of which are mere dust. Of the several million such particles that enter our atmosphere every day, perhaps one or two are as big as a penny, and perhaps only once every several centuries a chunk weighing many tons arrives. This description of the variety makes space seem crowded, but it is, on the average, far more empty than the best vacuums we can achieve in the laboratory.

The planets revolve about the sun in the same direction at systematically increasing distances from it. The orbits of the planets are nearly perfect circles and are within a few degrees of the same plane. Mercury and Pluto are exceptions in that their orbits are elliptical and are inclined at a somewhat greater angle to the one plane. Except for Venus, all of the planets rotate in the same sense as they revolve. Most of the 32 satellites re-volve around their planets in one direction and in the same plane as the planetary revolution, but again there are exceptions, both in terms of retro-grade revolution and deviations of orbits from the general plane of the solar system. The four inner or terrestrial planets, including the earth, all have densities greater than 4.0 g/cm³, as contrasted with the five outer or "great" planets, which have densities less than 2.5 g/cm³. Despite their high density, the total mass of the terrestrial planets is much less than 1 percent that of the outer planets, and the mass of all the planets together is less than 1 percent that of the sun. On the other hand, the angular momen-tum of the sun, which is a measure of the difficulty of stopping its rotation, is only 0.5 percent of the total system; that of the outer planets is 99.3 per-cent and that of the inner planets is 0.2 percent. The sun rotates only once in 25 days (Table 1.1).

Planetary materials can be conveniently classified as earthy, icy, and

Table 1.1
Characteristics of the Principal Planets of the Solar System[a]

	Sun	Terrestrial planets			
		Mercury	Venus	Earth	Mars
Distance from sun (millions of km)		58	108	150	229
Mass (units of 10^{27} g)	1,984,000	0.3244	4.861	5.975	0.6387
Density (g/cm³)	1.41	5.33	5.15	5.52	4.0
Volume (units of 10^{27} cm³)	1,410,000	0.06	0.94	1.08	0.16
Radius (km)	696,000	2,400	6,100	6,371	3,400
Maximum surface temperature (°C)	5,500	350	430	60	30–40
Number of moons		0	0	1	2
Gases in atmosphere	Many	None	CO_2, N_2	Many, primarily N_2 and O_2	CO_2, H_2O

[a] Data from Mason (1966) and Ringwood (1966).

gaseous. Hydrogen and helium are the chief gaseous materials; because their freezing points are near absolute zero ($-273°C$), they remain gaseous under all environmental conditions. On the other hand, compounds containing carbon, oxygen, nitrogen, and hydrogen in various combinations, such as water (H_2O), carbon dioxide (CO_2), methane (CH_4), and ammonia (NH_3), freeze at higher temperatures and exist primarily as solids in the outer planets.

The outer planets range in composition from that of Jupiter, which is about 90 percent gaseous and 10 percent icy with almost no earthy material, to that of Neptune, which is about 20 percent earthy, 70 percent icy, and 10 percent gaseous. The terrestrial planets are earthy and are made up almost entirely of compounds that are solid at most temperatures and pressures with which we are familiar. It seems probable that the general chemical makeup of the solid part of the inner planets is like that of the earth. The difference in properties, such as density and composition, between the

Great planets				
Jupiter	Saturn	Uranus	Neptune	Pluto
778	1,430	2,860	4,490	5,910
1,902	569.4	86.88	102.8	0.3?
1.35	0.71	1.56	2.47	2?
1,410	802	55.1	41.6	0.15
69,900	57,000	26,000	22,000	2,900?
-138	-153	-184	-200	Approximately absolute zero
12	10	5	2	0
CH_4, NH_3 (outer atmosphere); H_2 (inner atmosphere)	CH_4, NH_3 (outer atmosphere); H_2 (inner atmosphere)	CH_4, NH_3 (trace)	CH_4, NH_3 (trace)	?

Table 1.2
Bode's Law

Base	Add	Divide sum by 10	Actual distance (a.u.)	Planet
4	0 (3×0)	0.4	0.39	Mercury
4	3 (3×1)	0.7	0.72	Venus
4	6 (3×2)	1.0	1.00	Earth
4	12 (3×2^2)	1.6	1.52	Mars
4	24 (3×2^3)	2.8	—	(Asteroids)
4	48 (3×2^4)	5.2	5.20	Jupiter
4	96 (3×2^5)	10.0	9.54	Saturn
4	192 (3×2^6)	19.6	19.18	Uranus
—	—	—	30.06	Neptune
4	384 (3×2^7)	38.8	39.52	Pluto

(Distance from sun to earth $= 1$ a.u.)

inner and outer planets is so marked that it has even been suggested that they may have had different origins. However, there are so many regularities in relationships that there seems to be little serious question that all the planets originated at about the same time and as parts of a single dynamic system. Among the remarkable regularities is that described by "Bode's Law," which is illustrated in Table 1.2.

Bode's Law is an empirical relationship that gives the distances in astronomical units (1 a.u. is set as equal to the distance from the sun to the earth) between the sun and the various planets and may be expressed as

$$\text{Distance} = 4 + (3 \times 2^n)/10,$$

where $n = -\infty, 0, 1, 2, \ldots, 7$. The calculated distance values obtained from this relationship agree fairly well with the actual distances; the only striking anomaly is the planet Neptune, which occurs between two calculated distances. Also, Bode's Law predicts a planet between Mars and Jupiter; the asteroids, a swarm of particles and small bodies up to a few hundred kilometers in diameter, with a total mass of about 0.1 percent that of the earth, actually occupy the predicted position.

In addition to that in the belt of asteroids, there is a considerable amount of earthy, icy, and gaseous material scattered throughout the system. A great many comets, probably made up chiefly of tiny earthy and icy particles, are under the sun's gravitational influence, although their orbits may be highly irregular, and their paths may take them to distances of 100,000 a.u. from the sun. It has been estimated that the total mass of such material is about equivalent to that of the earth, and there may be as many as 100 billion comets.

A few samples of extraterrestrial materials have reached the earth in the form of meteorites, and, except for lunar samples, they still are the only material obtained from space that we can see and touch directly. Meteorites are of two chief types—the "stonies" and the "irons," and many, if not all, may have originated from the fragmentation of a small body or bodies occupying the present asteroid belt position. The ages of the meteorites, except for a controversial group of stones called tektites, have been determined to range from 4.0 to 4.6 billion years, in broad agreement with the 5 billion or so years currently accepted for the solar system and the earth itself.

The landing of astronauts on the moon has given us the chemical composition of samples of the moon's surface. The Russian space probe Venera made a landing on Venus and telemetered information on its elevation above the Venusian surface, as well as on the pressure and composition of the atmosphere, until it crashed.

In addition to the information obtained "on the spot" from space probes, there is a vast body of data obtained by astronomers from spectral analysis of energy received from stars and planets. When the light from the sun, emanating from its extremely hot subsurface, passes through the cooler outer materials, certain wavelengths are absorbed in various amounts by interaction with certain elements. These wavelengths are represented by dark lines on spectrographic film when sunlight is spread out into its component colors. From knowledge of these wavelengths it is possible to determine the elements that have done the absorbing and to calculate their relative percentages. The intensity distribution of the entire spectrum can be used to obtain the temperature of the radiating source. The same techniques can be used to estimate the temperatures of other bodies, including the moon.

Because we see the planets by reflected sunlight, we can compare their spectra with that of direct sunlight, determine the additional absorption that has been caused by reflection, and then obtain information on planetary atmospheres. It has been most heartening to find that spectral analysis is corroborated by the space probe analysis.

Analysis of light emitted or reflected naturally has recently been supplemented by radar waves which can be sent from earth to the planets. Enough of the reflected energy returns to permit spectral analysis of the radar waves in a manner similar to that used for light. Incredibly accurate measurement of the time required for the roundtrip of the radar waves has made it possible to determine interplanetary distances with high accuracy. In addition, radar techniques permit measurements of planetary diameters.

ORIGIN OF THE SOLAR SYSTEM

Most investigators agree that the planets, moons, asteroids, and comets either were formed during the creation of the sun itself or consist of materials derived from it at a later time. Emmanuel Kant and Pierre Laplace, in the late eighteenth century, postulated that the solar system was derived by the shrinkage and condensation of a great nebula or rotating lens of gaseous and particulate material. As the nebula contracted under the inward force of gravity, rings of gaseous and particulate material were left behind revolving in circular orbits. The circular, or near-circular, orbital paths were a result of the combined inward motion of the materials toward the center of the lens because of gravitational attraction and the tangential motion of the particles owing to their inertia. As the core of the original lens contracted to form the sun, the rings about the core condensed under the influence of gravity to form the planets. Each planet, in turn, threw off one or more rings from which the moons condensed and circled around the planetary cores. The condensation hypothesis is shown in Figure 1.1.

Such a concept accounts for many of the features of the system, such as the revolution of the planets in a plane and the rotation of almost all the planets in the same sense as their revolution, as well as the revolution of the moons in nearly the same plane as that of the planets. However, when analyzed by classical mechanics, the general concept has always had one major difficulty that has not been overcome. As the nebular mass shrank and condensed, the rate of rotation of the innermost part should have increased in order to conserve momentum. Thus, the sun should be rotating on its axis approximately once every few hours instead of its actual 25 days. Another difficulty inherent in the condensation hypothesis is that the planets should contain more mass than they now have; approximately one-third of the mass of the sun should have been left behind to form the planets. Actually the planets contain only about one thousandth of the mass of the sun.

Another approach to the explanation of the solar system is based on the dragging of material from the sun by the gravitational attraction of some other body, such as a passing star. According to this hypothesis, huge streamers of gaseous material were torn out of our sun and the passing body during this collision or near-collision. As the "foreign" body receded from the sun, some of these streamers of material under the influence of the sun's gravity were captured in orbits about the sun. Eventually, the planets condensed from these streamers. As early as 1745, Buffon suggested this idea; his "drag" was a comet. Buffon's basic concept has been examined and re-examined in this century, especially by T. C. Chamberlain and F. R. Moulton and by Sir James Jeans. Figure 1.1 also shows the essence of this concept. However, all careful analyses of the problem have shown apparently insuperable difficulties in accounting for the various details of this collision hypothesis. For example, the probability of a collision, or near-collision, between our sun and a passing body large enough to disrupt enough of the sun's and the passing body's mass to form the planets is very small. Indeed, it is likely that at the most only a few such collisions have occurred during the history of our galaxy, about 10 billion years. Also, the collision theory predicts that the planets should move about the sun in elongated, narrow, elliptical paths, whereas they actually have nearly circular orbits.

The condensation hypothesis offers an explanation for the circular orbits, and it fits into the latest ideas concerning stellar evolution. As men-

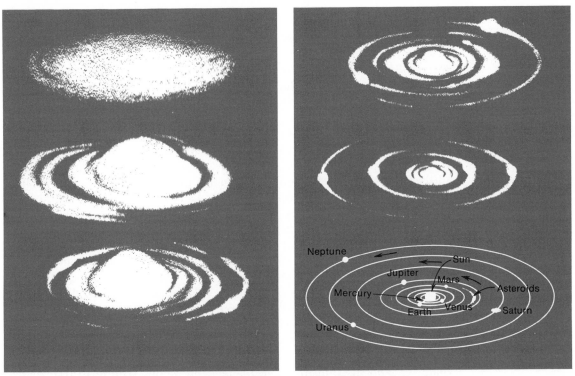

Drawings of Laplace's nebular hypothesis

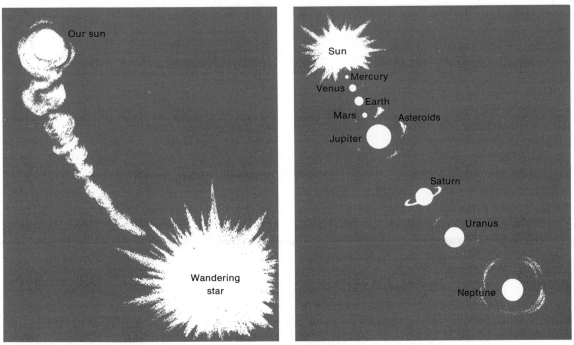

Sir James Jeans' tidal theory

Figure 1.1 Drawings showing essential features of the nebular and tidal hypotheses of the origin of the solar system (modified from Whipple, 1964).

tioned earlier, however, the concept has been plagued by the problem of momentum conservation. A possible solution to the problem of the conservation of momentum has developed recently from the study of the magnetohydrodynamics of the solar system. This work has been stimulated by the information on radioactive belts in the earth's vicinity that has been accumulated by observations from the many artificial satellites. Magnetohydrodynamics involves the study of the effects of magnetic fields on fluid systems. It appears that the problem of transfer of momentum from the sun, impossible to solve by Newtonian mechanics, may be accomplished by the magnetic and electric fields that are set up by streams of radioactively generated charged particles. Details of the complicated background are not appropriate here, but current hypotheses of the origin of the solar system are variants of the Kantian or Laplacian suggestion of origin from a cloud of gas and finely particulate material.

This cloud is assumed to have been initially chemically homogeneous, with an elemental composition similar to that of the sun. These assumptions are supported by the similarity of relative abundances of elements in the sun and in the stony meteorites and are consistent with present ideas of stellar evolution. As a result of gravitational and magnetohydrodynamic forces, particles and gases condensed first into droplets, then into small bodies, or planetesimals, perhaps as much as several hundred kilometers in diameter, and these in turn into planets and moons. The terrestrial planets are thought to have been too small to hold their gaseous and many of their icy constituents by gravitational effects as they became heated by the sun and by radioactive elements and were simultaneously subjected to stripping of their lighter elements by intense radiation from the sun. Thus they consist largely of nonvolatile compounds of relatively high density. The outer planets, being more massive, had sufficient gravitational pull to retain the icy and gaseous constituents. Also, being farther from the sun, their surfaces were not subjected to the same intense radiation that may have stripped the inner planets. Therefore they tend to resemble the overall composition of the sun much more than do the terrestrial planets. However, the reasons for these compositional differences between inner and outer planets are not completely understood.

ORIGIN OF THE EARTH

The origin of the earth is probably representative of the origin of the other terrestrial planets, although the range in their densities is large, and there are important differences in their compositions. The earthy and gaseous materials of the cosmic dust cloud accumulated, and as this occurred, the temperature began to rise as the planetary nucleus grew; heat was derived from decay of radioactive elements, from conversion of the kinetic energy of the particles into heat as they smashed into the accreting mass, and from conversion of potential energy as gravitational forces caused the proto-earth to shrink and consolidate.

It is thought that chondritic meteorites are nearly representative of the composition of much of the original material that accreted; note how closely the estimates of the composition of the earth as a whole (except for iron) agree with the composition of the average chondrite (Table 1.3). If it is indeed true that the agglomerated solid particles were all compositionally much the same, then the present gross internal zonation of earth composition required a tremendous movement of materials upward and downward during the very early stages of the earth's evolution.

The earth as a whole has 35 percent or more iron; today most of the iron has accumulated in the core, whereas the crust averages only about 6 percent. Of the elements of the entire earth, potassium is thirteenth in abundance and is estimated to make up only 0.1 percent or less of the earth mass; yet it is 2 percent of the continental crust—an enrichment factor of 20 or more. Furthermore, the differences between continental crust and oceanic crust for some elements are almost as great as those between oceanic crust and the overall earth composition. Potassium changes from 2.1 in continental crust to 0.8 in oceanic crust to 0.1 in the whole earth; for magnesium the corresponding values are from 2.3 to 4.5 to 13. According to Birch (1965), about 500 million years elapsed between the "initially cool unsorted conglomerate" and the development of a liquid iron core. He puts the initial accretion of the earth at about 5 billion years ago.

The formation of the core may have been a nearly cataclysmic event, occurring in a short time interval when internal heating had raised the temperature of the earth enough to permit migration of iron downward toward the earth's interior. Furthermore, core formation would result in a temperature rise of several thousand degrees centigrade, as the gravitational energy of the iron was converted into heat, as well as in a redistribution upward of radioactive materials responsible for further heating. Birch calculates that an internal thermal pattern qualitatively like that of today would have been established within the first billion years of the earth's history. He obtains a slow general temperature increase in the mantle since that time, accompanied by a nearly constant gradient in the outer mantle. It may be that as long as 4–4½ billion years ago, after core formation (Figure 1.2), the earth was a body closely resembling in internal zonation the one we know now. The greatest age that has been determined for surface rocks is about 3½ billion years, so the record available to us extends through much of post-core time.

It is also probable that the loss of the original atmosphere was attendant upon core formation, although this conclusion must be considered somewhat speculative at the moment. Several gases, such as xenon and neon, are deficient in the present earth; their concentrations are much lower than those expected from differentiation of material of the average composition of the solar system (Figure 1.3). Various lines of evidence (cf. Rubey, 1951) suggest that most of the present atmosphere, as well as all the water in the surface environment (*hydrosphere*), has originated from degassing of the earth. We are uncertain whether most of this degassing took place during core formation or very soon after core formation, or

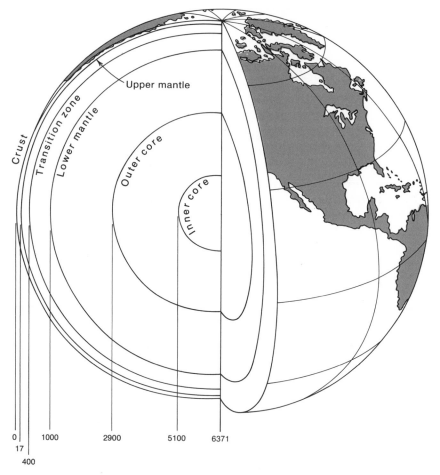

Distance from earth's surface, km

Figure 1.2 Cutaway of the earth showing the symmetrical internal zonation (redrawn from Sanders, 1967, *Chemical & Engineering News*, Vol. 45, pp. 1A–49A. Copyright 1967 by the American Chemical Society).

whether degassing has been taking place in important amounts throughout the earth's history.

ORIGIN OF THE MOON

Our own satellite continues to be one of the greatest puzzles of the whole solar system. Its low average density of about 3.33 g/cm³, not much greater than earth-surface rocks, indicates that it does not have a dense core. Yet if it is an agglomeration of primitive material, why was it not heated sufficiently by decay of radioactive elements and conversion of gravitational energy to heat, melt, and differentiate? One answer to this

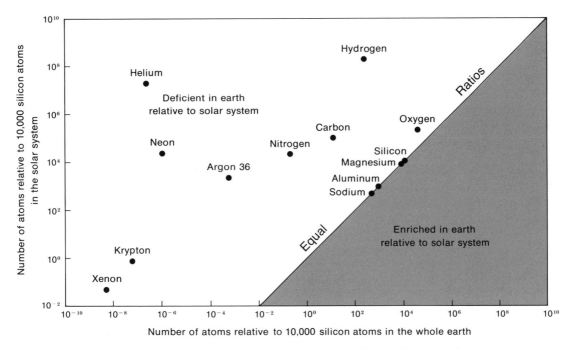

Figure 1.3 Comparison of the abundances of some elements in the earth versus those in cosmic material. Note the deficiencies in the earth of the typical icy and gaseous material of the outer planets (after Rasool, 1967. Copyright 1967 by the American Association for the Advancement of Science).

question is that moon formation perhaps took place after most of the rest of the solar system, and hence after significant decay of the radioactive elements in the cosmic dust that might otherwise have heated and melted it, but recent determinations of the age of moon rocks give ages of 3½ to 4½ billion years. An idea, suggested long ago by George Darwin, consistent with a late origin and also with the radioactive ages of moon rocks, is that the moon was born from the earth. Today we think the most likely time for such an event was when the earth's core was formed. As the density of the core increased, the earth's rotation would have been speeded up, and not only might this increase in angular velocity have helped cause the loss of the primordial atmosphere, but it also might have caused solid materials to be thrown out from the earth's equatorial regions to form the moon. If so, the low average density of the moon compared to that of the earth perhaps could be explained, for the density of the crustal materials of the earth, because of the partial differentiation of the earth, would have been lower than the earth average, even when the moon was lost.

Other hypotheses propose that the moon was formed independently of the earth. For example, it has been suggested that the moon is a planetary body that was formed in another part of the solar system and subsequently passed near enough to the earth to be captured and held in orbit by the gravitational force existing between the two bodies. Still other hypotheses

Table 1.3
Chemical Composition in Weight Percent of the Earth as a Whole Compared with the Chemical Composition of the Earth's Crust, the Sun, and Meteorites

Element	Earth as a whole			Earth's continental crust (Taylor, 1964)	Earth's oceanic crust[a] (Taylor, 196
	Mason's estimate (1966)	Washington's estimate (1925)	Niggli's estimate (1928)		
Fe	35	39.76	36.9	5.63	8.56
O	30	27.71	29.3	46.40	43.8
Si	15	14.53	14.9	28.15	24.0
Mg	13	8.69	6.73	2.33	4.5
Ni	2.4	3.16	2.94	0.0075	0.015
S	1.9	0.64	0.73	0.026	0.025
Ca	1.1	2.52	2.99	4.15	6.72
Al	1.1	1.79	3.01	8.23	8.76
Na	0.57	0.39	0.90	2.36	1.94
Cr	0.26	0.20	0.13	0.01	0.02
Co	0.13	0.23	0.18	0.0025	0.0048
P	0.10	0.11	0.15	0.105	0.14
K	0.07	0.14	0.29	2.09	0.83
Ti	0.05	0.02	0.54	0.57	0.90
Mn	0.22	0.07	0.14	0.095	0.15
H				0.14[d]	0.2
He					
C					

[a] Based on the assumption that the average composition of basalt represents the composition of the oceanic crust.
[b] Based on chemical composition of silicate phase of chondritic meteorites.

propose that the moon formed at the same time as the earth and in much the same way (condensation from the gaseous nebula) or that the moon is just one of the several bodies that once circled the earth and that the largest of these bodies captured the others to grow into the moon as we know it.

Information about the moon is growing every day but even the spectacular accomplishment of reaching the moon and being able to bring back samples for distribution to hundreds of scientists has not yet given definitive answers to the major problems. Chemical analyses of samples collected at several places on its surface have shown that the rocks there resemble earth-surface rocks like the lavas of the Hawaiian Islands more closely than other likely materials, such as chondritic meteorites. On the other hand, there are unique features of moon rock compositions, such as high titanium content, that leave their origin uncertain.

Measurements of the moon's gravitational field show marked local positive anomalies ("Mascons") that have been interpreted by some scientists as caused by deeply buried meteorites or asteroids of higher density than

Sun (Cameron, 1966)	Meteorites			
	Average iron (Brown and Patterson, 1948)	Average silicate[b] (Mason, 1966)	Average chondrite (Mason, 1966)	Carbonaceous chondrite[c] (type I)
0.00032	90.78	9.88	27.24	18.75
0.078		43.7	33.24	41.93
0.0027		22.5	17.10	10.58
0.0021		18.8	14.29	9.56
0.00007	8.59		1.64	1.02
0.0017			1.93	6.09
0.00012		1.67	1.27	1.11
0.00014		1.60	1.22	0.87
0.00017		0.84	0.64	0.56
0.000014		0.38	0.29	0.24
0.000004	0.63		0.09	0.04
0.000019		0.14	0.11	0.18
0.000004		0.11	0.08	0.06
0.000004		0.08	0.06	0.04
0.000007		0.33	0.25	0.16
86				0.21
13				3.97
0.045				5.53[e]

[c] Average of Orgueil and Ivuna meteorite analyses given by Wiik, 1956.
[d] Mason, 1966, Table 3.31.
[e] Organic matter.

typical lunar substances. This suggestion would simultaneously account for the large "Maria" as being the result of the impact when these bodies buried themselves in the moon. Other scientists disagree violently with this interpretation. At the moment, the moon has probably caused more scientific difficulties per metric ton than any other body in the universe.

METEORITES

Meteorites have been a major source of data helpful in our attempts to deduce the origin and history of the earth and solar system. Most of the meteorites apparently have come from the belt of asteroids between Mars and Jupiter. Because many of them seem to have been derived by fragmentation of larger bodies, up to several hundreds of kilometers in diameter, they may reveal to us the same kinds of information that would be available if the earth were similarly shattered and the fragments available to escapee scientists. Presumably the parent bodies of meteorites agglomerated at the same time as the earth and also went through a stage of heating and internal differentiation to form a metallic core and mantle of oxide and silicate minerals. Many of our conclusions concerning the earth's interior are based on the fact that the materials with the chemical compositions of meteorites, if disposed appropriately inside the earth, would cause the gravitational, seismic, and heat flow characteristics we observe at the earth's surface.

As indicated before, meteorites are subdivided into the preponderant stony or silicate-rich meteorites and into the iron meteorites that have important concentrations of nickel. The stony meteorites have been subdivided into the chondrites and the achondrites. The chondrites are so named because of a characteristic texture, consisting of an aggregate of rounded grains. Some investigators consider these grains as actual original particles that agglomerated to form the parent body of the meteorite. According to Ringwood (1966), the so-called "type 1" carbonaceous chondrites, containing about 5 percent carbon (Table 1.3), are the most primitive group of meteorites. Their composition suggests that they were formed by "accretion of the dust phase of the solar nebula into a small parent body which was subjected to a very mild degree of metamorphism."

The implication drawn is that the chondrites were derived from a parent body too small to have differentiated into a core and mantle and that their compositions are representative of "cosmic dust." On the other hand, the achondrites and the irons seem to have been derived from a body big enough to have melted and differentiated into these two types. Note in Table 1.3 that the chondritic composition could reasonably give rise to the irons and the silicates. Experimental and theoretical studies have attempted to define the size of the parent body or bodies for the irons and achondrites. The figure cited, a few hundred kilometers in diameter for the maximum size of the parent body, is based on a compromise between the requirement of a mass this large or larger to produce enough internal heat to melt the body and chemical and textural evidence that would rule against larger diameters.

STRUCTURE AND CHEMISTRY OF THE EARTH

The gross structure of the earth is shown in Figure 1.2. Surface features such as ocean basins and continents are too small to show in cross section. Three estimates are given (Table 1.3) of the overall composition of the earth. Despite the complex web of evidence that the investigators had to consider, it is worthwhile noting that even though there have been great increases in knowledge, changes in estimates have been small over the past 40 years.

Earth Shells

Some of the characteristics of the various earth shells are given in Table 1.4. Our direct observations have been limited to the upper 8 km of the crust, which in total make up only 0.4 percent of the earth's mass. The deepest drill hole is a penetration of less than 0.14 percent of the distance from the surface to the center! Yet we have fair confidence in the current conclusions concerning the structure and composition of the earth. Some of these conclusions are reviewed briefly here.

Size, Shape, and Density

The shape of the earth has been known with fair accuracy for a long time. It is a slightly oblate sphere, bulging a bit at the equator because of the centrifugal force of its rotation. The equatorial radius is 6,378 km, 21 km longer than from pole to center. A cross section at the equator is not quite a circle, but the deviation is very small.

The earth's mass is a huge 5.98×10^{27} g; its mean density of 5.52 g/cm³ is about twice that of average crustal rock. Measurements of gravitational force at various places on the earth's surface show that the density increase is virtually radially symmetrical but give no information as to whether the gradient is uniform, irregular, or discontinuous.

Internal Structure and Composition

Most of the interpretations of the internal structure of the earth came from studies of earthquake-wave velocities. When a liquid or a gas is subjected to a shock, energy is transmitted through it in waves; but because liquids and gases have no shear strength, only compressional waves in which particle displacement is parallel to wave propagation are formed. In solids, shear waves with particle displacements at an angle to the direction of wave propagation also are set up.

The important aspect, at this introductory stage, is that earthquakes set up compressional and shear waves that travel through the earth. Wave velocities are dependent on rock properties such as compressibility, density, and resistance to shear. Rocks can be tested in the laboratory to see the effects of increasing temperature and pressure on their physical properties and the resultant effects on wave velocities. Based in part on the results of these experiments, various models can be constructed for the nature and properties of the material within the earth. Inevitably, because wave velocities are not unique functions of a single property such as density, density profiles within the earth are not uniquely determined by velocity distributions.

Figure 1.4 shows velocities of P (compressional) and S (shear) waves as a function of depth within the earth. B, C, and D are interpretations of the

Table 1.4
Characteristics of the Earth Shells[a]

Shell	Chemical characteristics	State
Atmosphere	N_2, O_2, H_2O, CO_2, inert gases	Gas
Biosphere	H_2O, C, N, Ca, SiO_2	Solid and liquid matter, often colloidal
Hydrosphere	H_2O and dissolved salts	Liquid and solid
Crust	Silicate rocks with an average composition of igneous rock; oceanic crust low in SiO_2; continental crust high in SiO_2	Solid
Mantle	Silicate materials enriched in Fe and Mg and less siliceous than crustal rock; probably olivine and pyroxene and their high-pressure equivalents	Solid
Core	Iron-nickel alloy, perhaps with Si	Outer core, liquid; inner core, solid
Whole earth		

[a] Modified from Mason (1966).

density, pressure, and temperature relations. The salient features, as velocities change with depth, are as follows. Velocities beneath the continents, for a depth of about 40 km, increase irregularly downward, approximately as would be expected from the effects of increasing temperature and pressure on various mixtures of rock materials we can observe at the surface. Beneath the bottoms of the oceans, a similar increase is observed, but only for a depth of about 6 km. Then at the 40- and the 6-km depths, respectively, there is an abrupt increase in velocity over a short vertical distance, the Mohorovičič discontinuity. The reason for this velocity discontinuity is not completely known, but at the present state of our knowledge, it seems to be the result of a change in both the composition of the materials and the mineral phases present.

The P or compressional waves are transmitted through liquids and gases, as well as through solids; the S or shear waves pass only through solids. The absence of shear-wave transmission below about 2900 km is the main evidence for a liquid core, and the presence of a core is dramatically

Thickness (km)	Volume (units of 10^{27} cm³)	Mean density (g/cm³)	Mass (units of 10^{27} g)	Mass percent
			0.000005	0.00009
3.80 (mean)	0.00172	1.023	0.00176	0.024
40 (continental crust) 6 (oceanic crust) 17 (mean)	0.008	2.8	0.024	0.4
2,883	0.899	4.5	4.016	67.2
3,471	0.175	11.0	1.936	32.4
6,371	1.083	5.52	5.975	100.00

illustrated by the absence of S waves and the abrupt drop in P-wave velocity (Figure 1.4A). The P-wave velocity increase at about 5100 km indicates an inner core that probably is solid.

Core

The conclusion that the core of the earth is dominantly metallic iron is long-standing. This conclusion finds support in the fact that the average density of the earth is 5.52 g/cm³, whereas the average density of crustal material is only about 2.7 g/cm³; thus the interior of the earth must contain denser materials than its outer regions. The choice of metallic iron for core composition over other dense materials is based on indirect evidence that mainly has been accumulated during the last two decades.

For example, iron meteorites provide evidence that parts of other bodies, presumably similar in origin to the earth, in the solar system were composed of metallic iron, and thus the earth probably is also partially made

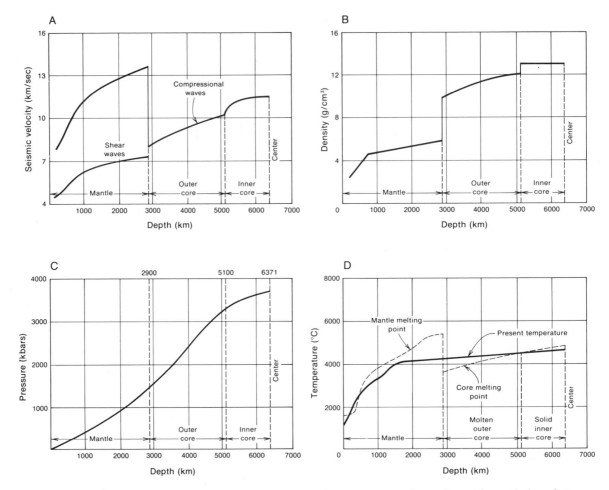

Figure 1.4 Comparison of a number of properties of the earth's interior. A and B redrawn from Clark and Ringwood, 1967, *Chemical & Engineering News*, Vol. 45, pp. 1A–49A. Copyright 1967 by the American Chemical Society. C and D after Takeuchi et al. (1967).

up of such material. Also, experimental studies of various materials, in which *P-T* conditions presumably comparable to those in the center of the earth (about 4500°C and 3700 kbars) have been achieved, have strongly reinforced the concept of a metallic iron core. Of all the materials that have been suggested as possible components of the core, only iron has properties of density and *P*-wave velocity that correspond to those previously deduced for the core (Birch, 1965). In addition, the melting point of iron, under the estimated pressure conditions of the core, is just about at the deduced temperature of the core, which would make the assumption of the presence of *both* solid and liquid in the core reasonable. Ringwood (1966) estimates that an alloy of iron, nickel, and silicon, in the ratios of $Fe^{84}/Si^{11}/Ni^5$, would fit the evidence much better than pure iron.

Mantle

Just as experiments have shown that almost without doubt the core is dominantly iron, they also have shown that the mantle, which makes up two thirds of the mass of the earth, has properties consistent with a silicate composition and cannot be composed of metals. This interpretation is in harmony with the preceding discussion of meteorites; the stony meteorites have properties appropriate for mantle material. An estimate of the approximate composition of the mantle, adapted from Ringwood (1966), is, in weight percentage (compare with silicate meteorites, Table 1.3),

Si	35.5
Mg	41.3
Fe	12.6
Al	3.7
Ca	4.6
Na	2.3

A major controversy rages concerning the compositions and structures of the individual compounds that are presumed to be in the mantle and the relationships of mantle materials to rocks of the crust. On the basis of our knowledge to date, it is thought that the elements listed above are perhaps combined as iron-magnesium silicates in the upper mantle and as oxides in the lower mantle. The chemical differences within the upper mantle are thought to be small, although important for interpretations of its behavior. Changes of wave velocities with depth appear to be consistent with those expected from differences in physical conditions. Uranium, thorium, tantalum, barium, strontium, zirconium, hafnium, beryllium, and the rare earths are thought to be relatively more concentrated in the upper mantle.

Crust

The composition of the crust has received a great deal of attention for many years. A fair proportion of the crust can be sampled directly. It is strongly enriched, relative to the mantle and to the earth as a whole, in silica, alumina, and potassa. Figure 1.5 shows abundances of the elements in the crust plotted as a function of atomic number and emphasizes that only oxygen, silicon, aluminum, iron, calcium, sodium, potassium, and magnesium exceed 1 percent by weight. The crust is so heterogeneous that we cannot classify the various parts without a long and detailed discussion. Two chief kinds of crust can be distinguished: continental and oceanic (Table 1.3). The composition of continental crust averages much higher in silica, alumina, and potassa than oceanic crust, and continental crust has a lower average density than oceanic crust. The crust will be considered in more detail in Chapter 2.

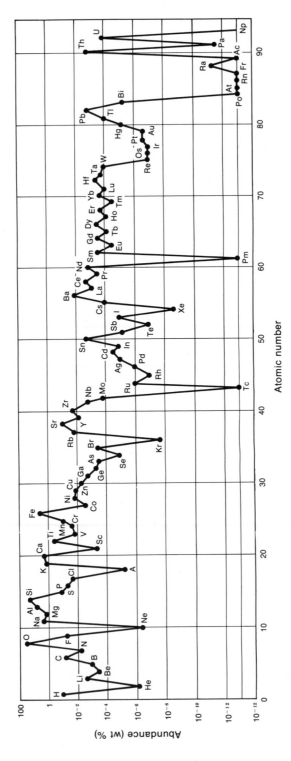

Figure 1.5 Crustal abundance of the elements 1–93 (Parker, 1967).

DISCUSSION OF EARTH STRUCTURE

A discussion by Press (1968) of earth density relations and his interpretation of the nature of the earth's interior is an excellent summary of our current state of knowledge and controversy. He concludes that the radius of the earth's core should be increased by about 20 km over most previous estimates. This would be an increase of less than 1 percent of a representative previous estimate of 3470 km and gives us a feeling for the accuracy of any current values. Press' models indicate that there is a chemical difference between the liquid outer core and the solid inner core, with the inner core made up of an iron-nickel alloy (20–50 percent nickel). The liquid core is thought to be an iron-silicon alloy with up to 25 percent silicon. He places the inner core density between 13.3 and 13.7 g/cm^3 and that of the top of the outer core between 9.4 and 10.0 g/cm^3.

Press interprets the lower mantle as having a downward increasing ratio of iron to iron plus magnesium, with a variation from top to bottom of a factor of 2. He equates the transition zone between upper and lower mantle to a compositional change as well as to changes in the structures of the minerals, and he finds that the upper mantle may well be inhomogeneous, representing various mixtures of eclogite (of the same composition as basaltic rocks at the surface) and of pyrolite (a composition higher in Fe and Mg than basalt, proposed by those who favor a chemical break at the bottom of the crust).

According to Press, the upper mantle, at least, is continuing to differentiate from a bulk composition lower in aluminum and higher in FeO and MgO than typical basaltic rocks at the surface by a process of partial melting, which would produce material of basaltic composition and leave behind a lower alumina residue. These basaltic liquids then are postulated to move upward in the mantle, sometimes solidifying within the mantle, sometimes intruding the lower crust beneath the continents, and sometimes breaking through the thin crust of the ocean floor to pour out as flows or to build up into islands like the Hawaiian chain. Also, the chemistry of differentiation of the upper mantle as specified is very likely to be modified as we learn more about this region. It is only quite recently that we have learned that the upper mantle is inhomogeneous laterally in terms of seismic wave velocities and heat flow; better knowledge of the areal extent and ranges of variations may produce drastic interpretive problems.

One of the important aspects of Press' conclusions is that a compositional change in the transition zone would make it impossible to have convection currents involving the entire mantle; such currents would tend to homogenize the mantle compositionally and would force an interpretation of its properties based entirely on structural and compositional changes of individual minerals, while keeping the bulk composition constant.

No matter what the details, it does seem that even today there may be continuing significant differentiation of the earth, with lighter, lower-melting materials moving upward and with concomitant sinking of the denser, more refractory residues. Just how the concentrations of the extreme differentiates became lumped together to form the continents is a major current problem, but there is little question that much of the H_2O, K_2O, SiO_2, and to a lesser extent Al_2O_3 have somehow been transported from an initial distribution throughout the earth's interior to form seven thin rafts of continental crust. These are now crowded into half of the northern hemisphere, leaving the top of the mantle almost bare on more than half of our globe.

MAGNETIC PROPERTIES

The earth is a huge magnet and its magnetic field is like the one that would result from having a bar magnet buried in its core. The magnetic axis is nearly coincident with the rotational axis; the angle between them is $11\frac{1}{2}°$. The north magnetic pole lies at 75°N latitude and 101°W longitude. Also, the field is somewhat distorted from that expected from a simple bar magnet; for example, the south magnetic pole is more than 2000 km from a point antipodal to the north magnetic pole.

The earth is a strong magnet in the sense that its field is easily detected even by crude instruments such as a cheap compass, and measurements of its intensity can be made routinely to within 1 ppm. The intensity of the field and the positions of the poles are changing significantly on a geologically short time scale. The intensity has diminished 6 percent in the last 150 years; at this rate (which well might not continue!) it would drop to zero in 2000 years. There is good evidence from the reversal of the magnetic orientation of magnetic minerals in rocks (see Chapter 2) that the earth's field has actually reversed many times in the last few million years. The north magnetic pole may change position as much as 1 km/year, but the movement may be dominantly precessive around the rotational axis over periods of millions of years.

The current hypothesis for the origin of the magnetic field is that temperature gradients in the liquid core cause vertical convection of the iron, and it is movement of this good electrical conductor that generates the magnetic field.

MAJOR SURFACE FEATURES

Mountains, valleys, and ocean basins are hardly distinguishable when we look at the entire globe, but much of our interest in the earth's history is pointed toward unraveling the nature of the surface environment through time. Consequently it seems worth elaborating somewhat on the major features of the earth's surface.

The average depth of the oceans is about 4 km. They cover 70 percent of the earth's surface. The continents average 0.84 km above sea level in elevation and have most of their land area in the northern hemisphere. Figure 1.6 shows the worldwide distribution of elevations of land and sea floor. Major features are labeled. The section shows that the continents are not entirely defined by their shorelines; they exist as platforms high above the general ocean floor, and the platforms (continental shelves and slopes) may extend far out beneath the sea.

Perhaps the most remarkable feature of the surface has been known for years; most areas larger than a few hundred square kilometers have elevations consistent with their behaving as crustal blocks in floating equilibrium in a substratum. The continents "float high" because they are thick plates of relatively light material; the ocean basins exist because the oceanic crust is thin and dense. Figure 1.7 shows a schematic interpretation of crust-mantle relations. This floating or *isostatic* equilibrium has some interesting consequences that will be explored later; for example, if material is eroded from mountains and deposited in an adjacent basin, the basin block should tend to sink and the mountain block to rise, relative to the medium in which they are floating—just as unloading sand from a raft into a boat would cause the raft to rise and the boat to sink in the supporting water.

The near-achievement of isostatic equilibrium of crustal elements is somewhat surprising in terms of our everyday concept of the behavior of materials. The earth, down as far as the core, gives every evidence of being solid, and we do not ordinarily consider that one solid can float in another.

Figure 1.6 Distribution of the area and elevation of the earth's solid surface between land and sea.

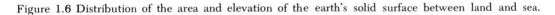
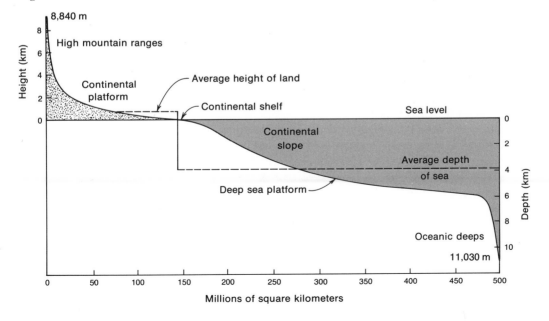

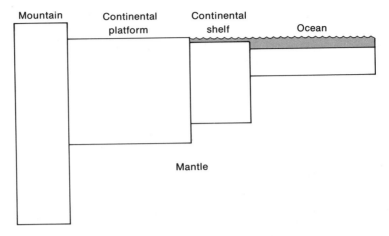

Figure 1.7 Schematic diagram showing isostatic equilibrium between various crustal blocks. The model is oversimplified in that the densities of the crustal blocks are all assumed to be the same. In actuality, each block has an irregular density gradient, and it is not at all certain that the density of the mantle is everywhere the same.

But rather simple calculations show that a cube of almost any kind of rock having a height of several kilometers would fail by crushing at the base under its own weight, just as a rope payed out over a cliff will eventually snap under its own weight. In large masses, and over long periods of time, rocks yield even to small stresses.

If the force of gravity on a given mass is measured at the earth's surface, it may differ significantly from place to place. But if the readings are corrected for (1) the differences in the distance of the various stations from the center of the earth, (2) the centrifugal force owing to the earth's rotation, and (3) the local irregularities of mass distribution because of topography, and, finally, if the assumption is made that the place of measurement is situated on a floating crustal block, the differences between readings, or *anomalies,* tend to disappear. There are a few regions in which there is a marked *residual anomaly* after the preceding corrections (for example, paralleling the south side of the arc of the Indonesian Archipelago), but the overall picture is one of nearly complete isostatic adjustment.

The Mohorovičič discontinuity between crust and mantle, indicating an abrupt change in properties of the rocks, permits us to think, as a first approximation, of light solid crustal blocks floating on a yielding denser mantle. The picture evoked is a little too simple but correlates well with gross relief features; mountains, which float higher, should project deeper, and it is generally true that the "Moho" is depressed beneath them.

Because rocks yield to long-applied stress differences, the areas of residual anomalies are of great interest. A negative anomaly results when, for whatever reason, the force of gravity calculated for a given station is greater than that observed. The lower observed value must result from a "deficiency in mass"; that is, the rocks contributing to the gravitational attraction at the station have an effective density less than that assumed in the isostatic model used for the calculation.

The principles involved in the interpretation of residual anomalies are illustrated in Figure 1.8. A negative anomaly can be explained on the basis that there are forces acting downward to prevent a relatively light crustal block from rising into an equilibrium position. This situation would result in a greater mass of low density crustal material beneath the station than assumed in an isostatic model and a lower than calculated gravitational force. A positive anomaly is given the inverse interpretation. One possible interpretation of major anomalies, but not the only one, is that the materials of the upper mantle are circulating in thermal convection cells, with the consequence that rising currents generate stresses that tend to "push" crustal blocks upward and laterally, and descending currents create stresses that "pull" the crust downward.

The fact of isostasy tends to force us immediately to the concept of an active crust; the continuing processes of erosion and deposition must be followed by compensating vertical shifts in crustal blocks. Furthermore, rates of erosion are so fast, relative to the great lengths of geologic time through which erosion and deposition have been active, that the continents should have been reduced to featureless plains long ago unless there have been mechanisms at work in addition to buoyant restoration and sinking.

ORIGIN OF CONTINENTS

The areal distribution of continents and oceans has always been a puzzle. The gross radial symmetry of the earth might lead us to expect a crust of

Figure 1.8 Illustration of the causes of gravity anomalies (after Strahler, 1963).

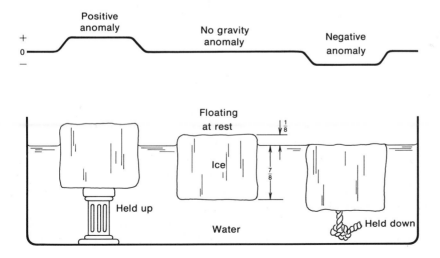

uniform thickness spread evenly around the sphere; the oceans, in turn, might form a thin shell, overlain by the atmosphere.

Various hypotheses for the present distribution, which will be considered in detail later, have been suggested. One hypothesis is that very early in the history of the earth a uniform thin crustal shell was formed, but it was disrupted during core formation, perhaps with the loss of material to form the moon. As a matter of interest, the Pacific Ocean was thought to represent the "scar" left behind after crustal materials escaped to make the moon. Even barring such an event, the high radioactivity and consequent heat production in the early earth would cause the surface region to be a place of great activity as material within the earth heated, melted, rose, cooled, and sank.

In addition to the problem of the origin of discontinuous continental crust is that of the permanency of the present continents. The obvious near-perfect fit of the east coast of South America against the west coast of Africa triggered initial speculations at the turn of this century that the continents had split apart and were moving relative to each other. Recently, the discovery of great fractures, or rifts, in the Atlantic and Pacific ocean basins, where active crustal movement apparently is taking place, has renewed interest in continental drifting, as have studies of records of the earth's magnetic field preserved in ancient rocks which indicate that the earth's magnetic pole has changed markedly and differentially relative to the continents through time. Finally, there is independent evidence of thickening and perhaps of marked change of continental areas through time, regardless of whether or not there has been relative lateral shifting.

The geologic history of the relations between continents and ocean basins is one of the most active areas of research today; although *few definitive conclusions* can be drawn at the moment, it is abundantly clear that the upper few hundred kilometers of the earth is still an active region, and that the earth, despite its great age, has not yet settled into an appropriate tranquillity.

DIFFERENTIATION OF THE EARTH

Few events that took place during the very early days in the history of our planet are well known, but on the basis of our present knowledge a reasonable model for the differentiation of the earth can be constructed. The various earth shells can now be considered as the product of the chemical and physical differentiation of an initial nearly homogeneous body with a composition, except for some of the more volatile elements such as hydrogen and argon, within the range of the stony meteorites. The energy required for differentiation was derived from gravitational energy as the initial particles condensed into the earth mass and from the energy of breakdown of radioactive elements. As the temperature rose, iron, nickel, and silicon were reduced from higher valence states to the metals by carbonaceous

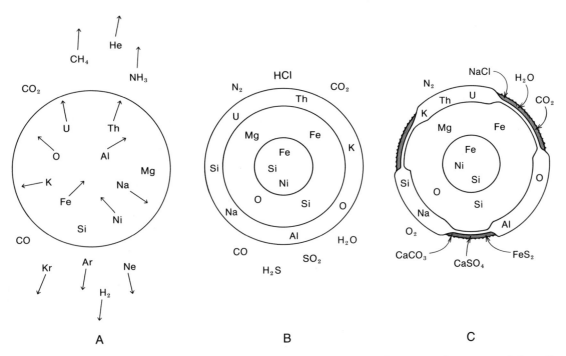

Figure 1.9 Three hypothetical stages of differentiation of the earth. A shows core formation and loss of original atmosphere. In B, the core and crust have formed and a reducing atmosphere consists of gases released by the internal differentiation. C represents an essentially modern earth.

materials and by reducing gases such as hydrogen and moved down to form the core. On the other hand, much of the SiO_2, Al_2O_3, and K_2O moved upward, along with some iron oxides, CaO, MgO, and Na_2O, to form the crust, where they crystallized on cooling to make various silicate and aluminosilicate minerals. The crust also became a region of concentration of the rare earth elements and many of the radioactive species such as uranium and thorium, as well as others such as technetium, which decays so rapidly that detectable concentrations no longer exist. The mantle developed a relatively uniform composition, with some tendency for higher values of the crustal elements in the upper mantle. During the period of active differentiation surface temperatures were probably very high, and many volatile substances were lost to space. Figure 1.9 shows diagrammatically an interpretation of the distribution of elements in the early earth, at approximately the time of core formation, and after formation of the oceans and the development of an oxygenated atmosphere.

The condensation and differentiation of the earth were probably completed in their major aspects within 1 billion years, and for the past 4 billion years surface conditions apparently have not differed drastically from today's, except for the progressive increase of oxygen in the atmosphere.

SUMMARY

The currently popular hypothesis is that the planets, their satellites, the asteroids, and a great many comets originated about 5 billion years ago as a result of coalescence from a nebula comprised of gases as well as icy and earthy particles. The inner planets agglomerated from dominantly earthy particles, and the outer planets from icy particles and gases.

As the earth coalesced, gravitational and radioactive heating brought the planet to or near the melting point, and it differentiated to form a metallic iron-rich core; a thin crust relatively high in oxygen, silicon, aluminum, and potassium, as well as radioactive elements; and an intermediate zone or mantle with oxygen, silicon, magnesium, and iron as major elements. It is possible that the moon was derived from debris thrown off at the same time the core formed and the primordial atmosphere was lost.

The crust averages 40 km thick beneath the continents and 6 km thick beneath the oceans and is separated from the mantle by a thin zone of rapid change of properties called the Mohorovičič discontinuity. Except locally, the crust floats in isostatic equilibrium in the mantle; large-scale vertical movements have clearly taken place. It also seems likely that the ocean basins are currently rifting and spreading, perhaps with concomitant major lateral movements of the continents. The origin of the continental distribution is still an active field of research and controversy; the complementary problem of the origin of the ocean basins is in similar flux. It is generally agreed that the crust, including the atmosphere and oceans, is a result of complex and, in many instances, highly selective processes of differentiation of a relatively homogeneous primordial mass.

REFERENCES

Berry, L. G., and Mason, B., 1959, *Mineralogy:* W. H. Freeman and Company, San Francisco.

Birch, F., 1965, Speculations on the earth's thermal history: *Bull. Geol. Soc. Am.,* 76, 133–154.

Brown, H., and Patterson, C., 1948, The composition of meteoritic matter. II. The composition of iron meteorites and of the metal phase of stony meteorites: J. Geol., 55, 508–510.

Cameron, A. G. W., 1966, Abundances of the elements: in Handbook of physical constants, S. P. Clark, Jr., ed., *Geol. Soc. Am. Mem.,* 97, 8–10.

Clark, S. P., Jr., and Ringwood, A. E., 1967: in Chemistry and the solid earth, H. J. Sanders, ed., *Chem. Eng. News,* Oct. 2, 1A–49A.

Garrels, R. M., 1951, *A Textbook of Geology:* Harper & Row, New York.

Holmes, A., 1965, *Principles of Physical Geology,* 2nd ed.: The Ronald Press Company, New York.

Mason, B., 1966, *Principles of Geochemistry,* 3rd ed.: John Wiley & Sons, Inc., New York.

Niggli, P., 1928, Geochemie und Konstitution der Atomkerne: *Fennia, 50,* no. 6, 1–24.

Parker, R. L., 1967, Composition of the earth's crust: in Data of geochemistry, 6th ed., M. Fleischer, ed., *U.S. Geol. Surv. Profess. Paper, 440–D,* D1–D19.

Press, F., 1968, Density distribution in the earth: *Science, 160,* 1218–1221.

Rasool, S. I., 1967, Evolution of the earth's atmosphere: *Science, 157,* 1466.

Ringwood, A. E., 1966, Chemical evolution of the terrestrial planets: *Geochim. Cosmochim. Acta, 30,* 41–104.

Rubey, W. W., 1951, Geologic history of sea water: an attempt to state the problem: *Bull. Geol. Soc. Am., 62,* 1111–1147.

Sanders, H. J., 1967, Chemistry and the solid earth: *Chem. Eng. News,* Oct. 2, 1A–49A.

Strahler, A. N., 1963, *The Earth Sciences:* Harper & Row, New York.

Takeuchi, H., Uyeda, S., and Kanamori, H., 1967, *Debate about the Earth:* Freeman, Cooper & Company, San Francisco.

Taylor, S. R., 1964, Abundance of chemical elements in the continental crust: a new table: *Geochim. Cosmochim. Acta, 28,* 1273—1285.

Washington, H. S., 1925, The chemical composition of the earth: *Am. J. Sci., 9,* 357–363.

Whipple, F. L., 1964, The history of the solar system: in *The Scientific Endeavor:* Rockefeller University Press, New York, 69–107.

Wiik, H. B., 1956, The chemical composition of some stony meteorites: *Geochim. Cosmochim. Acta, 9,* 279–289.

2 | Conditions of the Crust

The crust of the earth on which we live is the only part we have directly observed. In terms of the earth as a whole, it is almost too thin to notice, comprising only 0.3 percent of the earth's radius; it averages about 40 km in thickness beneath the continental surface and 6 km beneath the ocean floor. The deepest bore hole has penetrated only about one fifth of the continental crustal thickness. We are microbes on the skin of the earth; let us see in this chapter how the microbes have attempted to learn about the epidermis of their host.

We have given the overall chemical composition of the crust in Chapter 1, Table 1.3; it is particularly high in silicon, aluminum, and potassium with respect to the total earth. The elements of the crust are organized into minerals, and the minerals are associated in various proportions to make up rocks (Appendix A provides some fundamental information concerning major mineral species). One of the first results of the study of rocks was the recognition that chemical and mineralogical variability is not continuous; rocks tend to occur in masses of various shapes and sizes in which the ratios of the constituent minerals are fairly constant, and there is a clear-cut break from one mass to the next.

IGNEOUS ROCKS

Most of the crustal rocks are interpreted as having resulted from the crystallization of molten material; these are the igneous rocks. Lavas are igneous rocks that have flowed out upon the surface of the earth; intrusive or plutonic rocks are those that crystallized below. Figure 2.1 shows that lavas usually have the tabular shape expected from their origin; intrusive rocks range from tabular bodies that cooled in fractures (sills and dikes) to irregular masses of many cubic miles (batholiths) that formed owing to invasion and displacement or assimilation of pre-existing rock by molten material that later cooled.

Although an individual lava or intrusive body tends to have a given chemical and mineralogical composition, a plot of the compositions of many bodies shows that they apparently are all genetically related, for when the chemical compositions of a group of igneous bodies are plotted, the points for the chemical constituents of individual bodies fall on smooth curves. Figure 2.2 shows how the minerals of igneous rocks vary in SiO_2 content and also provides an illustration of a classification of major rock types. A few minerals and their proportions are all that we need to know to describe the major aspects of the composition of most of the crust. On the other hand, if we begin to look at the details of the variations in composition of these minerals, if we examine all the minerals, and if we con-

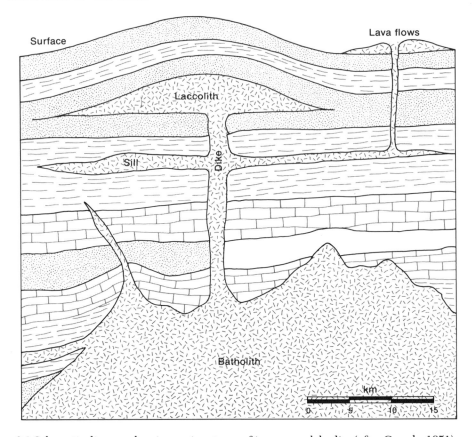

Figure 2.1 Schematic diagram showing various types of igneous rock bodies (after Garrels, 1951).

sider all the different kinds of rocks that are exceptions to the generalization, the picture becomes almost infinitely complex.

The term *basalt* (see Figure 2.2) is commonly applied to lavas as well as to material of similar composition of intrusive origin. Most crustal rocks of basaltic and andesitic composition are extrusive, whereas those of granitic or granodioritic composition are intrusive. Granites and granodiorites are typically continental, whereas basalts and andesites are oceanic. However, basaltic lavas are also found as tabular bodies covering large areas of the continents.

SEDIMENTARY AND METAMORPHIC ROCKS

Although igneous rocks make up most of the crust, about 80 percent of the continents' surface is covered with sedimentary rocks. They are the debris that has accumulated from eons of operation of surface processes. Sedimentary rocks can be grouped into four chief categories: sandstones, lutites, carbonates, and evaporites.

Sandstones are defined as rocks composed of particles ranging in size from a few to 0.06 mm in diameter. Sandstones show a wide range of chemical composition. Quartz is usually the major mineral; feldspars, especially K-feldspar, are common. Sandstones are composed of individual mineral grains or rock fragments that have been derived from the weathering of all kinds of rocks and tend to be accumulations of minerals that are chemically and physically durable. The particles are transported by streams, waves, wind, or ice and accumulate in tabular or lenticular bodies in stream beds, lakes, or the ocean.   ·

Lutites are rocks composed of particles smaller than those of sandstone, commonly in the range of a few microns, and represent finer debris carried by the erosional agents of running water, ice, and wind. The more exclusive term *shale*, which denotes lamination or fissility in addition to grains smaller than those of sandstones, is also applied to these fine-grained rocks. The clay minerals are dominant constituents of lutites but fine-grained quartz and feldspar are common.

Carbonate rocks are made up of the carbonate minerals calcite, aragon-

Figure 2.2 Mineralogical composition of some important igneous rock types as a function of SiO_2 content (not including glasses and minor minerals). Olivines, pyroxenes, and amphiboles are shown in areas that represent rock types in which a particular mineral group is commonly found (after Clark, 1966). 1, Diorite, gabbro (intrusive); andesite, basalt (extrusive). 2, Quartz diorite (intrusive or extrusive). 3, Granodiorite (intrusive). 4, Granite (intrusive); rhyolite (extrusive). 5, Alkali granite (intrusive).

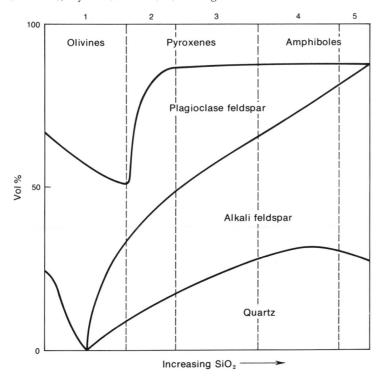

ite, and dolomite. Calcite- or aragonite-rich carbonate rocks are termed limestones; dolomite-rich rocks are called dolomite. Many carbonate rocks have formed as a result of the precipitation of calcite or aragonite by organisms and incorporation of the skeletons of these organisms into sedimentary deposits. Inorganic precipitation of carbonate minerals may also lead to the formation of carbonate rocks; however, the importance of this process remains to be evaluated. Table 2.1 summarizes some key relationships among the sedimentary rocks.

Sandstones and shales are initially deposited as particles of sand and mud. With burial and time, cementation and a certain amount of mineral reconstitution take place, and the coherent material is regarded as rock. When post-depositional changes become drastic, the rock is said to be metamorphosed, and new names are given to the altered products. Sandstone is altered to quartzite, limestone to marble, shale to slate, and so on. These altered products are called metamorphic rocks. Generally speaking, the gross chemical composition of a rock tends to be maintained during alteration, so for our purposes and to avoid unnecessary repetition of terms we shall use the term *sedimentary rock,* to include the initial unconsolidated

Table 2.1
Some Characteristics of Sedimentary Rocks

Rock	Some sediment equivalents	Some metamorphic equivalents	Origin	Grain size	Major minerals	Wt % of sedimentary rock mass
Conglomerate	Pebbles Cobbles Boulders	Conglomerite	Clastic particles	>2 mm	Rock fragments	<1
Sandstone	Sand	Quartzite	Clastic particles	$\frac{1}{16}$–2 mm	Quartz Feldspar Rock fragments	15–20
Lutite	Silt Mud Clay	Slate Schist	Clastic particles	<$\frac{1}{16}$ mm	Clay minerals Quartz Feldspar	70–80
Limestone	Reef Lime mud	Marble	Biochemical precipitate, inorganic precipitate, clastic particles (shell fragments)	Variable, up to cms	Calcite Dolomite Aragonite	≃5–15
Evaporite	Salt	None	Precipitate from evaporating water	Variable, up to cms	Halite Gypsum Anhydrite	≃5

sediments, the resultant sedimentary rock, and the metamorphic equivalent. We exclude from sedimentary rocks those metamorphic rocks of enigmatic origin which cannot be recognized to be of sedimentary origin on the basis of diagnostic textures, structures, and/or chemical compositions.

Evaporites are chemically precipitated rocks resulting from evaporative concentration of a water body. When the medium evaporating is sea water, the chief minerals formed are gypsum, anhydrite, and halite. Thick deposits containing these minerals are quite common; however, they constitute only a small percentage of all sedimentary rocks.

HEAT FLOW AND TEMPERATURE WITHIN THE EARTH

The earth is still a very energetic globe. One of the manifestations of its internal energy is the continuous emission of heat from the interior. In bore holes, shallow as they are, we find increasing temperatures with depth. The probability that the liquid part of the core is mostly molten iron indicates that temperatures within the earth are several thousand degrees Centigrade.

The average temperature gradient measured in drill holes is about 30°C/km, and most measurements fall in the range 20–50°C/km. Exceptionally high gradients are found locally, as in volcanic or hot-spring areas, where the temperature rise may be measured in degrees per meter, but these areas do not contribute significantly to the heat budget of the total earth. The temperature gradient diminishes downward. At the bottom of the crust, temperatures are estimated to be about 400–600°C. For the continents this would give an average gradient of about 12°C/km, a value considerably less than that measured at shallow depths. Corroboration of this conclusion of a decreasing gradient is found by extrapolating the average near-surface gradient of 30°C/km to the center of the earth; the predicted temperature would be 192,000°C above the surface temperature! Figure 2.3 shows one estimate of temperatures beneath the continents and oceans as a function of depth. This figure illustrates two points well: Although the positions of the isotemperature lines are currently hotly disputed, the steep gradient near the surface diminishes rapidly with depth and becomes a degree or two per kilometer at a few hundred kilometers; and whatever the exact numbers, the gradient beneath the continents appears to be different from the gradient beneath the oceans.

The conclusion that there is probably a difference in the temperature gradient beneath the continents and ocean basins comes as a result of studies of heat flow. From temperature gradients and knowledge of the thermal conductivities of rocks, it is possible to calculate the heat released per unit area of the earth's surface. For a linear gradient heat loss can be calculated from the equation

$$q/t = \frac{KA(T_2^\circ - T_1^\circ)}{d},$$

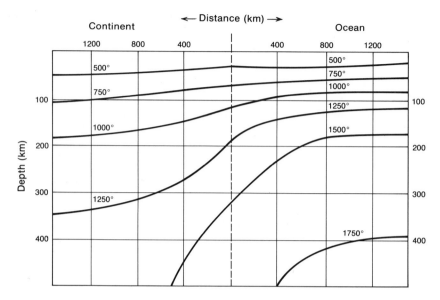

Figure 2.3 An estimate of the distribution of temperature isotherms (isotemperature lines) as a function of depth beneath the continents and oceans (after MacDonald, 1964). Comparison of this figure with Figure 2.6 shows important differences between the investigators' interpretations and provides a measure of the uncertainty of our knowledge of temperature with depth.

where q is the heat in calories, t is time in seconds, K is the thermal conductivity of the rock, A is the cross-sectional area of interest in centimeters squared, T_2° and T_1° are the temperatures in degrees centigrade at two points in a drill hole, and d is the distance between them in centimeters. As a specific example, for a 30°C/km gradient in a particular granite, the heat flow per second per square centimeter is

$$q/\text{sec} = \frac{(7.9 \times 10^{-3})(1^2)(30)}{10^5} = 2.4 \times 10^{-6} \text{ cal/cm}^2/\text{sec},$$

or $2.4\,\mu$ cal/cm²/sec.

Several thousand measurements of heat flow have now been made. Figure 2.4 shows one interpretation of the measurements plotted on a global basis. The values range between 0.7 and $3.0\,\mu$ cal/sec and average about 1.4. Many striking local anomalies have been averaged out. Furthermore, there is no apparent difference between *average* oceanic and continental heat-flow values. This observation came as a rude shock to investigators a few years ago when enough measurements had been made to be fairly certain of this conclusion. Notice, however, in Figure 2.4 that there are areas of relatively high heat flow on continents and in ocean basins. At any rate, it is apparent that the earth is losing heat through the crust with remarkable uniformity; if any area of continental dimensions is considered, it is delivering just about as much heat as any other comparable area, regardless of the complexities of the underlying materials!

It is a little difficult to think in terms of microcalories; $1.4\,\mu\,cal/cm^2/sec$ represents approximately 44 $cal/cm^2/year$, or 2.2×10^{20} cal/year for the whole earth. This is enough heat, if it were not dissipated, to raise the temperature of the oceans 1°C in about 6000 years.

For a long time it was thought that the earth was continuously cooling, and in fact heat flow had been used to calculate the original temperatures within the earth and the cooling history required to produce the present condition. Now it appears that approximately the present temperature distribution may be long-lived indeed and that the interior may be heating or cooling slightly. The source of heat is radioactive elements.

Uranium, thorium, and potassium are the chief contributors to heat production, and these elements show greatest abundance in the rocks of the continental crust. Because it was thought that the mantle was homogeneous below the Mohorovičič discontinuity and that heat flow *into* the crust would be uniform, everyone was confident that the additional heat in the continental crust contributed by the decay of uranium, thorium, and potassium would produce continental heat-flow values 50 percent or more greater than those through the relatively "dead" oceanic crust, yet the average values are about the same. The explanation seems to be that the upper mantle is *not* homogeneous but contains more radioactive material

Figure 2.4 Representation of heat flow values over the earth. Contour lines are in microcalories per square centimeter per second. Lines are dashed over regions of little or no data (Lee and Uyeda, 1955).

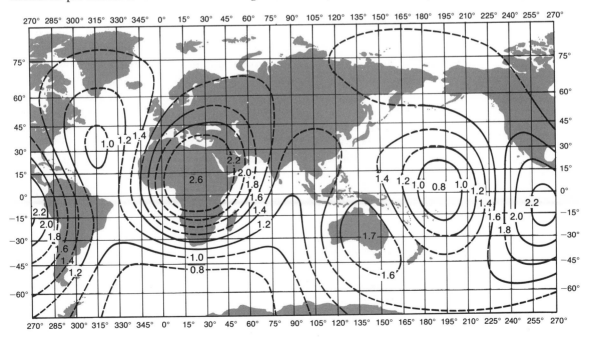

beneath the oceans and is depleted beneath the continents. This relationship has important consequences, as yet incompletely assessed, concerning the evolution of the crust from the mantle and bears on the problem of continental drift. If the continents are moving *on* the mantle, why is the mantle depleted in radioactive elements immediately below them? One would assume that it would take a very long time for the required depletion by differentiation, in which case any continental movement would require simultaneous movement of subjacent mantle.

CRUST-MANTLE RELATIONSHIPS AND PRESSURE-TEMPERATURE CONDITIONS

In addition to the apparent inhomogeneities of the upper mantle in terms of radioactive elements, the explanation of the nature of the crust-mantle contact itself, the Mohorovičič discontinuity, is a major problem. In some places the discontinuity is characterized by an abrupt change in compressional earthquake-wave velocities from about 6.5 km/sec to about 8.0 km/sec within a vertical distance of several kilometers. Elsewhere the velocity change is not so abrupt, and, in fact, near some continental margins there is no break in the velocity gradient. Beneath the oceans the velocities are consistent with a thin layer of sediments (~1 km), an even thinner layer (0.2 km) of rock with properties like those of a granite or rhyolite, and about 5 km consistent with the properties of basalt. In the continental crust there is much uncertainty in generalizing the structure from earthquake evidence. The upper few kilometers are consistent with the behavior expected from average igneous rocks (granodiorite-like material). The range of velocities in the lower part of the crust is considerable and suggests complex relations involving chemical and physical changes in the crustal material. Wave speeds, as indicated in Chapter 1, are functions of density and incompressibility, so that changes with depth can result from a chemical change of material or simply from a physical change in a given material. Figure 2.5 is adapted from Poldervaart's concept (1955) of the nature of the crust in various regions.

The interpretation of the nature of continental crust is based on deduction of pressure and temperature conditions, followed by testing of known rocks to see if their properties fit the required earthquake-wave velocities, the densities deduced from isostatic adjustment, heat flow, and other parameters.

In most of the interpretations of the crust, the rock densities range from about 2.5 g/cm³ for sedimentary rocks to 2.7 g/cm³ for granitic rocks, and up to 3.2 g/cm³ for basalts, with an average density for a typical continental block of about 3.0 g/cm³. Consequently, the pressure at the bottom of the continental crust is that of a 40-km rock column. Each square centime-

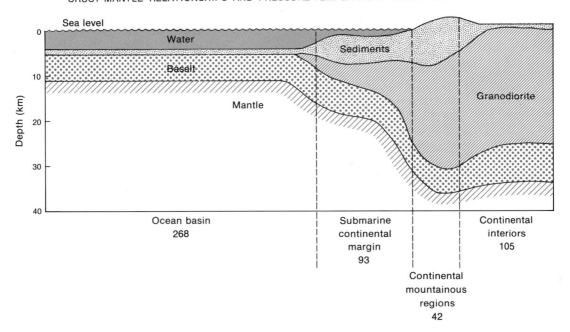

Figure 2.5 Schematic cross section of the earth's crust. The approximate areas of crust beneath the four major segments of the earth's surface are given in units of 10^6 km^2 (after Poldervaart, 1955).

ter of the mantle is supporting 4×10^6 cm^3 of rock with a weight of 12.0×10^3 kg, or a pressure of 12,000 kg/cm^2.[1]

Beneath the oceans, the Moho lies under about 4 km of sea water, 1 km of sediment, and 5–6 km of basalt. Assuming sea water, sediment, and basalt densities of 1.03, 2.5, and 3.2 g/cm^3, respectively, the pressure at the base of the oceanic crust is about 2.6 kbars. Figure 2.6 shows calculated temperature-pressure-depth curves below continental and oceanic crust. One general requirement of crustal material is that it does not melt at depth; the transmission of shear waves through the crust proves the rocks are solid. This requirement is not very stringent in a dry system; almost all minerals melt at P-T conditions that are higher than those encountered in the crust and upper mantle. However, the presence of H_2O lowers melting points drastically, as shown by the positions of the melting points for albite feldspar in Figure 2.6. Our lack of knowledge concerning the "wetness" of rocks at depth in the earth prevents us from drawing melting curves as a function of pressure and temperature alone. It may be that regions of relatively high H_2O content in the mantle are sites of partial melting and generation of lavas.

A more useful condition to be fulfilled is possession of proper earth-

---

[1] The units most commonly used to express pressures are bars or kilobars (1000 bars). One bar equals 1.0197 kg/cm^2 = 0.969 atm; for most purposes these units can be considered interchangeably.

quake-wave velocities. Figure 2.7 shows compressional-wave velocities as a function of pressure and temperature for granite and basalt. This figure shows that the wave speeds of granite and basalt permit discrimination between these two chief types of materials reported in the crust. Also, the spread of continental wave-velocity values, indicated by the ×'s, shows that the materials of the continents are heterogeneous.

The big jump in speeds at the Moho has been explained by one school of thought as being the result of polymorphism of the same minerals that exist in the lower crust. It is well known, for example, that quartz changes with pressure and temperature to a whole series of denser forms. Also, many minerals can be expected to react under deep crustal conditions either to make new species by reconstituting into new phases or by reacting with neighboring minerals to give new products. The most widespread concept is that the lowermost part of the crust is basalt, made up of olivine, pyroxene, and plagioclase feldspar, and that the basalt changes to a garnet-pyroxene rock at the Moho, keeping its original chemical composition. This rock (eclogite) should be both denser and more rigid than the parent basalt. Small bodies of eclogite have been found at the earth's sur-

Figure 2.6 Calculated temperature-pressure-depth curves beneath continental and oceanic crust. The pressure scale is based on an average density of rock equal to 3 g/cm³ (after Clark and Ringwood, 1964).

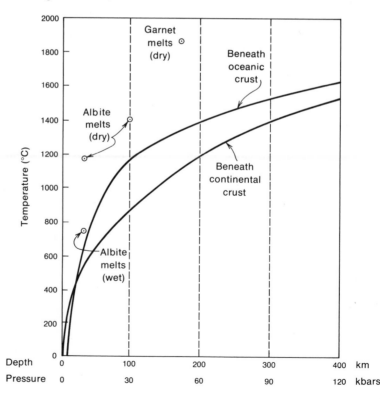

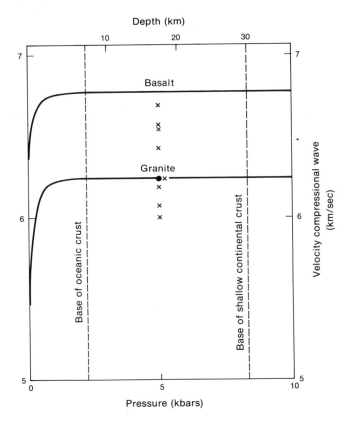

Figure 2.7 Calculated curves for compressional-wave velocities in the crust. Measured velocities in continental crust are on the order of 6 km/sec; in the oceanic crust, on the order of 7 km/sec. The ×'s are actual values chosen to show the range of continental crustal velocities. The solid circle shows the average of a large number of measurements from several continents (adapted from Press, 1966).

face, and their properties match those required for the upper mantle quite well. Eclogites have densities of about 3.4 g/cm³ as opposed to 2.9–3.2 g/cm³ for basalt, and compressional waves travel at about 8 km/sec through them. In essence, the change from basalt to eclogite involves the recrystallization of basalt, with the elimination of lower-density/lower-velocity feldspar and the formation of higher-density/higher-incompressibility/higher-velocity garnet. Figure 2.8 shows the results of laboratory work on the transformation of basalt to eclogite. The P-T conditions required are appropriate for those expected at the base of the continental crust, although the depth at the base of the oceanic crust would seem to be too low for the transformation.

The proponents of a distinct change in chemical composition at the Moho, who engage in heated controversy with the isochemical group, really are not so far removed from them. One of the favorite mixtures for upper-mantle composition is called pyrolite, a rock composed of three

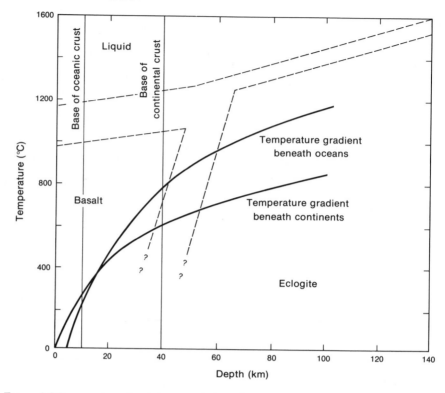

Figure 2.8 Temperature-depth (pressure) phase diagram of basalt and eclogite (Yoder and Tilley, 1962). Estimated temperature gradients beneath oceans and continents are also shown. Note that the temperature at the base of the continental crust is apparently suffi-cient for transformation of basalt to eclogite.

parts peridotite (pyroxene and olivine) and one part basalt. Such a mixture would be higher in FeO and MgO than basalt and lower in SiO_2, CaO, and Al_2O_3. The important point is that there is some agreement on the general chemical nature of the upper-mantle material. The principle at stake is whether the Moho appears automatically, so to speak, with the in-creasing P and T of depth, or whether the crust and mantle are chemically disconnected there.

When the Moho was originally discovered, the break was thought to be a major chemical and mechanical discontinuity. Seismic and heat-flow data obtained during the past decade, however, suggest that the nature of the contact is more complicated; chemical differences between the lower crust and the upper mantle may be minor and the contact may not represent a zone of mechanical decoupling.

UPPER CRUST

We are all familiar with the irregular cracked surface of the Earth, under-lain by the debris of billions of years. Here is the zone of intersection of the atmosphere with the rocks, where rain falls and accumulates into rivers, or sinks into the ground and percolates through pores and fractures. Rocks are broken up and form soils, which are eventually swept into the sea; other rocks rise to take their places, with the same eventual result.

The conditions of the shallow crust are hard to describe, because they are so varied and the rock units so discontinuous. But let us look at some of the processes that transfer materials within the upper few thousands of meters of the solid earth. In Chapters 4, 5, and 6, surface processes will be considered separately.

Porosity

Below a depth of 5 or 6 km, the pressure exerted by the overlying rock column is so great that open spaces in rocks are eliminated and porosity is only a small fraction of 1 percent of the rock volume. Transportation of liquids or gases in important quantities can take place only with the development of forces sufficient to break the rocks, permitting movement along fractures, or by ionic or molecular diffusion in microscopically thin films along mineral-grain boundaries.

In the upper few kilometers of the crust confining pressures are less, and most rocks have the strength to maintain openings. The maximum size of these openings diminishes from huge caves near the surface to grain boundary films at depth.

Igneous rocks almost never have significant intergranular porosity. In many fine-grained extrusive rocks the grains are typically so tightly intergrown that the rock density is almost exactly that of the individual minerals. Intrusive igneous rocks are almost as tightly put together; a granite may have 0.1 percent pore space, most of which occurs where the intergrowth of the original grains left small irregular cavities. Consequently transport of water, which is the chief agent of material transfer in the shallow crust, takes place in igneous rocks through major cracks and fractures. The degree to which igneous rocks are fractured near the surface is extremely variable. In some places closely spaced fractures subdivide the rock into fragments a few centimeters in diameter; elsewhere the fracture spacing may leave great blocks many meters across. When the rock is heavily fractured, water can pass through it easily and rapidly, but even in such rocks the percentage of open space is small so that they do not contain much water per unit of rock volume.

On the other hand, the sedimentary rocks, which make up 10–20 percent of the entire crust and underlie about 80 percent of the continental surfaces, range from materials with no significant intergranular porosity to those with 30–40 percent pore space. The clastic sedimentary rocks—those deposited mechanically as grains—contain from 30 to 90 percent by volume of water at the time of deposition; as they are compacted by the weight of subsequent deposits, the water is squeezed out until the grains make sufficient contact with each other to form a load-supporting structural framework. In the absence of any later cementing material, sand-

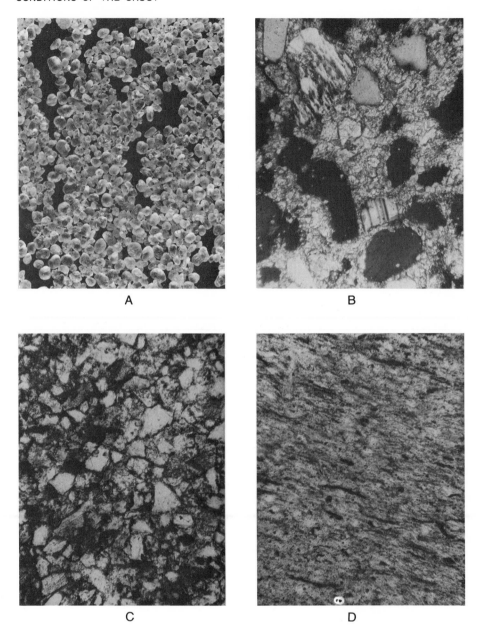

Figure 2.9 Photomicrographs of some sandstones and shales. A, Sand, uncemented, quartz-rich sediment exhibiting good porosity and permeability. B, Sandstone, quartz, and feldspar fragments in a fine-grained chlorite and quartz matrix—poor to moderate porosity and permeability. C, Sandstone, fine-grained, quartz grains embedded in hematite-quartz cement—poor porosity and permeability. D, Shale, very fine-grained clay minerals; black areas are concentrations of carbonaceous material—moderate porosity and poor permeability. (Courtesy of E. C. Dapples, Department of Geological Sciences, Northwestern University, Evanston, Illinois.)

stones and shales average about 30 percent porosity. Average actual porosities are about 15 percent for sandstones and shales, but differences between sandstones and shales, or from one part of a rock to another, are large. Figure 2.9 shows photomicrographs of thin sections of sandstones and shales and serves to illustrate the textural complexity of these rocks.

The typical history of a sedimentary rock comprises its original deposition and compaction in sea water; burial to depths ranging up to several kilometers; uplift above the sea, with little or great distortion and fracturing; and eventual exposure at the surface of the earth. During this history the original pore water may be in part retained, apparently with selective loss of water and salts with increasing compaction, or it may be completely displaced by fresh waters percolating downward from the surface or by waters moving from adjacent rocks. The original porosity may be increased by solution of grains by moving waters, or it may be eliminated by precipitation and cementation from such waters.

At any rate, rain falling on the earth's surface tends eventually to reach the sea. Part of it is carried in streams; part of it sinks into the ground and circulates downward wherever it can, its movement dominated by those rocks through which it can move easily and continuously. Figure 2.10 shows an idealized situation involving an upland, a stream, and a valley in which the subsurface material is uniformly porous and in which the force causing water movement is the height of the zone of saturation under the upland above that in the valley. The figure can be used to indicate the sequence of events due to the pulsating drive of percolating rain followed by

Figure 2.10 Movement of underground water from the surface to a stream. Dashed lines show the path traveled through homogeneous materials by a water particle. The path is the resultant of the force of gravity (F_g) and the "pulling" force (F_s) of the stream. Only the force of gravity is operating in the unsaturated zone (Garrels, 1951).

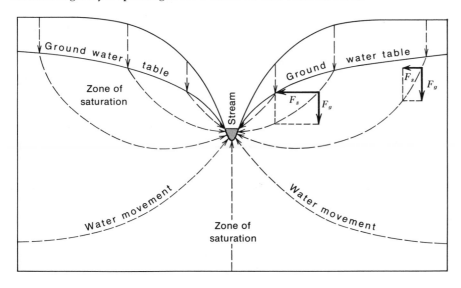

dry weather. The pore spaces of the rocks tend to fill up to the ground surface during the rain, when addition is faster than flow through the rock; this increase in potential energy beneath the upland tends to disappear between rains, causing continuous flow. The surface of saturation (water table) is clearly a function of the amount and frequency of rainfall and of the water-carrying characteristics of the rocks. In arid areas, evaporation may exceed rainfall, and no zone of saturation may exist; in areas of heavy continuous rain the saturation surface remains at the topographic surface. As a generalization, the zone of saturation in the United States, exclusive of the desert areas, is encountered at an average depth of 6–12 m and appears as a subdued image of the topography.

The most effective water-carriers are the sandstones. The pore spaces are relatively abundant and large, which makes it easy for water to pass through them, and sandstones commonly occur as continuous thin sheets over many thousands or tens of thousands of square kilometers.

Permeability

A measure of the ease of transmission of water through a rock is called its permeability. Movement of water is driven by a pressure gradient, commonly gravitational, and is retarded by friction with the grains of the rocks. The permeability can be tested by determining the rate at which the water will flow through a given cross-sectional area of a rock sample of a given length. The equation is analogous to that for heat flow (p. 41),

$$q/t = \frac{K(P_2 - P_1)A}{d},$$

where q is the quantity of water delivered in cubic centimeters, t is time in seconds, $P_2 - P_1$ is the pressure drop across the sample in atmospheres, A is its cross-sectional area in square centimeters, d is its length in centimeters, and K is the coefficient of permeability. A permeability of 1 darcy is possessed by a rock that transmits 1 cm^3 of water in 1 sec through an area of 1 cm^2 down a length of 1 cm under a pressure gradient of 1 atm. Few rocks have permeabilities as high as 1 darcy; the mdarcy (darcy/1000) is in common use. Table 2.2 lists permeabilities for some sedimentary rock types. The differences in permeability of several orders of magnitude between sandstones and shales is the major reason sandstones are the dominant carriers of underground water.

Figure 2.11 illustrates a situation that occurs in the Great Plains of the United States and is fairly typical of what obtains in areas underlain by slightly inclined sedimentary rocks. Rain infiltrates the sandstone in its catchment area, and the underground waters (ground water) move downward through the sandstone beneath an impervious layer. Where a drill hole pierces the capping rock, the pressure in the sandstone is released and the water flows upward. If the permeability of the sandstone were 10 mdarcies, the head of water 500 m, and the distance from catchment area to dis-

Table 2.2
Average Permeabilities of Some Sedimentary Rocks[a]

Sedimentary unit and age	Average permeability (mdarcies)
Woodbine sandstone (Cretaceous)	1500
Leduc limestone (Devonian)	800
Smackover limestone (Jurassic)	737
Stevens sandstone (Miocene)	140
Bradford sandstone (Devonian)	50
Spraberry silty shale (Devonian)	0.5
Asmari limestone (Tertiary)	<0.5
Shale (Cretaceous)	0.004
Slate (Precambrian)	0.0013
Chert (Precambrian)	0.00019

[a] Data from Levorsen (1956) and Davis and DeWiest (1966).

charge 200 km, the flow per year per square centimeter would be (100 m $H_2O = 10$ atm):

$$q/3.15 \times 10^8 = \frac{(0.01)(50)(1)}{2 \times 10^7}$$

$$q = 7.9 \text{ cm}^3 \text{ H}_2\text{O}.$$

In other words, the water would be expected to move only 8 cm during a year. Such a rate is not atypical; ground-water flow rates of several centimeters to a few tens of meters per year are usual, and rates as high as 1000 m/year or more are attained in unconsolidated gravelly sediments. Even a slow rate of movement of 8 cm/year becomes significant on a geologic time

Figure 2.11 Water movement in slightly dipping heterogeneous sedimentary rocks. Rain infiltrates the pervious layer where it intersects the surface. The water filters into the pervious rock beneath the impervious layer, and if a drill hole pierces the capping rock and penetrates the water-saturated layer, the pressure is released and the water flows upward (Garrels, 1951).

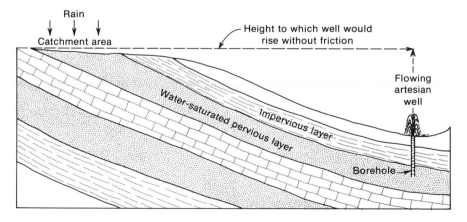

scale. A liter of water would be transmitted through each square centimeter of rock in 125 years, and 8000 liters in a million years. Ground waters in rocks are highly variable in their chemical composition, but about 0.1 g/liter of dissolved solids in sandstone waters is not uncommon, and therefore about 800 g of dissolved material could be transported through each square centimeter in a million years.

In the history of sedimentary rocks, the tendency is to displace the original sea water with rain water, but in a highly irregular way, controlled by the permeability of the rock, its geometry, its relation to topography and to adjacent rocks, the climatic conditions of the region, and a host of other variables. But for any given mass of rocks, it is apparent that sandstones will be the chief water-carriers, along with some of the limestones. Shales will transmit little if any water, even over geologic periods. The older a rock, the more chance it has had to be flushed out, with accompanying effects of solution and alteration of mineral grains, as well as cementation and pore filling.

About 0.32×10^{20} g, or 32 trillion metric tons, of water flow down the rivers into the oceans each year. All this water has at least a short period of contact with surface materials, and perhaps 25–35 percent percolates through these materials before draining into streams. Most of the percolating water moves close to the surface; with depth the amount diminishes rapidly. The total amount of water in the pores of rocks is estimated to be as much as 20–25 percent of that in the oceans; some of the molecules have not seen the surface of the earth since the day of their entrapment; others have been residents of many different rocks at many different times.

In addition to a gravity drive of ground water, for which energy is derived ultimately from the sun, water circulation is driven by the earth's internal energy, where variations in the temperature gradient set up pressure differences. The geysers and hot springs of Iceland and of the Yellowstone National Park are extreme examples of heat-driven circulation. There are many other areas, without obvious surface expression, where a steepened thermal gradient drives water upward through the rocks.

Diffusion

Bulk flow of water with its contained solutes is the chief medium of material transfer in the upper crust. Differential movement of water and solutes by ionic and molecular diffusion is of secondary importance in moving large quantities of materials long distances but is one of the dominant factors in controlling reactions between waters and rocks. For example, if an aqueous solution is moving through a fractured rock and the solution is not in equilibrium with the rock, the rock usually tends to be altered at the solution-rock interface. The alteration process involves transport of ions from the solution through the altered material to react with fresh rock and back-transport of the soluble reaction products into the external solution. Figure 2.12 illustrates the process schematically.

Fissure through which a solution of
constant composition is flowing

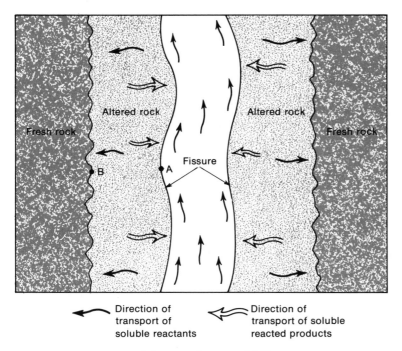

Figure 2.12 Schematic diagram of diffusion process.

The simplest situation occurs when a flowing solution maintains a constant composition at the outer boundary of alteration, and the alteration reaction fixes the solution composition at the inner boundary with fresh rock, as at A and B in Figure 2.12. In this situation a fixed concentration difference is maintained, and the diffusion of a given ion is described by the equation analogous to those for heat flow and water flow:

$$q/t = \frac{D(C_2 - C_1)A}{d},$$

where if q is expressed in equivalents of ions, t is in days, D is the diffusion coefficient of the species in question in square centimeters per day, $C_2 - C_1$ is the concentration difference in equivalents per cubic centimeter, A is the effective cross-sectional area in square centimeters, and d is the distance in centimeters over which diffusion is occurring. The effective cross-sectional area is that fraction of the actual cross-sectional area that behaves as if it were a straight solution-filled pore. In most rocks it is about two thirds of the total porosity. Table 2.3 lists diffusion coefficients for a number of aqueous species.

As an example to provide some feeling for the time required for mate-

Table 2.3
Diffusion Coefficients for Some Individual Ions in Solutions of Infinite Dilution at 25°C and in Which Flow Rate Is in Terms of Equivalents of Ions Per Day[a]

Ion	D (cm^2/day)
H^+	8.05
K^+	1.69
Na^+	1.15
Ca^{2+}	0.68
Sr^{2+}	0.68
Fe^{2+}	0.63
Mg^{2+}	0.61
OH^-	4.60
SO_4^{2-}	1.79
Cl^-	1.75

[a] After Garrels, Dreyer, and Howland (1949).

rial transport by diffusion, consider the alteration of a large feldspar grain in a sandstone to kaolinite. The reaction involved is

$$H_2O + 2KAlSi_3O_8 + 2H^+ = Al_2Si_2O_5(OH)_4 + 4SiO_2 + 2K^+.$$

feldspar   kaolinite

An example of this reaction is the near-surface alteration of some sedimentary rocks. In Wyoming, in sedimentary rocks that are made up chiefly of granitic rock fragments, there are cubic kilometers of the rock in which all the original feldspar grains have been kaolinized by ground water but retain their original grain form. The kaolinized grains are porous, because of the removal of potassium and silica in solution. The time required for alteration of half of a 1-cm feldspar grain at room temperature is about 10 years and represents the transport of about 0.6 g of potassium from the feldspar grain into the pore-water solution.

In some places, where water has moved through fractures in rocks, diffusion-controlled alteration extends for a few feet to a few tens of feet outward from the fracture and may represent tens of thousands or hundreds of thousands of years of solution flow in the fracture during which diffusion of ions into and out of the wall rock was taking place.

P-T Gradients in the Shallow Crust

In the upper few kilometers of the crust, as was pointed out, rocks have the strength to maintain openings, and the rocks and pore waters can be considered two separate interpenetrating systems. Pressures in the water-filled pores usually approach those of a water column, whereas those of the rocks are usually higher by a factor proportional to the ratio of rock density to water density.

A 100-m column of water exerts a pressure of about 10 bars; consequently the hydrostatic pressure increases about 100 bars/km of depth. The pressure of the rock column depends on the density of the rocks at a particular place. Sedimentary rocks have densities of the order of 2.5 g/cm³; metamorphic and high-silica igneous rocks, such as granite and granodiorite, have densities of about 2.7 g/cm³; and for low-silica igneous rocks, such as basalts, densities are a little over 3 g/cm³. Therefore the lithostatic pressure increases by about 250–300 bars/km. Figure 2.13 shows hydrostatic and lithostatic pressure-temperature-depth relations for various

Figure 2.13 Relations of hydrostatic and lithostatic pressures, temperature, and depth for various thermal gradients. The density (ρ) of H₂O as a function of temperature and pressure and the vapor pressure curve of H₂O are also shown. C.P. is the critical point (after Goguel, 1953).

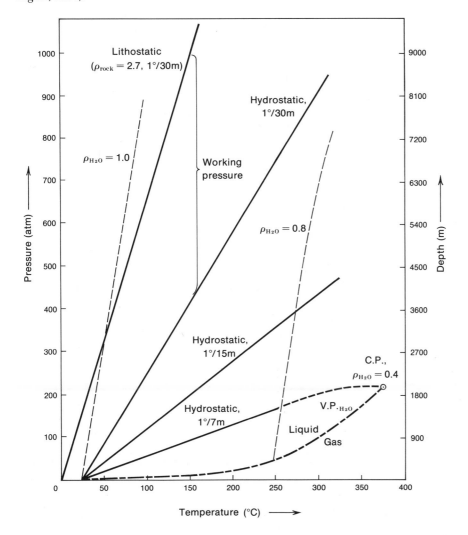

thermal gradients. The boiling curve for water is plotted as well and shows that for shallow depths, water remains liquid under almost all thermal gradients likely to be encountered and that production of steam or the presence of geysers requires somewhat unusual conditions. Furthermore, down to depths of several kilometers, the density of liquid water lies between 0.8 and 1.0 g/cm³.

The pressure difference, for a given temperature, between the typical lithostatic and hydrostatic pressure curves, is labeled "working pressure" on the diagram. This difference in pressure is the maximum that can be achieved underground as $P_2 - P_1$ in the permeability equation (p. 52). In other words, if the pore-water pressure exceeds lithostatic pressures, the overlying rock column would be lifted and fractured. Note that this working pressure increases with depth, so that the maximum heads of water pressure for driving water through rocks increase linearly with depth.

Although the usual situation encountered in the shallow earth is the expected hydrostatic pressure owing to the weight of the water column, water pressures approaching the lithostatic head are observed in a few places. This leads to a situation in which the weight of the overlying rock column is almost entirely supported, and lateral sliding of large masses of rock under the influence of very small forces may occur at depth. The presence of nearly horizontal, thick, plastic, impermeable shale layers is probably necessary for high water pressures to obtain; in other types of rocks the water would escape through fractures.

SUMMARY

The crust of the earth forms a thin rind of silicon-, aluminum-, potassium-, and sodium-rich rocks overlying the mantle. Igneous rocks, such as basalts and granites, probably compose 60–80 percent of the crust; the rest consists of sedimentary and metamorphic rocks. Our ideas concerning the nature of the crust at depth are derived from a meager number of direct observations of materials collected from bore holes and indirectly from deduction of *P-T* conditions, followed by testing of rock materials to see if their properties fit the necessary earthquake-wave velocities, heat flow properties, densities, and other parameters.

The reason for the sharp increase in earthquake-wave velocities at the crust-mantle contact, the Mohorovičič discontinuity, is a subject of hot debate even today. Two concepts of the nature of the Moho in vogue now are (1) the crust-mantle contact represents a phase change at which the basaltic rock of the lower crust changes to eclogite at the Moho, and (2) the contact represents a chemical change at which basalt gives way to a rock richer in iron and magnesium at the Moho; pyrolite (three parts peridotite, one part basalt) is a possible mixture for the upper-mantle material. The important point is that whatever the nature of the crust-mantle contact, the Moho does not appear to separate two distinct systems; there probably is considerable coupling between the crust and the mantle.

The crust is an active region of the earth; surface rocks are broken down and the debris is carried to the oceans by streams where it accumulates as thick wedges of sedimentary rock in slowly sinking basins. New rocks rise to replace the destroyed ones. Materials are transported long distances through the shallow crust by pressure-induced bulk flow of fluids through the pores of permeable sedimentary rocks. Differential movement of water and solutes by diffusion is important over distances of a few millimeters to meters. The water content of rocks generally diminishes with depth as temperature and pressure rise. In the shallow crust, waters in pores and fractures form a nearly continuous hydraulic system, and pressures are usually close to hydrostatic. Temperature rise in the shallow continental crust averages about $30°C/km$; under these conditions water is a liquid with a density near 1 g/cm^3.

In this chapter we have tried to describe the crust and to indicate controls of some of the fundamental processes that are continuously operating to change it. Also, we have presented a few of the many problems of current interest. We shall return to these problems, and others, in Chapter 12 but not until we have looked in detail at sedimentary rocks.

REFERENCES

Clark, S. P., Jr., 1966, Composition of rocks: in Handbook of physical constants, S. P. Clark, Jr., ed., *Geol. Soc. Am. Mem.*, 97, 1–5.

Clark, S. P., and Ringwood, A. E., 1964, Density distribution and constitution of the mantle: *Rev. geophys.*, 2, 35–88.

Davis, S. M., and DeWiest, R. J. M., 1966, *Hydrogeology:* John Wiley & Sons, Inc., New York.

Garrels, R. M., 1951, *A Textbook of Geology:* Harper & Row, New York.

Garrels, R. M., Dreyer, R. M., and Howland, A. L., 1949, Diffusion of ions through intergranular spaces in water-saturated rocks: *Bull. Geol. Soc. Am.*, 60, 1800–1828.

Goguel, J., 1953, Le regime thermique de l'eau souterraine: *Ann. Mines, 10,* 1–31.

Lee, W. H. K., and Uyeda, S., 1965, Review of heat flow data: in *Terrestrial Heat Flow*, W. H. K. Lee, ed.: American Geophysical Union, Washington, D.C., *Geophysical Monograph Series* No. 8, 87–190.

Levorsen, A. I., 1954, *Geology of Petroleum:* W. H. Freeman and Company, San Francisco.

MacDonald, G. J. F., 1964, Dependence of the surface heat flow on the radioactivity of the earth: *J. Geophys. Res.,* 69, 2933–2946.

Poldervaart, A., 1955, Chemistry of the earth's crust: in Crust of the earth, A. Poldervaart, ed., *Geol. Soc. Am. Spec. Papers*, 62, 119–144.

Press, F., 1966, Seismic velocities: in Handbook of physical constants, S. P. Clark, Jr., ed., *Geol. Soc. Am. Mem.*, 97, 195–218.

Yoder, H. S., Jr., and Tilley, C. E., 1962, Origin of basalt magmas: An experimental study of natural and synthetic rock systems: *J. Petrol.*, 3, 346–532.

3 | Isotopes and the Geologic Time Scale

In this chapter we shall review briefly the history of the development of the geologic time scale and shall also discuss stable and radioactive isotopes. As we shall see, the discovery of radioactivity provided a way of assigning actual years for the duration of eras and periods for which only relative time spans could previously be assigned and also provided a new tool for the correlation of rocks and events in distant parts of the earth. In addition, the discovery focused attention on the existence of isotopes of the elements, some of which are radioactive and thus useful in age dating but most of which are not. It is convenient, while emphasizing the uses of radioactive isotopes in age dating to discuss other aspects of isotopes that are important in deciphering the earth's history, such as the role of radioactive isotopes in generating heat within the earth or that of stable isotopes in the measurement of paleotemperatures.

We are just beginning to exploit the possibilities of the uses of isotopes in geologic studies, for they have opened up a realm of chemistry comparable in extent to all of classic chemistry. In the ensuing chapters we shall need to refer to isotopes frequently; here we shall provide a minimum discussion and review as background for that use.

EARTH HISTORY

Time is determined by a sequence of events. Man does not find it difficult to place himself within the temporal framework of his own existence. However, to understand ancient events is somewhat more strenuous, owing to the long intervals of time involved and the fact that these "happenings" are not directly observable. The geologist interprets ancient events by observing the effects that they have had on natural materials preserved in sequences of rocks. To understand the earth's history, it is necessary to have some sequential framework in which to place our observations; that is, the relative or absolute ages of the rock units must be known.

Prior to the discovery of radioactivity, only relative ages of rock units could be determined. For example, an igneous intrusive is certainly younger than the rocks it injects; younger sedimentary rocks lie above older rocks unless there has been a major disturbance of the rock sequence. Now ancient events as recorded in the rock record can be assigned ages in years. With a time scale and the assumption that natural laws are invariant, the geologist is armed with the tools necessary to interpret the earth's history.

In the preceding chapters, the age of the earth has been given as approximately 5.0 billion years. This figure is generally accepted by earth scientists, and a questionnaire circulated to them would probably reveal that only a few expect the estimate to change drastically in the future. Whether this faith is justified is an interesting question; it certainly appears highly suspect if we look at the present value in historical perspective.

If we were to plot estimates of the earth's age versus the date on which the estimate was made, we would find that every generation of students has been given a greater age for the earth than was conveyed to the preceding generation. We would have to conclude that our children will be taught that the earth is more than 10 billion years old. Why, then, is there confidence in today's estimate of the age of the earth?

Historical Background

For most of historic time, the age of the earth has been determined by religious beliefs or philosophic considerations. In many cultures the age has been considered to be infinite. According to fundamentalist Christian doctrine, the earth is very young; one of the most celebrated of the specific estimates is that of Archbishop Ussher of Ireland, who, in 1658, after careful study of the *Old Testament,* pronounced that the earth was created at 9 A.M. on October 23, 4004 B.C.!

During the Renaissance there were individuals (e.g., Leonardo da Vinci) who studied rocks and fossils and became convinced that the earth must be very old—at least relative to a few thousand years.

Even as late as the 1820's, speculation concerning the age of the earth as deduced from the rock record was tied closely to religious and philosophic beliefs. For example, one of the prevailing geologic doctrines at this time and in the late eighteenth century was that *all* rocks, including those we now call igneous, were deposited from an ancient ocean that once covered the whole earth and upon recession of the sea were left in their present state. This idea is called the "Neptunist" concept and was championed by Abraham Werner of Freiburg, Saxony. Of course, because of the Biblical record of the flood, the concept had substantial religious appeal, which probably accounted for its widespread acceptance.

The Neptunist philosophy met a slow death because of its compatibility with theologic concepts; however, it soon became apparent to geologists of the early nineteenth century that the doctrine could not stand up to the pressure of field observations of the rock record. For example, basalt rocks were shown to have crystallized directly from a melt. In the late eighteenth century, James Hutton of Scotland, often called the Father of Geology, developed the concept of Uniformitarianism. He saw that streams and waves deposit sand and mud and recognized that the sandstones and shales in the hills are simply hardened and uplifted deposits of ancient streams or waves. Everything he saw in the record of the rocks seemed to him to be explicable on the basis of processes like those of the present. Furthermore, his studies indicated an earth history consisting of an endless series of cycles of deposition, uplift, erosion, and deposition. He saw no evidence for a beginning to the history of the earth and no suggestion of an end. His Uniformitarian concept that "the present is the key to the past," as envisioned today, refers to the time invariance of natural physical laws and is simply an extension of general scientific methodology to interpretation of the rock record. But, nevertheless, Hutton's ideas, which were popularized by John Playfair and extended by Charles Lyell, necessitated an earth age of much more than a few thousand, or hundred thousand, years.

In the second half of the nineteenth century, Lord Kelvin, a renowned Scottish physicist, attempted to calculate an age for the earth on the basis of its thermal history. Radioactivity was unknown at the time, so he calculated the time required for the earth to cool from an original extremely high temperature to its present state and considered the time required as a maximum for the earth. He obtained a "best estimate" of the order of 100 million years. Even this estimate was at variance with observations of the rock record and, perhaps even more significantly, did not provide the geologic time required by Charles Darwin's concepts of organic evolution. The origin of species through natural selection as envisioned by Darwin necessitated a much older age of the earth.

Thus, as geologic studies progressed rapidly in the late nineteenth century, many investigators became dissatisfied with Kelvin's estimate as being too short to account for the numerous cycles of erosion and deposition they could read from the rocks, as well as seeming too short to permit the observed evolutionary changes of organisms, but there was no way for them to mount a convincing quantitative argument.

The discovery of radioactivity by Becquerel in 1896 produced almost immediate repercussions on the age problem; for one thing, Kelvin's estimate necessarily was much too short, for the heat produced by the decay of radioactive minerals had not been considered in his calculations. Not only did the discovery of radioactivity upset previous age estimates, it supplied a method for determining the ages of minerals based on the degree of disintegration of radioactive elements contained in them.

THE GEOLOGIC TIME SCALE

After Hutton had laid the foundations for studies of the rock cycle, geologists began to study the sequences of sedimentary rocks. They found in many instances that fossils, particularly those of marine organisms, were guides to the relative ages of the layers. Darwin's theory of evolution, coupled with modern concepts of genetics and individual variation, forms a basis for determining the relative time equivalence of sedimentary strata deposited on different continents.

Because organic evolution is accompanied by the appearance of new plants and animals and the disappearance of others, certain organisms or groups of organisms were found to have existed only over short time intervals. If they had a wide geographic distribution, they could be used to determine that sedimentary layers were deposited during the same short time interval, even though the layers occurred on different continents. Correlation—determination of the equivalence of sedimentary strata—can also be accomplished by comparing the physical attributes of stratigraphic units in two, or more, different regions. Distribution of organisms in time and space also is controlled by their environment, but judicious selection of organisms with widespread environmental and geographical distributions and short evolutionary histories has enabled geologists to correlate strata from widely separated regions.

Geologists took advantage of the obvious principle that younger sedimentary rocks must lie upon older ones and so could assign relative ages to layers without the help of fossils. Erosional breaks, or unconformities, between sedimentary sequences also aid in correlation. Unconformities range from local and minor breaks in the rock record, shown perhaps only by a slight discontinuity in the fossil record and slight erosion of the older sequence, to major breaks, perhaps in some instances worldwide, in which the older rocks may have been folded and metamorphosed, intruded by masses of igneous rock, uplifted several kilometers, and eroded severely. These great angular unconformities commonly reveal their presence by tremendous biological changes between the rocks below and those above, indicating a time break of millions of years.

It was thought for a long time that many of these great unconformities were indeed worldwide and represented a time of high continents, with the sediments being swept into the deep sea and lost. Today this degree of simultaneity seems unlikely. At any rate, the four eras of geologic time are, from youngest to oldest, the Cenozoic, Mesozoic, Paleozoic, and the Precambrian or Cryptozoic. The Cenozoic, Mesozoic, and Paleozoic comprise the Phanerozoic eon.

The eras are subdivided in turn into periods, based on widespread angular unconformities of lesser extent than those separating eras. However, in some regions they are records of major crustal disturbance—the fossil record of mountains—and have been satisfactory markers. The most recent two periods have been further subdivided into epochs.

Table 3.1 gives the time scale with the various eras and periods. The subdivisions start with the beginning of the Paleozoic and include only the most recent 600 million years of a 4.5-billion-year history—about 14 percent. Only with the determination of many age dates by radioactivity methods can the Precambrian record be worked out. Fossils are almost absent, and although there have been some recent finds of soft-bodied forms in very old rocks, the high degree of metamorphism and contortion, plus the sporadic occurrence of older rocks, has made correlations from one area to another almost impossible. It begins to appear that there are perhaps four major subdivisions of the Precambrian; a large body of rocks about 1 billion years old, another about 2 billion, a third about 2.7 billion, and a fourth of the order of 3.3 billion. Our information diminishes sharply in an exponential fashion with increasing age of the rocks investigated.

The sedimentary rocks of the past 600 million years are estimated by us to have a mass about equal to the Precambrian mass and in general are not as complexly contorted or metamorphosed; the history of the time interval they represent is more completely preserved in these strata. From the Paleozoic on, a relatively continuous fossil record of the evolutionary history of plants and animals is found in the sedimentary beds. Abundant fossils of shelled organisms first appear in Cambrian rocks. No definite shell-bearing fossils have been found in rocks older than Cambrian. The distinctive evolutionary events of the various geologic periods are shown in Table 3.1.

The great divisions of the time scale were developed early in the study of rocks; in general, rocks of Cenozoic age still tend to have a "new" look about them. Porosity is high and the mineralogy is much like that of modern sediments; they may be poorly cemented, and some fossil shells retain vestiges of their original colors. Precambrian rocks of sedimentary origin, on the other hand, are generally contorted and metamorphosed; fossils, if once present, have been destroyed and porosity is very low.

However, bitter experience has shown that use of the degree of chemical and physical change from original sediments as a criterion of age is an extremely dangerous practice. It is true only as the broadest kind of generalization. In some areas rocks 2 billion years old, by the accident of their geography, are less altered than others that have survived only a few million years.

Hutton's "No vestige of a beginning . . ." still holds in one sense for the results of studies of the earth's surface. Presumably there was a beginning of the cycles of erosion, deposition, and uplift, when the first sediments were formed by weathering of primordial crust, but no one has yet discovered remnants of that crust. The oldest rocks we know are of sedimentary origin.

One of the most remarkable aspects of the rock record is the obvious similarity of erosional and depositional processes throughout the past 3 billion years. If we do our best to eliminate postdepositional changes, it emerges that the compositions, textures, and the structures of the most ancient sedimentary rocks can be matched in detail by comparable features of modern sediments. It is perhaps this aspect of the rock record that has kept the concept of Uniformitarianism as a basic principle for almost 200 years.

RADIOACTIVE ELEMENTS AND AGE DATING

The discovery by Becquerel that uranium disintegrates spontaneously into other elements caused a revolution in the physical sciences. Let us review briefly some of the results of the investigations he triggered that are pertinent to earth problems.

We shall ignore all of the recent advances in particle physics and treat atoms as if they consisted only of a nucleus containing protons of unit mass and unit positive charge and of neutrons of unit mass and no charge, orbited by electrons of unit negative charge and negligible mass. The chemical properties of the elements are controlled by the configuration of the electron cloud around the relatively tiny nucleus. The number of protons equals the number of electrons and is the basis for assignment of atomic number. The atomic weight equals the sum of protons and neutrons. According to this picture, the atoms should all have whole-number atomic weights, and this conclusion is true to a close approximation.

Because chemical properties depend almost entirely on the number of electrons, and hence the number of protons, atoms of a given element might and do exist with identical numbers of protons but with different numbers of neutrons. Such atoms of the same element with different masses are called isotopes.

Table 3.1
The Geologic Time Scale[a]

Periods and epochs	Maximum known thicknesses (km)	Duration (millions of years)	Total from present (millions of years)	Distinctive life
PHANEROZOIC				
CENOZOIC ERA				
Quaternary	1.8	1 or 2	1 or 2	
Recent				Modern man
Pleistocene				Stone-age man
Tertiary				
Pliocene	4.6	9 or 10	12	Great variety of mammals Elephants widespread
Miocene	6.4	13	25	Flowering plants in full development Ancestral dogs and bears
Oligocene	7.9	15	40	Ancestral pigs and apes
Eocene	9.1	20	60 ⎫	Ancestral horses, cattle, and elephants
Paleocene	3.7	10	70 ⎭	appear
MESOZOIC ERA				
Cretaceous	15.5	65	135	Extinction of dinosaurs and ammonites Mammals and flowering plants slowly appear
Jurassic	13.4	45	180	Dinosaurs and ammonites abundant Birds and mammals appear
Triassic	9.1	45	225	Flying reptiles and dinosaurs appear First corals of modern types

[a] Adapted from Holmes (1960).

It was many years after the discovery of the spontaneous disintegration of uranium that its decay scheme was worked out; by chance it turned out to be one of the most complicated of the radioactive elements. In a magnificent series of early experiments, carried on in large part at the Cavendish Laboratories in England under the direction of Lord Rutherford, it was discovered that three kinds of emanations from radioactive materials were taking place: alpha particles, consisting of two protons and two neutrons (the nucleus of a helium atom); beta particles, identified as electrons; and gamma rays, electromagnetic radiation of high frequency and great energy. Later it was found that uranium itself consists of isotopes, one with 92 protons and 146 neutrons and hence an atomic weight of 238 ($_{92}U^{238}$), and another with 92 protons and 143 neutrons and an atomic weight of 235 ($_{92}U^{235}$). The ratio of $_{92}U^{238}$ to $_{92}U^{235}$ in nature is about 138:1. A third isotope, $_{92}U^{234}$, is less than 0.006 percent of all uranium.

Periods and epochs	Maximum known thicknesses (km)	Duration (millions of years)	Total from present (millions of years)	Distinctive life
PALEOZOIC ERA				
Permian	5.8	45	270	Rise of reptiles and amphibians Conifers and beetles appear
Carboniferous				
Pennsylvanian	6.1	35	305	Coal forests
Mississippian	7.9	45	350	First reptiles and winged insects
Devonian	11.6	50	400	First amphibians and ammonites Earliest trees and spiders Rise of fishes
Silurian	10.4	40	440	First spore-bearing land plants Earliest known coral reefs
Ordovician	12.2	60	500	First fish-like vertebrates Trilobites and graptolites abundant
Cambrian	12.2	100	600	Trilobites, graptolites, brachiopods, molluscs, crinoids, radiolaria, foraminifera Abundant fossils first appear
PRECAMBRIAN				
Late Precambrian		1000	2000	Scanty remains of primitive invertebrates: sponges, worms, algae, bacteria
Earlier Precambrian		1500	3500	Rare algae and bacteria back to at least 3000 million years for oldest known traces of life

Table 3.2
Decay of $_{92}U^{238}$

Element	Symbol	Emission	Original name	Half-life
Uranium	$_{92}U^{238}$	Alpha	Uranium I	4.5×10^9 years
Thorium	$_{90}Th^{234}$	Beta	Uranium X_1	24 days
Protactinium	$_{91}Pa^{234}$	Beta	Uranium X_2	1.2 min
Uranium	$_{92}U^{234}$	Alpha	Uranium II	248,000 years
Thorium	$_{90}Th^{230}$	Alpha	Ionium	80,000 years
Radium	$_{88}Ra^{226}$	Alpha	Radium	1,622 years
Radon	$_{86}Rn^{222}$	Alpha	Radon	3.8 days
Polonium	$_{84}Po^{218}$	Alpha	Radium A	3.0 min
Lead	$_{82}Pb^{214}$	Beta	Radium B	26.8 min
Bismuth	$_{83}Bi^{214}$	Beta	Radium C	19.7 min
Polonium	$_{84}Po^{214}$	Alpha	Radium C'	1.6×10^{-4} sec
Lead	$_{82}Pb^{210}$	Beta	Radium D	21 years
Bismuth	$_{83}Bi^{210}$	Beta	Radium E	5.0 days
Polonium	$_{84}Po^{210}$	Alpha	Polonium	138.4 days
Lead	$_{82}Pb^{206}$	—	Lead	Stable

Table 3.2 shows the decay scheme of $_{92}U^{238}$. A brief discussion of the first few steps should suffice to indicate the consecutive processes involved when an individual atom of $_{92}U^{238}$ decays. First an alpha particle is emitted from the nucleus, reducing the mass by 4 units and the nuclear charge by 2 units, producing the nucleus of thorium 234 ($_{90}Th^{234}$). Two electrons also are lost from the outer part of the electron cloud to balance the loss of protons from the nucleus. However, the energy change is so small in this loss, compared to the violent expulsion of the alpha particle from the nucleus, that the process can be ignored in an energy-mass balance. Next, a beta particle (electron) is expelled from the nucleus, and the reaction can be considered to be

$$\text{neutron} = \text{proton} + e^- \uparrow.$$

This process increases the nuclear charge by 1 and produces $_{91}Pa^{234}$, with the same mass as $_{90}Th^{234}$ but a higher atomic number. Then another beta particle is expelled, and $_{92}U^{234}$ results. The final product of the long series of disintegrations and new elements is $_{82}Pb^{206}$.

If we were to crystallize today a block of pure $_{92}U^{238}$, consisting of a very large number of atoms, and then analyze it several million years hence, we would find all the elements of the reaction series in proportions determined by the relative rates of the decay reactions. At the end of a few million years more, the proportions of the various daughter elements would be found to have stayed the same, except for the accumulation of $_{82}Pb^{206}$, the stable end product. For any large number of uranium atoms, the fraction that disintegrates per unit time is constant; that is, the rate of loss of uranium is proportional to the amount present:

$$\frac{dU}{dt} = -k'U \quad \text{or} \quad \frac{dU}{U} = -k'dt,$$

where t is time and k' is the decay constant. Integration of this expression from time zero to time t yields a negative exponential relation:

$$\log \frac{U}{U_0} = -kt,$$

where U_0 and U are the number of uranium atoms present at time zero and time t, respectively.

Thus a plot of the logarithm of the fraction remaining versus time for a particular radioactive element yields a linear plot (Figure 3.1). The time required for decay of half of the original uranium atoms is known as the half-life and is a convenient way to describe the process. The half-life of $_{92}U^{238}$ is 4.51 billion years, by coincidence about the same as the current estimates of earth age. This means that half the $_{92}U^{238}$ atoms present at the beginning of the earth have decayed; half the remainder will decay in the next 4.51 billion years, leaving one quarter the original amount.

Each radioactive element has a characteristic half-life; Figure 3.1 shows several of major interest today.

Age Dating of Minerals

The principle of determining the age of a mineral from the degree of decay of its radioactive elements is simple, but the practice has been extremely difficult. It was recognized a few years after the discovery of radioactivity that if one could find a uranium mineral such as uraninite (UO_2) that would crystallize originally containing only U and O and no Pb that its lead content would be a measure of the time interval since formation. Let us examine some of the conditions that must be fulfilled if a trustworthy age is to be obtained.

First of all, there should be some assurance that the rate of nuclear disintegration has not changed with time. Perhaps the 50 percent or so of the original uranium-238 atoms that remain on the earth today behave somewhat differently from those that already have decayed. Our measurements of decay rates have been made in the past 60 years; this is a brief time span to use to probe 4 billion.

Happily, a test of the constancy of the energy and frequency of nuclear

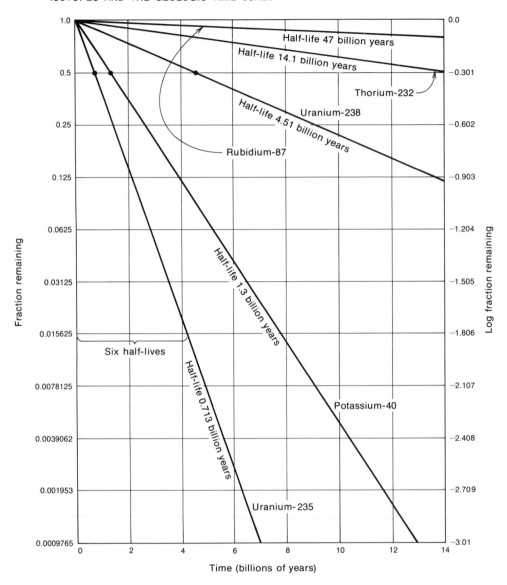

Figure 3.1 Plot of the fraction remaining of a number of radioactive elements as a function of time. Note that the fraction remaining after six half-lives is only about 1.5 percent of the original amount.

disintegration through time actually exists. Long ago mineralogists noticed concentric light and dark rings of microscopic dimensions in mica. Eventually it was shown that these haloes have been caused by tiny grains of uranium minerals included in the mica. Particles, shot out during disintegration, literally drill holes in the mica, and in fact the tracks of individual particles can be counted.

The lengths of the tracks can be related to the energies with which the particles have been shot out of the nuclei of the uranium atoms in the radioactive mineral; the various rings are the result of radiation damage at various discrete distances from the radioactive mineral source and can be correlated with the disintegration of daughter elements in the chain of nuclear changes from uranium to lead. Examination of many pleochroic haloes shows that the alpha particle energies for a given ring are always the same, showing no dependence on the geologic occurrence of the micas. This evidence suggests that the processes of disintegration of uranium atoms have not been affected by the different pressures and temperatures to which the micas have undoubtedly been subjected or by their differences in age.

Second, the original lead content, if any, of the mineral must be known. This is indeed difficult to ascertain in many instances and tends to limit the applicability of the method to a few minerals that are almost free of lead when they crystallize, although reliable values can be obtained by some experienced investigators from original lead-bearing minerals.

Third, the age determined is the time of crystallization of the mineral studied and not necessarily that of the rock that includes it. The particles of a sedimentary rock may have been deposited 300 million years ago, but the individual minerals, carried bodily to the site of sedimentation, may have crystallized during the cooling of an igneous rock 2 billion years old. For example, potassium-bearing minerals, such as illite (a micaceous mineral), are being deposited in the Atlantic deeps today; however, the ages of these minerals as determined by radioactive dating are 200–400 million years. These minerals are preponderantly eroded from older rocks on the continents and transported to the ocean by running water and wind. As a corollary, the metamorphism of a rock may cause the minerals to recrystallize, in which case accumulated lead and other decay products may be mobilized and escape, and the age determined is the time of metamorphism and not that of original crystallization of the rock.

Fourth is a relation already implied by the discussion of the effects of metamorphism: The disintegrative process must have been permitted to continue through time without addition or removal of material. For example, one of the elements in the chain of uranium decay is the gas radon. If it has opportunity to escape, the amount of lead found in the mineral will be less than that expected for the actual time elapsed.

Fifth, crude application of the U/Pb ratio is not satisfactory. Lead is the final product of the disintegration of $_{92}U^{238}$, $_{92}U^{235}$, and $_{90}Th^{232}$. Therefore the lead present is the sum of any original lead plus that from several radioactive sources. However, the leads of various origins can be identified fairly well if the lead isotopes can be separated and determined. The mass spectrograph, invented by Aston in 1923 and put into an instrument suitable for general use by Nier in the late 1930's, permits accurate determination of isotope mass ratios. The method can be applied for all isotopes, and age determinations by a variety of methods are accomplished by many scientists today.

Uranium-Lead Dating

Most of the early determinations of mineral ages were done by the uranium-lead method. It continues to be an important contributor to our fund of dated rocks, and even though some other methods are more widely applicable, the method will be developed in some detail here because it·illustrates the general theoretical approach to methods of dating. Also, the discussion will provide background for the use of uranium and lead isotopes for purposes other than direct age dating.

Most of the suitable uranium-rich minerals occur sparingly in igneous rocks of granitic composition, and it can usually be demonstrated that they crystallized at the same time as the other rock minerals when the melt solidified. Therefore the age obtained is that of the solidification of the intrusive rock. Spectrometric analyses are made for $_{92}U^{238}$, $_{92}U^{235}$, and the lead isotopes $_{82}Pb^{206}$, $_{82}Pb^{207}$, and $_{82}Pb^{204}$. Also, a test may be made to see if the minerals are in radioactive equilibrium, i.e., whether all decay products between U and Pb are present in their proper proportions. If they are, it indicates that at least for the most recent 100,000 years or so there has been no loss of decay products.

Then the ratio of $_{92}U^{238}/_{82}Pb^{206}$ is determined, and sometimes that of $_{92}U^{235}/_{82}Pb^{207}$. The age of the sample can be calculated by substituting in the uranium decay equation (p. 71). Because each uranium atom is replaced by a lead atom when it disintegrates, the equation can be written

$$\log \frac{U}{U+Pb} = -kt. \tag{1}$$

This equation can be rearranged to obtain

$$\frac{U}{Pb} = \frac{10^{-kt}}{(1 - 10^{-kt})}. \tag{2}$$

For example, the half-life of $_{92}U^{238}$ is 4.51 billion years, so when half an initial mass m of uranium has disintegrated to $_{82}Pb^{206}$, substitution in equation (1) yields the disintegration constant

$$\log \frac{0.5m}{0.5m + 0.5m} = -k \times 4.51$$

$$-0.3 = -k \times 4.51$$

$$k = 0.067.$$

Thus for any measured ratio of a uranium isotope to its lead isotope decay product, substitution of the value for U/Pb and that for k permits obtaining the age t in billions of years from equation (2). Figure 3.2 represents solutions of equation (2) for the ratios $_{92}U^{238}/_{82}Pb^{206}$ and $_{92}U^{235}/_{82}Pb^{207}$. If, for example, the ratio of $_{92}U^{238}/_{82}Pb^{206}$ is 2.0, reference to Figure 3.2 shows that the age of the mineral should be 2.6 billion years.

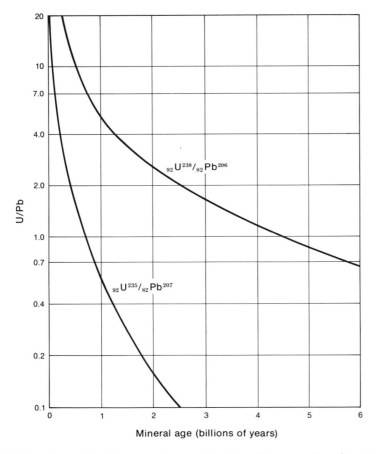

Figure 3.2 Uranium to lead isotope ratios as a function of the age of a radioactive mineral, assuming that no lead was originally present and that no lead or uranium has been lost.

This can sometimes be checked by the $_{92}U^{235}/_{82}Pb^{207}$ ratio, using the decay curve based on the $_{92}U^{235}$ half-life of 0.713 billion years. In the example cited, $_{92}U^{235}/_{82}Pb^{207}$ should be a little less than 0.1.

Potassium-Argon and Rubidium-Strontium Methods

Perhaps the most widely used age-dating method today utilizes the decay of the radioactive isotope of potassium $_{19}K^{40}$ to $_{18}Ar^{40}$. Although the $_{19}K^{40}$ isotope is much less abundant than the stable $_{19}K^{39}$ and totals only a few parts per million in most igneous rocks, the K-Ar method has been used to determine the ages of minerals only a few million years old. This is a truly remarkable analytical feat, because in 10 million years only 0.6 percent of the original $_{19}K^{40}$, with its half-life of 1.3 billion years, has decayed.

The widespread use of this method stems in part from the great range of ages, from a few million to several billion years, for which it can be employed and from the fact that potassium is an essential constituent of common rock-forming minerals such as micas and feldspars.

The rubidium-strontium method, utilizing the decay of $_{37}Rb^{87}$ to $_{38}Sr^{87}$, is gaining in popularity as techniques improve. The very long half-life tends to restrict its use to old rocks, and the abundance of $_{37}Rb^{87}$ is low.

Ages have now been obtained for many samples in which both $_{19}K^{40}$-$_{18}Ar^{40}$ and $_{37}Rb^{87}$-$_{38}Sr^{87}$ dates have been obtained for each sample. The validity of an age determined for one mineral of a rock by one isotope ratio is almost impossible to assess, but if dates on two individual minerals agree, or on one mineral by two different methods, concordance breeds reasonable confidence, and discordance demonstrates interferences of some kind.

An example of the use of K-Ar and Rb-Sr dating is shown in Figure 3.3, which also illustrates some of the complexities involved in age dating. The figure shows the variation of mineral ages as a function of distance from an igneous rock body that was intruded into metamorphic rocks about 54 million years ago. Notice that mineral ages in metamorphic rocks near the contact with the igneous intrusion are much younger than ages obtained farther away. The daughter products of the decay of radiogenic potassium

Figure 3.3 Schematic diagram showing the variation in ages of minerals found in metamorphic rocks as a function of distance from an igneous body intruded 54 million years ago. The true age of the metamorphic rocks is about 1700 million years, although metamorphic events in this region have occurred at 1300 and 1000 million years and in the Tertiary (after Hart, 1964).

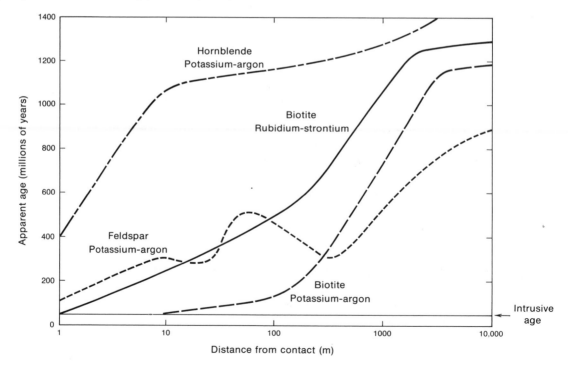

and rubidium, respectively, $_{18}Ar^{40}$ and $_{38}Sr^{87}$, escaped by diffusion from the minerals hornblende, biotite, and feldspar because of the heat produced during intrusion of the igneous mass, thus resulting in young ages for these minerals. The mineral ages obtained far away from the intrusive body should represent the approximate age of metamorphism; however, notice that the ages obtained from different minerals and even from the same mineral by different isotope techniques do not agree. These different ages are due to the complex history of Precambrian metamorphic events in this region and to uncertainties in the decay constants for $_{19}K^{40}$-$_{18}Ar^{40}$ and $_{37}Rb^{87}$-$_{38}Sr^{87}$.

Many ingenious and complicated graphical and mathematical procedures have been developed in an attempt to determine the true age of a system which has suffered loss during its history of radiogenic isotopes or their daughter products. For example, even though there has been loss of lead from a crystal of uraninite, both $_{82}Pb^{207}$ and $_{82}Pb^{206}$ will be lost in amounts proportional to their concentrations at the time of loss. A graphical method (concordia plot) based on the assumption that the lead isotopes are not fractionated by natural processes permits determination of true ages despite loss as a result of metamorphism or percolating ground waters.

Carbon-14 Dating

One of the most widely used methods for ages up to about 40,000 years is the one involving radioactive carbon. Whereas in the preceding examples the radioactivity of the atoms presumably originated only at the initial time of synthesis of the elements themselves, $_6C^{14}$ is continuously formed from $_7N^{14}$ by cosmic-ray-induced neutron bombardment in the upper atmosphere;

$$_7N^{14} + \text{neutron} = {_7N^{15}}$$

$$_7N^{15} = {_6C^{14}} + \text{proton}.$$

The $_6C^{14}$ formed disintegrates back into $_7N^{14}$ by loss of a beta particle from the nucleus.

The $_6C^{14}$ becomes mixed into the entire atmosphere and is taken up by living organisms, notably in the tissues of plants and in the calcium carbonate shells of organisms. Thus the ratio of $_6C^{14}$ to the most common isotope $_6C^{12}$ is a measure of the time since any carbon-fixing organism removed the CO_2 from the atmosphere. With the $_6C^{14}$ method, it is possible to make satisfactory measurements of about six half-lives, that is, on specimens in which only about 1.5 percent of the original $_6C^{14}$ remains.

The method has been used to determine the ages of trees and has been checked by counts of annual growth rings. Archeologists have benefitted particularly in their studies of prehistoric cultures; students of the Great Ice Age have a tool that takes them back before the most recent retreat of

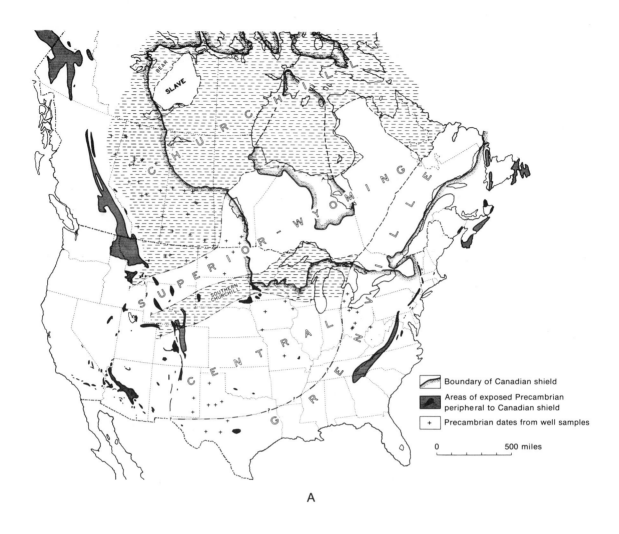

A

Figure 3.4 Map (A) and histograms (B and C) showing Precambrian provinces and the ages of Precambrian rocks in various parts of North America. Note the central belt of ages in the 2500-million-year range and the peripheral maxima at about 1800 million and 1000 million. C summarizes the histograms of B, and demonstrates the presence of clear-cut maxima. A and B courtesy L. L. Sloss and L. H. Nobles.

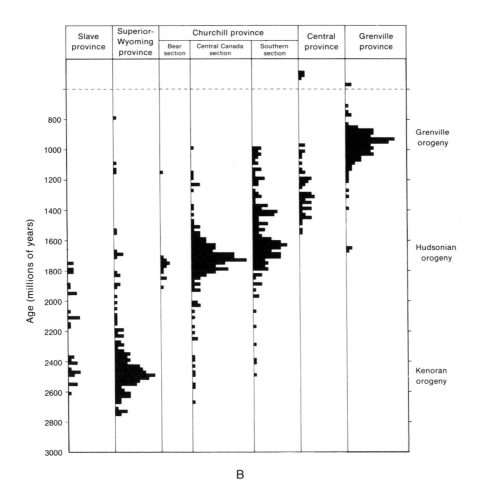

B

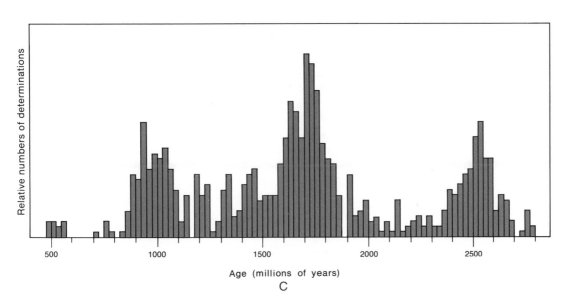

Age (millions of years)

C

79

the ice. Ancient peats found at various depths below present sea level have been dated using the $_6C^{14}$ method. These peats were formed during low levels of the sea within the last 10,000 years. From these age and depth data, a sea level curve for the past 10,000 years of earth history has been constructed, illustrating the rate of rise and elevation of sea level during this interval of time.

Two important changes have taken place in the initial assumptions used in obtaining C-14 ages. One came from the discovery that trees 200 years old have approximately the same $_6C^{14}/_6C^{12}$ ratio as modern trees— in fact all samples up to about 200 years old give about the same apparent age. This puzzle was solved when it was realized that in the last 200 years the amount of "dead" CO_2 poured into the atmosphere by the burning of fossil fuels actually has diminished the $_6C^{14}/_6C^{12}$ ratio in the atmosphere.

The other change is in the original assumption that the cosmic-ray flux which produces C-14 has been approximately constant through time.

Tree ring counts on trees more than 5000 years old show definite discrepancies between the C-14 age and the tree ring age, indicating temporal variations in the cosmic-ray flux.

SOME RESULTS OF RADIOACTIVE AGE DATING

It is hard to keep up with the flood of rock ages now appearing in the scientific literature. Figure 3.4 shows histograms summarizing the results of many age determinations on Precambrian rocks from North America and a map showing the parts of the continent from which they came. The pattern exhibited, with oldest rocks in the center and progressively younger rocks outward, is one basis for the suggestion that the continent has "grown" through time.

The age determinations on these rocks, as was indicated before, should not necessarily be accepted as the time of formation of the rocks. An age determination commonly means only that an "event" took place, which in many instances can be correlated with a time of metamorphism. However, the age determined ordinarily represents a minimum age of the rock in question. It may well be that the minima and maxima on the histogram are correlative with times of widespread crustal disturbance.

The distribution of ages for North America agrees fairly well with the subdivisions of geologic time that had been made previously based on correlations and unconformities. Interpretation of the age belts in terms of crustal history is treated in Chapter 12. Here it suffices to note that the interval between peaks on the histogram is of the order of 300–600 million years, suggesting that the broad concept of eras as representing repetitions of major cycles of events on a continental or global scale has some validity, at least as opposed to a concept of the earth's history as a continuum, in which the only major change with time is the place in which events occur. This latter view would be consistent with classic Uniformitarianism and has many strong proponents.

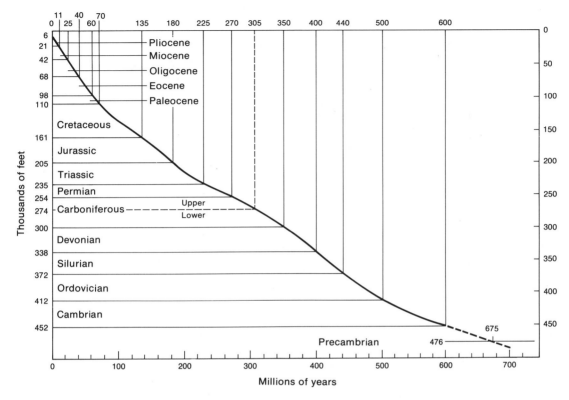

Figure 3.5 Correlation between maximum known thickness of sedimentary layers deposited and the absolute ages of the layers (Kummel, 1961).

One result of all the radioactive age work is the reasonably firm establishment of the duration of the various periods and eras in terms of actual years, as listed in Table 3.1. It is a remarkable tribute to the geologists, who had estimated the relative lengths of the time divisions from evolutionary changes, thickness and type of rock deposited, and other such criteria, that there has been little change in their estimates of relative duration. Figure 3.5 shows the good correlation between the maximum thicknesses of rocks deposited during the various periods, which have been used as a measure of their durations, and the duration of the periods in years as determined by radioactive methods.

LEAD ISOTOPES AND THE AGE OF THE EARTH

The oldest rocks so far found at the surface of the earth are about 3.5 billion years old. The best estimate of the age of the earth itself is 4.5–5.0 billion years. The method for determining the time of earth origin is an outgrowth of U-Pb age determinations.

The isotopes of lead are $_{82}Pb^{204}$, $_{82}Pb^{206}$, $_{82}Pb^{207}$, and $_{82}Pb^{208}$.

$_{82}Pb^{204}$ has no known radioactive parent element; the others are products of the decay of $_{92}U^{238}$, $_{92}U^{235}$, and $_{90}Th^{232}$, respectively. Consequently the present-day abundance of $_{82}Pb^{204}$ presumably is the same as it was originally, but the amounts of the other isotopes have increased from some initial value through time by decay of their parent elements. The relative abundances of the other isotopes are usually expressed in terms of their ratio to the unchanging $_{82}Pb^{204}$.

We can get maximum ages for the crust of the earth by treating it as if it were a single radioactive mineral and by making the simplifying assumption that at the time of crust formation there were no $_{82}Pb^{206}$ and $_{82}Pb^{207}$ present and that *all* the $_{82}Pb^{206}$ and $_{82}Pb^{207}$ now present have been derived from decay of $_{92}U^{238}$ and $_{92}U^{235}$. Then we can use equation (1) or (2) to obtain a time for the formation of the crust. The relative abundances of the isotopes (which must be compared on an atomic basis, not on a weight ratio, because decay produces one atom of lead for one of uranium) are $_{92}U^{238}/_{82}Pb^{206} = 10{:}18.9 = 0.53$ and $_{92}U^{235}/_{82}Pb^{207} = 0.0725{:}15.73 = 0.0046$. Substituting these values into equation (1), with the appropriate decay constants, yields

$$\log \frac{U^{238}}{U^{238} + Pb^{206}} = \log \frac{10}{10 + 18.9} = -0.067t$$

$$-0.46 = -0.067t$$

$$t = 6.9 \text{ billion years,}$$

and

$$\log \frac{U^{235}}{U^{235} + Pb^{207}} = \log \frac{0.0725}{0.0725 + 15.73} = -0.42t$$

$$-2.34 = -0.42t$$

$$t = 5.6 \text{ billion years.}$$

These discordant results presumably stem from the simplifying assumption concerning the lead isotopes; there must have been some of each at the time of crust formation.

One method of estimating the initial abundance of $_{82}Pb^{206}$ and $_{82}Pb^{207}$ is from their occurrence in iron meteorites. The iron meteorites contain very little uranium—just about enough to get age determinations on them. Consequently their lead isotope content must represent almost entirely lead present at the time of their formation. If we use their contents of $_{82}Pb^{206}$ and $_{82}Pb^{207}$ as a measure of that present in the primordial crust of the earth, we can subtract this "primordial lead" from that now present to obtain a new number for the amount of lead that has resulted from uranium breakdown. If we do so, the relative abundance of $_{82}Pb^{206}$ is $18.9 - 9.5 = 9.4$ and that of $_{82}Pb^{207}$ is $15.7 - 10.4 = 5.3$. Equation (1) thus becomes

$$\log \frac{U^{238}}{U^{238} + Pb^{206}} = \log \frac{10}{10 + 9.4} = -0.067t$$

$$t = 4.35 \text{ billion years,}$$

and equation (2) becomes

$$\log \frac{U^{235}}{U^{235} + Pb^{207}} = \log \frac{0.0725}{0.0725 + 5.3} = -0.42t$$

$$t = 4.45 \text{ billion years.}$$

The correction for initial lead has brought the ages almost into concordance. Further refinements of the type of calculation presented here yield the "best value" of 4.53 billion years.

Figure 3.6 illustrates one of the techniques that has been used to get the age of the crust. The ratios of the radiogenic lead isotopes to $_{82}Pb^{204}$, obtained from recent lead minerals, are shown on the ordinate. The ratios from deposits of known age also are plotted, and a smooth curve is drawn through them. These ratios presumably represent the relative abundances of the lead isotopes at various times in the past and illustrate the changes that have taken place because of the differences in the decay rates of $_{92}U^{235}$ and $_{92}U^{238}$. Then the points where the "best estimates" of primordial lead/lead ratios intersect the curves become the starting point of radioactive generation of lead in the crust. The figure shows that an age of 4.5 billion years is consistent with the ratios of all three radiogenic isotopes. The interval between formation of the earth and formation of the crust has been estimated between 100 and 500 million years.

Thus this convergent evidence begins to give us some faith in the validity of the current age estimate; furthermore, the agreement between earth age and meteorite ages suggests a time of origin for the solar system. Also, an age of 4.5–5.0 billion years is satisfactory to account for the present thermal state of the earth, when estimates like those of Kelvin are revised to take into account heating by radioactive elements.

GEOLOGIC TIME AND DAY LENGTH

It has been suggested that the rate of rotation of the earth is slowly decreasing because of tidal friction; thus the length of the day has been increasing. This increase is estimated to be about 2 secs/100,000 years. However, the period of revolution of the earth around the sun has apparently remained constant with time; hence, the number of days in a year has been decreasing. For example, a year at the end of the Precambrian should have contained about 420 days.

A paleontological method is also available for determining past day length. Growth features of some invertebrates exhibit daily, lunar, and annual cycles. For example, the concentric ridges on the modern mollusk *Pecten diegensis* show daily and tidal cycles. Fossil corals exhibit surface ridges, banding, and striations apparently due to yearly, monthly, and daily increments of shell growth, respectively.

Wells (1963) reasoned that fossils with both daily and yearly growth increments could be used as fossil "clocks." He collected corals from Pennsyl-

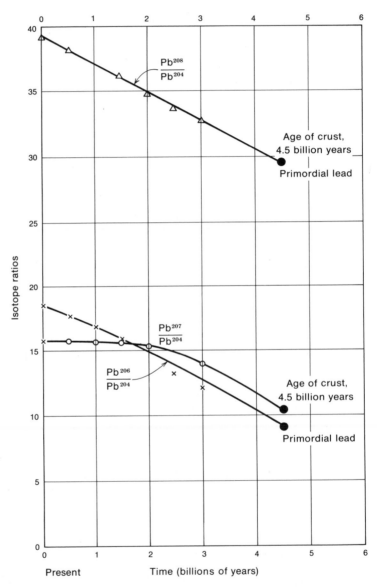

Figure 3.6 Ratios of $_{82}Pb^{208}$, $_{82}Pb^{207}$, and $_{82}Pb^{206}$ to $_{82}Pb^{204}$ from common leads of various ages plotted versus time to obtain an age for the earth's crust. Note that the extrapolations of the curves beyond the data for common leads to intersect the composition of primordial lead is subject to considerable interpretation.

vanian and Devonian strata to test this hypothesis. These corals exhibit surface textures consisting of a series of prominent ridges and fine striations encircling the coral. By counting the fine striations between the coarser ridges, Wells was able to estimate the number of days per year during the Pennsylvanian and Devonian periods. His estimates agree reasonably well with the geophysical results based on tidal retardation of the earth's rotational rate and radiometric ages. Although there is still some controversy concerning the significance of growth increments found in fossils, the agreement between the number of days per year obtained from fossil corals and from calculations based on tidal retardation supports the interpretation of the growth rings and indirectly validates the assumption of constancy through time of radioactive decay processes. Research concerned with the use of fossils as geologic "clocks" will undoubtedly progress rapidly in future years.

RADIOACTIVE HEAT GENERATION

The major radioactive elements U, Th, and K disintegrate slowly, but the age of the earth is sufficient to have produced an important decrease in their amounts. $_{92}U^{238}$, with a half-life the same as that of the earth, has been reduced by half; $_{92}U^{235}$, with a half-life of only 0.713×10^9 years, has been decaying for more than six half-lives, so only about 1 percent of the original amount is left. Most of the $_{90}Th^{232}$, with its 14×10^9 year half-life, still remains, but radioactive $_{19}K^{40}$ has been reduced to approximately 10 percent of the primordial amount.

Birch (1965) estimates that heat production had to drop from an early high to a value 3–5 times the present value before a continental crust could form; at higher values, crustal rocks would melt. Table 3.3 gives the heat production per gram of some of the elements.

The heat production per year does not seem very large; a gram of uranium would require more than 1000 years to heat a liter of water 1° C; but

Table 3.3
Radioactive Heat Production

Isotope	cal/g/year
$_{92}U^{238}$	0.71
$_{92}U^{235}$	4.3
$_{92}U$ (ordinary mixture of isotopes)	0.73
$_{90}Th^{232}$	0.20
$_{19}K^{40}$	0.21
$_{19}K$ (ordinary mixture of isotopes)	27×10^{-6}

one must remember that the same gram, if present at the beginning of the earth, would liberate as much heat during the earth's history as would the burning of half a ton of coal.

The radioactive heat production of a ton of granite, with about 4 g of $_{92}U^{238}$, 13 g of $_{90}Th^{232}$, and 4.1 g of $_{19}K^{40}$, is about 6 cal/year. Therefore it would take only about 7 tons of granite, or a 1-cm² column 26 km long, to produce the observed average crustal heat flow of about 40 cal/cm²/year. Clearly radioactive heat is responsible for most of the heat flow; in fact, it can be concluded that where the crust is 40 km thick, the material cannot be all granite.

STABLE ISOTOPES

Although this chapter is devoted chiefly to age determination and radioactivity, it is a convenient place to discuss some of the uses and potentials of measurements of stable isotope ratios. Because isotopes have almost identical chemical properties but different masses, they have a tendency to become separated only by mass-dependent processes, such as evaporation or ionic diffusion. Also, many organisms tend to use one isotope selectively in their metabolic processes.

Sulfur Isotopes

The two chief isotopes of sulfur are $_{16}S^{32}$ and $_{16}S^{34}$. $_{16}S^{32}$ is by far the more abundant; the average ratio in earth materials is 22.6:1. $_{16}S^{33}$ and $_{16}S^{36}$ also exist but make up only 1 percent of all sulfur. The mass difference of some 6 percent between $_{16}S^{32}$ and $_{16}S^{34}$ is sufficient for marked fractionation in many earth-surface processes.

Variations in the isotope ratio are currently expressed in parts per mil, (parts per mil equals parts per thousand) and are given as deviations from some arbitrary standard. In the case of sulfur, the ratio in meteorites is nearly constant and agrees with the earth average, so it has been utilized as the standard.

Deviations from the standard, in parts per mil, δ, are calculated from the relation

$$\delta S^{34} (\%_0) = \left(\frac{S^{34}/S^{32} \text{ in sample}}{S^{34}/S^{32} \text{ in standard}} - 1 \right) \times 1000.$$

Figure 3.7 is a summary diagram of the S^{34}/S^{32} relations for a number of sulfur-containing earth materials. It shows that sulfur in sulfates is "heavier" than that in sulfides. This difference is caused by organisms; sulfur-reducing bacteria selectively reduce $_{16}S^{32}$ when they change sulfate into sulfide. These bacteria live at the interface between sediments and overlying water and oxidize organic material as they reduce sulfate. The hydrogen sulfide generated usually reacts with minerals containing iron to

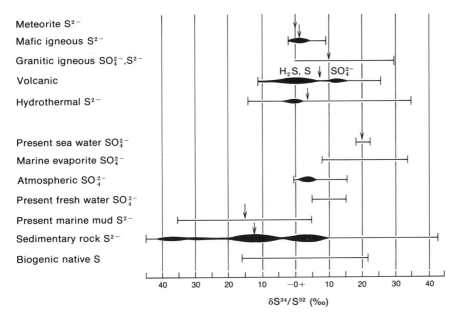

Meteorite S^{2-}
Mafic igneous S^{2-}
Granitic igneous SO_4^{2-}, S^{2-}
Volcanic
Hydrothermal S^{2-}

Present sea water SO_4^{2-}
Marine evaporite SO_4^{2-}
Atmospheric SO_4^{2-}
Present fresh water SO_4^{2-}
Present marine mud S^{2-}
Sedimentary rock S^{2-}
Biogenic native S

H_2S, S SO_4^{2-}

40   30   20   10   −0+   10   20   30   40

$\delta S^{34}/S^{32}$ (‰)

Figure 3.7 Variations in $\delta S^{34}/S^{32}$ in various types of earth materials (redrawn from W. T. Holser and I. R. Kaplan, 1966. Isotope geochemistry of sedimentary sulfates. *Chemical Geology,* Vol. I, Fig. 1).

produce iron sulfides, most of which eventually become pyrite, FeS_2, or is released to the atmosphere where it is oxidized to sulfate. Sulfate can be reduced inorganically under sedimentary conditions only at a vanishingly small rate; fractionation is therefore evidence of biochemical action.

The sulfate of the present-day oceans averages about $+20$/mil, and sedimentary sulfides about -20; therefore at present about half of the total sulfur of the sedimentary rock-ocean system is present as sulfate, and most of the other half resides in sedimentary rocks as iron sulfides. Figure 3.8 shows the isotopic composition of sedimentary sulfate deposits as a function of their geologic age. Because the precipitation of gypsum ($CaSO_4 \cdot 2H_2O$) or anhydrite ($CaSO_4$) from sea water causes little fractionation, we must conclude that the distribution of sulfide and sulfate sulfur between rocks and oceans has changed markedly from time to time.

Sulfur isotope ratios have provided information on many other aspects of geologic history; for example, they have been used to determine that certain economic deposits of copper sulfide gained their sulfur from sedimentary sources and that certain others have been derived from the mantle or the deep crust. Also, sulfur isotope ratios have been used to determine the sources of atmospheric sulfur. Sulfates from samples of rain and snow are isotopically light when compared with sea water sulfate. These data indicate that sulfate in the atmosphere comes from oxidation of naturally produced H_2S and of industrially derived sulfur dioxide, both of which are enriched in S^{32} relative to the standard, and not from sea water sulfate transported into the atmosphere.

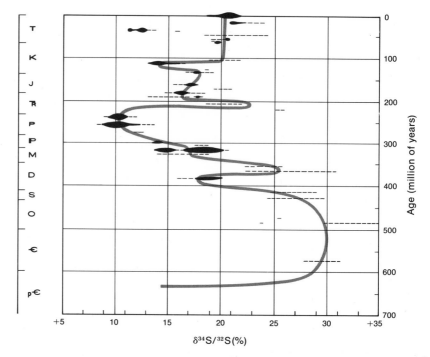

Figure 3.8 Variation with time of $\delta S^{34}/S^{32}$ in evaporite deposits (redrawn from W. T. Holser and I. R. Kaplan, 1966. Isotope geochemistry of sedimentary sulfates. *Chemical Geology*, Vol. 1, Fig. 5).

Oxygen has three isotopes, $_8O^{16}$, $_8O^{17}$, and $_8O^{18}$, with relative abundances of 99.8, 0.04, and 0.2%, respectively. The $_8O^{18}/_8O^{16}$ ratios in minerals and in waters have been used extensively for their classification. Figure 3.9 shows the isotopic variation of oxygen in nature. During evaporation the first H_2O is enriched in the light isotope, so atmospheric processes lead to variations of $_8O^{18}/_8O^{16}$ in various water bodies. At least as far back as Cambrian time, the oceans have been fairly constant in $_8O^{18}/_8O^{16}$ ratio or have had a small decrease in this ratio. Because minerals precipitated from various water bodies tend to be influenced by the isotopic composition of the medium, oxygen isotopes have been used to help decide whether various minerals are of marine or fresh water origin. One of the complicating factors is the role of organisms as intermediaries in the precipitation processes for many minerals; various parts of a single coral, for example, may be quite different in their isotope ratios.

Oxygen isotopes have been used as a thermometer; if isotopic equilibrium is established between two oxygen-containing minerals, the equilibrium constant and hence the $_8O^{18}/_8O^{16}$ ratios in the minerals are a function of temperature.

For a rock containing hematite and chert, perhaps from the ancient iron formations of Michigan, the reaction can be written

$$Fe_2O_3^{18}+SiO_2^{16}=Fe_2O_3^{16}+SiO_2^{18}.$$

The equilibrium constant is

$$K=\frac{{}^{a}Fe_2O_3^{16}\times{}^{a}SiO_2^{18}}{{}^{a}Fe_2O_3^{18}\times{}^{a}SiO_2^{16}}.$$

Isotope analysis of the minerals yields a value for K and hence a temperature from experimentally determined values of K. Clayton (1958) obtained temperatures for the Michigan iron formations of 150–300°C and interpreted them as representing the extremes to which the rocks had been heated after deposition. A surprising number of rocks have been found in which the minerals, as checked by two or more mineral pairs, are in isotopic equilibrium.

Oxygen isotopes also have been used to interpret the temperature history of the oceans during the past 400,000 years. Small, floating protozoans called Foraminifera, which secrete a $CaCO_3$ skeleton, live in the surface layers of the sea. When the organisms die their shells fall to the bottom and are incorporated into the sediments of the deep sea. Long cores of these sediments have been obtained. The forams have been separated from the sediments and the $_8O^{18}/_8O^{16}$ ratios of the calcareous tests determined as a function of depth in the cores. Shifts in the oxygen-isotope ratio occur with depth in the cores and have been interpreted as being the result of shifts in the surface ocean temperature. Figure 3.10 shows a plot of the sur-

Figure 3.9 Oxygen isotope variation in nature (Garlick, 1969).

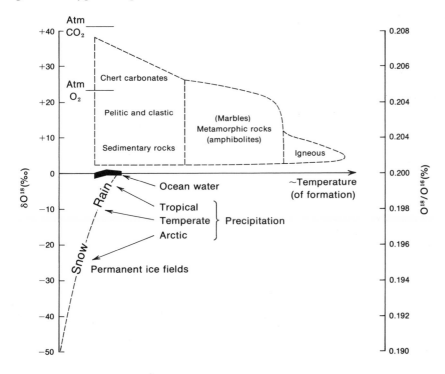

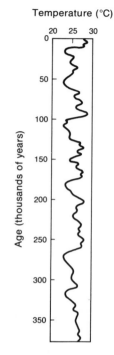

Figure 3.10 Generalized temperature curve for surface ocean water for the past 400,000 years as obtained from oxygen isotope data (Emiliani, 1964).

face ocean temperature obtained from $_8O^{18}/_8O^{16}$ isotopic ratios as a function of time. Surface ocean water temperatures have apparently varied 5–6°C in the past 400,000 years. The colder periods are thought to represent times of intense glaciation on the continents during the Pleistocene, whereas the warmer periods represent times of less active continental glaciation and temperate climate.

Carbon Isotopes

Much work has been done with the $_6C^{13}$ isotope in relation to the abundant $_6C^{12}$. Variations of the ratio of as much as 10–12 percent occur. Photosynthesis, during which CO_2 from the atmosphere is utilized, is attended by marked selection of $_6C^{12}$. Fossil organic matter has a nearly constant ratio. Figure 3.11 shows the carbon-isotope ratios of some natural carbon-bearing materials.

Among the problems that have been studied by the aid of $_6C^{13}/_6C^{12}$ ratios is the origin of petroleum. In the laboratory, progressive distillation of oily materials yields fractions with increasing $_6C^{13}/_6C^{12}$ ratios. In the field, attempts have been made to follow the migration routes of hydrocarbons through rocks to form oil accumulations by mapping out the spatial variation of the isotope ratio. Also, $_6C^{13}/_6C^{12}$ ratios of petroleums are compatible with an organic origin of this important energy source. In fact, carbon isotope ratios indicate that petroleum is probably derived from the lipid (fat and waxes) fraction of organisms.

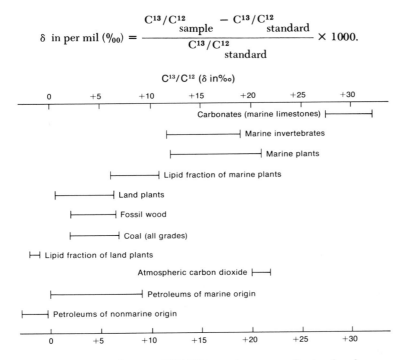

$$\delta \text{ in per mil (\%}_{00}) = \frac{C^{13}/C^{12}_{\text{sample}} - C^{13}/C^{12}_{\text{standard}}}{C^{13}/C^{12}_{\text{standard}}} \times 1000.$$

Figure 3.11 Variations in the ratio C^{13}/C^{12} in various types of natural carbonaceous materials (Silverman, 1964).

Much more could be written about stable isotopes, because many different kinds of studies have been made with isotopes of many elements. Their application to geologic problems is growing rapidly, and their usefulness is continuously increasing as we learn more and more about their occurrence.

SUMMARY

Prior to the discovery of radioactivity, only the relative age of a rock unit could be determined by criteria such as its stratigraphic position and the fossils contained within it. Regional and intercontinental correlations were made on this basis.

Estimates of the age of the earth ranged widely; in the late 1800's Kelvin's estimate of about 100 million years, based on calculations of the thermal history of the earth, was widely accepted, although many students of organic evolution were convinced that an age of many hundreds of millions of years was required.

With the discovery of radioactivity, it became possible to assign ages in years to many rocks and thus to estimate the actual duration of the various Eras and Periods of geologic time that had been established before. Good agreement has been found between the earlier estimates of relative duration and the current ones of absolute duration, suggesting that the basic assumption of radioactive age dating, that the rates of disintegration today are invariant with time, is valid.

In addition to permitting the development of an absolute time scale, the discovery of radioactivity dealt a death blow to Kelvin's 100-million-year-old earth by providing a new source of heat for the earth's interior. The cooling time required for an originally hot earth was increased many-fold over his model.

A tremendous amount of work during the last 20 years on the determination of rock ages by methods based primarily on the radioactive decay of uranium, thorium, potassium, and rubidium leads us to a "firm" estimate for the age of the crust of 4.5 billion years, and a current age of about 3.5 billion years for the oldest rock yet discovered. The general agreement of the age of the crust with ages of meteorites indicates a time of earth and solar system origin between 4.5 and 5.0 billion years ago. Studies of carbon-14 provide us with a detailed chronology for the past 40,000 years.

Stable isotopes, particularly those of sulfur, oxygen, and carbon, have become useful tools in the investigation of earth history. Sulfur isotopes provide information about the chemistry of the ancient ocean, the source of sulfur in metallic sulfide ore deposits and of sulfur in the atmosphere. Oxygen isotopes can be used to determine the temperature of ancient surface ocean water, the temperatures of formation of minerals, and the marine or fresh water origin of minerals. Carbon isotopes have been particularly useful in problems related to the genesis and migration of hydrocarbons.

REFERENCES

Birch, F., 1965, Speculation on the earth's thermal history: *Bull. Geol. Soc. Am.*, 76, 133–154.

Clayton, R. N., 1958, The relationship between O^{18}/O^{16} ratios in coexisting quartz, carbonate, and iron oxides from various geological deposits: *J. Geol.*, 66, 352–373.

Degens, E. T., 1965, *Geochemistry of Sediments:* Prentice-Hall, Inc., Englewood Cliffs, N. J.

Emiliani, C., 1964, Paleotemperature analysis of the Caribbean cores A254-BR-C and CP-28: *Bull. Geol. Soc. Am.*, 75, 129–144.

Garlick, G. D., 1969, The stable isotopes of oxygen: in Handbook of Geochemistry, vol. II, part 1, K. H. Wedepohl, ed.: Springer-Verlag, New York.

Hart, S. R., 1964, The petrology and isotopic-mineral age relations of a contact zone in the Front Range, Colorado: *J. Geol.*, 72, 493–525.

Holmes, A., 1960, A revised geological time-scale: *Trans. Edinburgh Geol. Soc.*, 17, 183–216.

Holser, W. T., and Kaplan, I. R., 1966, Isotope geochemistry of sedimentary sulfates: *Chem. Geol.*, 1, 93–135.

Kummel, B., 1961, *History of the Earth:* W. H. Freeman and Company, San Francisco.

Silverman, S. R., 1964, Investigations of petroleum origin and evolution mechanisms by carbon isotope studies: in *Isotopic and Cosmic Chemistry*, H. Craig, S. L. Miller, and G. J. Wasserburg, eds.: North-Holland Publishing Co., Amsterdam, 92–102.

Sloss, L. L, and Nobles, L. H., 1964, *Earth History:* Northwestern University Bookstore, Distributor, Evanston, Ill.

Wells, J. W., 1963, Coral growth and geochronometry: *Nature, 197,* 948–950.

Whipple, F. L., 1964, The history of the solar system, in *The Scientific Endeavor:* Rockefeller University Press, New York, 69–107.

4

Present Fluxes of Earth Materials

The continents are continuously being lowered by loss of material to the ocean basins. Interactions of many kinds between the atmosphere and the earth's land surface cause the rocks to break up or dissolve; then gravitational forces, in one guise or another, move the material downward until it rests on the sea floor. The chief agents of transport are mass movement, such as landslides and soil creep, streams, wind, waves, ice, and underground water.

Plants break up rock to form soil; rain infiltrates the soil or runs across the surface; rills collect into streams that in turn coalesce to form still larger ones. Each continent is frayed by a network of stream systems. As streams cut downward they create valley wall slopes, which accelerate mass movements as well as underground drainage into the streams. Because of their winding courses, streams also cut laterally on the outer banks of their curves and deposit at the inner sides. A major river such as the Amazon or the Mississippi has a wide flood plain; as it meanders back and forth it continuously reworks the valley alluvium, leaching soluble constituents from it. The overall erosional effect of streams is to funnel the load derived from the entire continent into a relatively few major rivers carrying suspended particulate matter plus dissolved constituents.

In desert areas, or in other places where surface materials are dry and loose, such as plowed fields during drought, wind can lift the finest particles high into the atmosphere and can move considerable debris as a surface sheet. Dust that gets into the upper atmosphere may be transported around the earth several times before sifting out.

Waves cut continuously at the shores of the continents. Where they encounter loose materials, or rocks so poorly cemented that the waves can disaggregate the rocks by their mechanical energy, erosion is rapid and has striking effects in developing cliffs and benches cut at water level. However, the most important effects of waves are in sorting and reworking the materials brought to the oceans by streams.

Erosion and transport by ice is of major importance today only in Antarctica and Greenland, where ice sheets thousands of feet thick bury most of the land surface. Snow constantly accumulates and then changes to ice, causing a continuous radial outward movement of the ice mass. As this occurs, the underlying rocks are scoured and quarried, and the debris is incorporated into the ice and carried to the sea.

Because of the very great age of the earth, the presence of continents would seem to be anomalous. Erosion must have started very early in the earth's history; energy from the sun causes pressure differences in the atmosphere and hence is responsible for wind and waves; it also evaporates water from the oceans and dumps it on the continents as rain or snow.

Rain, snow, and waves tend to dissolve or granulate rocks and render them susceptible to the tendency of gravitational forces to fill the depressions and lower the mountains.

Only in the past few years has it been possible to assess the quantitative importance of the various agents that are destroying the continents and filling the ocean basins. In the following pages we want to look at the distribution of water on the face of the globe, for it is water that does most of the erosional work, and it is water in the oceans that protects the materials that are carried into the ocean basins. Then we try to determine the present-day rate at which the ocean basins are being filled and, conversely, the rate at which the continents are being destroyed. The data are still so incomplete that we have included estimates by various researchers in the hope that the reader, by noting the ranges of values given, can form an opinion of the probable error of the estimates. Direct observation, even if we include the oldest written records of some thousands of years ago, constitutes but an instant of geologic time; we can determine only the present

Figure 4.1 Total masses of materials transported to and leaving the oceans by various agents. The mass values are given in units of 10^{14} g/year, except for the value of total dissolved solids in the ocean, which is in units of 10^{20} g. The net addition of materials to the oceans is about 250×10^{14} g/year. About 90 percent of this material is transported to the oceans by streams.

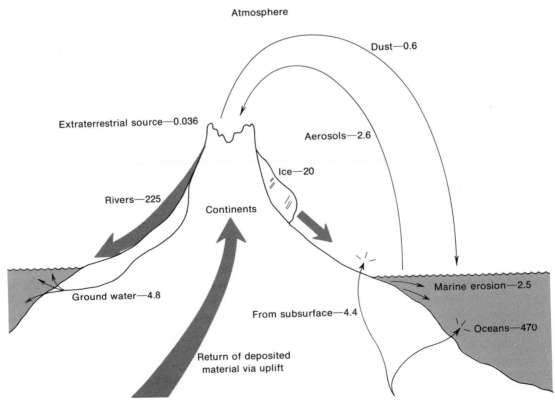

rates of processes. The sedimentary geologic record is an integration of many different rates through time; one of the most difficult problems is to determine the extent to which present-day rates are representative of the geologic past, or better, the ranges of rates that are required to explain the ages, types, and relative mass abundances of sedimentary rocks.

Because we necessarily must deal in large numbers in describing mass transfer on a global scale, we have tried to use units of 10^{20} g or of 10^{14} g. The unit of 10^{20} g is sometimes called the *geogram*, so we suppose that the 10^{14} unit is a *microgeogram*. In any case, they differ by an easily remembered factor of 1 million. The smaller unit is generally convenient for describing annual rates of material transfer on the global scale, whereas the larger one is suitable for such numbers as the mass of the oceans or the mass of the crust. After a little use of the units, they become familiar, if not comprehensible. A microgeogram is roughly equivalent to the amount of material that could be transported by 100,000 trainloads of 100-car trains. A geogram is even harder to put into reasonably familiar terms; perhaps we can visualize a block of ice 50 km (30 miles) on an edge, which would be about the right mass. A microgeogram would be a 0.5 km ice cube.

The discussion in the following sections and the values of mass transfer of materials into and out of the oceans are summarized pictorially in Figure 4.1; Table 4.11 gives estimates of the mass of individual major chemical species entering or leaving the oceans. The totals listed in this table do not necessarily agree with those of Figure 4.1 because only partial totals are listed. For example, the suspended load of streams is primarily composed of aluminosilicates; the mass of oxygen in the aluminosilicates is not included in the estimate of Table 4.11.

THE HYDROSPHERE

Ocean waters and sediment pore waters contain most of the present hydrosphere (Table 4.1). About 80 percent of the water in the hydrosphere is contained in the oceans and seas. The pores of sediments and sedimentary rocks hold nearly all the rest—about 24 percent of that in the ocean, or 20 percent of the total. Ice now locks up a little more than 1 percent of the total and may have accounted for 3 percent or so during the height of the Ice Age, but water storage is trivial in rivers, lakes, or the atmosphere. On the other hand, the rate of water circulation through the rain-river-ocean-atmosphere system is relatively fast (Figure 4.2); the amount of water discharged into the oceans each year from the land is approximately equal to the total mass of water stored at any instant in rivers and lakes.

We estimate the mass of the oceans to be about 30–50 percent of the mass of sedimentary rocks now in existence and about 5 percent of the

Table 4.1
Mass of the Present Hydrosphere[a]

	Total mass (units of 10^{20} g)	Percentage of total hydrosphere
Oceans	13,700	80.0
Pore waters in sediments	3,300	18.8
Ice	200	1.2
Rivers, lakes	0.3	0.002
Atmosphere	0.13	0.0008
Total hydrosphere	17,200	100.0

[a] Adapted from Horn (1966), Hutchinson (1967), Kalle (1943), Kossinna (1921), and Meinardus (1928).

mass of the earth's crust. Dissolved solids constitute about 3 percent of the oceanic mass; thus storage by the oceans accounts for roughly 1 percent of the materials currently involved in the sedimentary cycle. Another 0.25 percent or so can be attributed to solutes in the pore waters of rocks.

Figure 4.2 H_2O cycle. Annual transfer of H_2O as well as total mass of H_2O in the oceans, in the pore waters of sediments, in ice, and in the atmosphere are given in units of 10^{20} g.

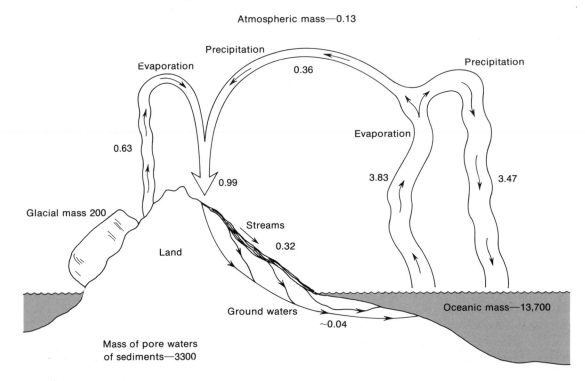

RIVER- AND GROUND-WATER FLUX

Introduction

Rivers are the major sources of dissolved and particulate materials for the oceans. River water is primarily a dilute solution of calcium bicarbonate. The composition of average river water and that of sea water are compared in Table 4.2. River water contains only about 130 ppm dissolved solids, whereas sea water has 35,000 ppm. Also, the proportions of the various dissolved species differ markedly in the two media. For example, silica is actually in higher concentration in stream water than in the oceans, but chloride in the oceans is 2500 times that in streams. Although, as we shall see, sea water must be derived by chemical differentiation and evaporation of stream water, the processes must be such as to affect every element differentially, showing that simple evaporation and concentration are entirely secondary to other processes.

The composition of river water given in Table 4.2, taken almost entirely from Livingstone's work (1963a), is the one we shall use henceforth in var-

Table 4.2
Major Constituents of River and Sea Water

Constituents	River water [a]		Sea water [b]	
	ppm	mmoles per liter	ppm	mmoles per liter
Cl^-	7.8	0.220	19,000	535.2
Na^+	6.3	0.270	10,500	456.2
Mg^{2+}	4.1	0.171	1,300	54.2
SO_4^{2-}	11.2	0.117	2,650	27.6
K^+	2.3	0.059	380	9.7
Ca^{2+}	15	0.375	400	10.0
HCO_3^-	58.4	0.958	140	2.3
SiO_2	13.1	0.218	6	0.1
NO_3^-	1	0.016	—	—
Fe^{2+}	0.67	0.012	—	—
Al	0.01[c]	—	0.001[c]	—
Br^-	—	—	65	0.8
CO_3^{2-}	—	—	18	0.3
Sr^{2+}	—	—	8	0.1
Dissolved organic C	9.6[d]	—	0.5[d]	—
Total	129.5	—	34,467	—

[a] Livingstone (1963a).
[b] Goldberg (1957).
[c] Sackett and Arrhenius (1959).
[d] Williams (1966).

ious calculations. However, the reliability of his estimate can best be assessed by comparison with other estimates.

Dissolved Solids in Rivers

Calculation of the total amount of dissolved solids transported to the oceans annually by streams requires estimates of total stream discharge and average chemical composition of the waters. Because most of the major rivers of the world, particularly in the humid tropics and arctic regions, have not been sampled seasonally, only rough estimates of discharge and mean salinity are available for many rivers. However, there is fair agreement among various investigators on total discharge. Alekhin and Brazhnikova (1963) estimate 0.36×10^{20} g/year; Livingstone (1963a) obtained 0.32×10^{20} g/year, whereas Holeman (1968) reports 0.30×10^{20} g/year. We have elected to use Livingstone's figure, partially because it is an intermediate value and also because it is convenient to use both his salinity and discharge data to get internal consistency. Furthermore, Gibbs (1967) indicates that the discharge of the Amazon River may be about double previous estimates, which would perhaps tend to increase Holeman's estimate. The Amazon contributes 10–18 percent of the total world runoff.

In Table 4.3 we have listed four estimates of the mean composition of the rivers of the world to give some historical perspective. Note that although there is a range from Murray's estimate of 258 ppm of total dis-

Table 4.3
Comparison of Estimates of the Mean Composition of River Waters of the World

Constituent	Salinity (258 ppm)[a]		Salinity (137 ppm) (based on Clarke's river content data and Murray's H_2O runoff estimate)[b]	
	ppm	Percent total dissolved solids	ppm	Percent total dissolved solids
HCO_3^-	152	58.7	71.5	52.3
Ca^{2+}	37.1	14.3	20.4	14.9
Na^+	6.3	2.4	5.8	4.2
Cl^-	3.3	1.3	5.7	4.2
SiO_2	19.5	7.6	11.7	8.6
Mg^{2+}	8.4	3.2	3.4	2.5
SO_4^{2-}	14.9	5.8	12.1	8.9
K^+	2.4	0.9	2.1	1.6
NO_3^-	5.1	2.0	0.9	0.7
Fe	8.6[c]	3.5	2.8[c]	2.0
Others	0.7	0.3	0.2	0.1

[a] Murray (1887).
[b] Clarke (1924).
[c] Conway (1942).

solved solids in 1887 to Livingstone's value of 120 ppm in 1963, the relative proportions of the various constituents are pretty well agreed upon by all four investigators. The lower values of Livingstone's recent estimate of total salinity stem largely from data on tropical rivers unavailable to previous investigators; waters of the streams draining areas of high rainfall have been found to be more dilute than previously estimated. In fact, Gibbs' work on the Amazon indicates that Livingstone's 120 ppm might eventually have to be revised downward a few percent.

Little global information exists on the dissolved organic constituents of streams; Williams (1966) indicated that 10 ppm might be an order-of-magnitude figure.

From the total discharge and mean composition values we derive an estimate of total dissolved materials carried to the ocean basins each year:

$$0.32 \times 10^{20} \times 130 \times 10^{-6} = 42 \times 10^{14} \text{ g/year.}$$

If organic material is excluded, the value is reduced to 39×10^{14} g/year.

Dissolved Solids in Ground Waters

The mass of dissolved constituents carried directly into the oceans by ground waters can be estimated only within an order of magnitude. Until recently when drilling into the sediments of the continental shelves revealed fresh water moving outward through the sedimentary layers from

Salinity (199 ppm) (Conway's estimate from Clarke's data)[e]		Inorganic salinity (120 ppm)[d]; total salinity including dissolved organics (130 ppm)	
ppm	Percent total dissolved solids	ppm	Percent total dissolved solids
104	52.3	58.4	48.7
29.8	14.9	15	12.5
8.4	4.2	6.3	5.3
8.3	4.2	7.8	6.5
17.1	8.6	13.1	10.9
5	2.5	4.1	3.4
17.7	8.9	11.2	9.4
3.1	1.6	2.3	1.9
1.3	0.7	1	0.8
4[e]	2.0	0.67	0.6
0.3	0.1	—	—

[d] Livingstone (1963a).
[e] Includes aluminum.

the land, direct ground-water addition to the oceans would have been assumed to be an insignificant fraction of that of streams. A maximum estimate can be made indirectly from the difference between total precipitation on and evaporation from the continents; the values thus derived for water that must reach the oceans through some terrestrial route usually average about 10 percent higher than estimates of river discharge. Conceivably this excess could be delivered by subsurface flow. If so, and if these ground waters have about the same total salinity as streams, approximately 4×10^{14} g/year of dissolved solids could be entering the ocean basins from subterranean flow. Both required assumptions are shaky; from the preceding discussion of stream discharge it is clear that a 10 percent difference between total precipitation minus evaporation and stream discharge could be accounted for by errors in either estimate. Also, we do not have good numbers for the dissolved solid content of those ground waters reaching the sea.

Suspended Sediment

To obtain the total flux of river-borne elements into the oceans, it is necessary to know the mass and composition of the suspended sediment load as well as of the dissolved load. Unfortunately, there is a notable lack of data dealing with the concentration, mineralogy, and chemistry of suspended sediment in the world's rivers. Many estimates have been made of the mass of suspended detritus carried to the oceans; here we have tabulated (Table 4.4) those of Conway (1942), Kuenen (1950), Strakhov (1967), Holeman (1968), Judson (1968), and our own.

Of the estimates listed, that of Holeman is the best documented in terms of the inclusion of many data accumulated within the past few years. One of the major problems in estimating worldwide addition of solid materials to the ocean basins is the overwhelming contribution from Southeast Asia. According to Holeman, about 80 percent of all the solids delivered come from that area. In contrast to dissolved materials, for which drainage area is an important consideration and for which the contributions per square kilometer do not differ drastically from one area to another, dis-

Table 4.4
Various Estimates of Suspended Sediment Carried to the Oceans Each Year

Author	Year	Suspended sediment (units of 10^{14} g)	10^6 g/km²
Conway	1942	214	140
Kuenen	1950	325	217
Strakhov	1967	127	85
Holeman	1968	183	120
Judson	1968	93 (corrected by estimate of effect of man's activities)	62
Authors' estimate	1969	183	120

Table 4.5
Suspended Sediment Concentration and Salinity of Several of the World's Rivers

River	Sediment (ppm)	Salinity (ppm)	Suspended sediment/salinity ratio	Source
Mississippi	670	166	4:1	Conway (1942)
	250	223	1.1:1	Edwards et al. (1956); Livingstone (1963a)
Nile	490	168	2.9:1	Conway (1942)
	595	161	3.7:1	Livingstone (1963a); Clarke (1924)
Rio Grande	42,600	881	48:1	Livingstone (1963a); Kennedy (1965)
Danube	350	151	2.3:1	Conway (1942)
Rhone	560	182	3.1:1	Conway (1942)
Amazon	90	36	2.5:1	Gibbs (1967)
Congo	ca. 23	ca. 78	ca. 0.3:1	Gibbs (1967); Livingstone (1963a)
Ob	ca. 65	ca. 130	ca. 0.5:1	Strakhov (1967); Livingstone (1963a)
La Plata	ca. 430	ca. 103	ca. 4.2:1	Strakhov (1967); Livingstone (1963a)
Yukon	ca. 980	ca. 208	ca. 4.7:1	Strakhov (1967); Livingstone (1963a)
Average of rivers of southeastern Asia	—	—	ca. 4.2:1	Strakhov (1967); Livingstone (1963a)

charge of solids shows a wide range. This relation is illustrated by Table 4.5, which lists a number of important rivers and gives the ratio of suspended sediment to dissolved solids. Note especially the differences in the ratios for the Mississippi and for the Nile as cited by various investigators.

We have accepted Holeman's estimate of the mass of solids delivered to the ocean basins each year, approximately 183×10^{14} g/year. This estimate, combined with our preceding decision to use 42×10^{14} g/year as the best value for addition of dissolved solids, yields a worldwide ratio of suspended/dissolved load of about 4:1 and a total flux of about 225×10^{14} g/year through streams to the ocean basins.

The chemical composition of the suspended load poses a problem almost as difficult as the estimation of the volume and composition of ground waters entering the oceans. The only chemical analyses extant of representative material from major rivers are data for the Nile and for the Mississippi. The analyses are cited in Table 4.6, along with a composite analysis made by averaging the two. The marked differences between the Nile

Table 4.6
Chemical Composition of Nile Mud and Mississippi Silt[a]

	Nile	*Mississippi*	*Composite*[b]
SiO_2	45.10	69.96	61.0
Al_2O_3	15.95	10.52	14.3
Fe_2O_3	13.25	3.47	9.0
MgO	2.64	1.41	2.2
CaO	4.85	2.17	3.8
K_2O	1.95	2.30	2.3
Na_2O	0.85	1.51	1.4
CO_2	?	1.40	—
H_2O	6.70	3.78	5.7

[a] Data from Clarke (1924).
[b] Calculated by adjusting the partial analyses to 100 percent and mixing 1:1.

muds and the Mississippi silt serve to illustrate the great differences even between major rivers. The Mississippi silt obviously includes coarser sediment, as evidenced by its high silica content reflecting quartz particles. The Nile muds, on the other hand, are extremely high in Fe_2O_3. The composite would take on much greater significance if data were available for the Amazon, as a tropical representative, and for the Mekong, which would reflect the large Southeast Asia contributions.

Atmospherically Cycled Constituents and Marine Storage Times

The dissolved materials transported by streams and ground water to the oceans remain as dissolved constituents in the oceans or interstitial waters of marine sediments, are chemically or biochemically precipitated, or are released into the atmosphere as sea-spray particles generated at the ocean-atmosphere interface. These atmospherically cycled salts must be subtracted from the total river flux of dissolved materials to arrive at a value for the net land-derived flux of dissolved solids into the oceans.

By using the net annual flux of H_2O to the continents from the ocean and estimating the mean composition of salts in ocean-derived rain falling on the continents, an estimate can be made of the amount of cyclic marine constituents in river waters. Table 4.7 shows the mean chemical composition of rain waters from various parts of the world. An estimate (Jacobs, 1937) of the mean composition of precipitation is also given. To arrive at a "best" estimate of the composition of rain waters and thus, presumably, the composition of marine cyclic salts, Jacobs' average estimate was modified by adding 0.30 ppm K^+ to achieve electrical neutrality and by assuming that the average pH of rain water is 5.7 and that the H_2O is in equilibrium with the P_{CO_2} ($10^{-3.5}$ atm) of the atmosphere. These are reasonable assumptions because median pH values of rain are just under 6, and the residence time of rain in the atmosphere is long enough to attain near-equilibrium with CO_2.

Table 4.8 compares the annual amounts of atmospherically cycled salts derived from the oceans with the amounts of these constituents carried to the oceans annually by rivers. Conway (1942) independently estimated that atmospherically derived Cl constitutes 44 percent of the annual river flux of chlorine; Livingstone (1963b) estimated that 46 percent of the annual river flux of sodium is atmospherically cycled. Both of these estimates are similar to the results shown in Table 4.8.

More than 99 percent of the dissolved solids in river water can be accounted for by the constituents chlorine, sodium, sulfate, magnesium, calcium, potassium, bicarbonate, and silica; the same constituents make up more than 99 percent of the dissolved materials in ocean water (Tables 4.2

Table 4.7
Mean Composition of Cyclic Constituents in Rain Water (ppm)

	Na^+	K^+	Mg^{2+}	Ca^{2+}	Cl^-	SO_4^{2-}	HCO_3^-	SiO_2	Reference
Northern Europe	2.05	0.35	0.39	1.42	3.47	2.19	—	—	After Carroll (1962)
Southeastern Australia	2.46	0.37	0.50	1.20	4.43	Tr	—	—	After Carroll (1962)
United States	0.90	0.23	—	—	1.13	2.02	—	—	Junge and Werby (1958)
Menlo Park, California	2.0	0.25	0.37	0.79	3.43	1.39	4.0	0.29	Whitehead and Feth (1964)
Africa	—	—	—	—	4.1	—	—	—	After Eriksson (1952)
Bermuda	7.23	0.36	—	2.91	12.41	2.12	—	—	Junge and Werby (1958)
Hawaii	1.7	—	—	—	1.7	—	—	—	Mordy (1953)
North Carolina and Virginia	0.56	0.11	0.14	0.65	0.57	2.18	—	—	Gambell and Fisher (1966)
Jacobs' estimate of average composition of precipitation	1.98	—	0.27	0.09	3.57	0.58	—	—	Jacobs (1937)
"Best" estimate assuming pH of average rain water = 5.7 and equilibrium with P_{CO_2} of atmosphere	1.98	0.30	0.27	0.09	3.79	0.58	0.12	—	
Milliequivalents Cations	0.086	0.008	0.023	0.004	—	—	—	0.121 total	
Anions	—	—	—	—	0.107	0.012	0.002	0.121 total	

Table 4.8
Comparison of Atmospherically Cycled Salts and Annual River Load

Constituent	Annual flux of atmospherically cycled salts (units of 10^{13} g)	Amount delivered to oceans annually by rivers (units of 10^{13} g)	Atmospherically cycled salts in rivers (%)
Na^+	7.33	20.7	35.0
K^+	1.11	7.4	15.0
Mg^{2+}	0.99	13.3	7.0
Ca^{2+}	0.33	48.8	0.7
Cl^-	14.0	25.4	55.0
SO_4^{2-}	2.15	36.7	6.0 [a]
HCO_3^-	0.44	190.2	0.2

[a] Holser and Kaplan (1966) suggest that this value may be as high as 30 percent.

and 4.9). A feeling for the huge amount of H_2O and dissolved solids carried to the oceans by streams can be attained simply by calculating the time it would take for river fluxes alone to attain oceanic amounts. Table 4.9 shows that these times vary considerably but only Cl should require

Table 4.9
Mass of Major Dissolved Constituents Delivered to the Oceans Annually and the Time Required for These Constituents To Reach Oceanic Amounts With and Without Correction for Atmospherically Cycled Salts (conventional residence times are given for comparison with storage times)

Constituent	Mass delivered by rivers to ocean annually (units of 10^{14} g)	Mass in ocean (units of 10^{20} g)
Fe^{2+}	0.223	0.0000137
Al	0.003	0.0000137
SiO_2	4.26	0.08
HCO_3^-	19.02	1.9
Ca^{2+}	4.88	6
K^+	0.74	5
SO_4^{2-}	3.67	37
Mg^{2+}	1.33	19
Na^+	2.07	144
Cl^-	2.54	261
Dissolved organic C	3.2	0.007
H_2O	325,000	13,550

[a] After Goldberg (1965). Residence time is expressible in terms of the same variables as storage time; however, residence time calculations may or may not contain corrections

more than 100 million years. A correction for atmospherically cycled marine salts can be applied to river fluxes and the same calculation made. The time necessary to attain oceanic amounts for river-derived sodium and chlorine is about doubled when the corrections for cyclic sodium and chlorine are made. This further illustrates the large amount of sodium and chlorine that is cycled through the atmosphere. The other major constituents are predominantly derived from the weathering of rocks, and the correction for atmospherically cycled salts affects only slightly the time required for them to reach oceanic amounts.

These times, uncorrected for cyclic salts, may be called *storage times;* storage time can be expressed by

$$\lambda = \frac{A}{dA/dt},$$

where A is the total amount of the constituent in the ocean and dA/dt is the amount in solution introduced by streams per unit time. Storage times for elements are given in Table 4.9.

The storage time of a particular constituent in the ocean, if we consider the ocean as a well-mixed, steady-state system, is indirectly a measure of the rate at which stream-derived dissolved constituents leave the ocean.

Storage time—time for river fluxes to attain oceanic amounts (units of 10^6 years)	Time for river fluxes to attain oceanic amounts correcting for atmospherically cycled salts (units of 10^6 years)	Residence time[a] (units of 10^6 years)	
		River input	Sedimentation
0.00006	0.00006		0.00001
0.0046	0.0046	0.004	
0.02	0.02	0.04	0.01
0.1	0.1		
1.23	1.24	1	8
6.8	8	10	11
10.1	10.7		
14.3	15.4	22	45
69.7	108	210	260
103	230		
0.0002			
0.042			

for atmospherically and rock-derived cyclic materials and for eolian particulate matter. Also, the amount of the constituent in suspension in the ocean or streams may enter the residence time calculation.

The sinks and processes by which these constituents are removed from the ocean will be considered in Chapter 11.

GLACIAL FLUX

Until recently the mass of material carried to the oceans by glacial processes was impossible to estimate because of our scant knowledge of the great ice-covered continent of Antarctica. As a result of the International Geophysical Year and the researches of an international group of scientists, much has been learned about Antarctic geography, geology, atmospheric physics, and oceanography within the last decade. We are now more aware of the large effect processes in the Antarctic have on the whole earth system.

The present flux of materials into the oceanic system by glaciers is second only to total river flux. Yevteyev (1959) estimated that about 0.69 km³ of chemically unweathered, fine-grained rock material is eroded annually from Antarctica. Because of the absence of river runoff and chemical weathering, all of this material is eroded from the continent by glaciers. Presumably the composition of this detritus would be very near that of average crustal rock. Therefore, if we assume a density of 2.7 g/cm³ for this material, about 19×10^{14} g of detritus is eroded by glaciers from Antarctica yearly.

Because the Antarctic continent makes up nearly 90 percent of the great modern icecaps, an estimate for the amount of detritus delivered directly by all glaciers to the oceans is $19/0.9 \sim 20 \times 10^{14}$ g/year. The composition of glacial debris may be approximated by that of average igneous rock and the flux of major constituents carried by glaciers to the oceanic system calculated (Table 4.11).

The total annual load of ice-borne materials is about 50 percent of the total dissolved load of rivers, and ice is second only to rivers as a source of materials for the oceans. This glacial source has varied with time because of changes in the magnitude of glaciation; e.g., during times of major continental glaciation during the last 1 or 2 million years of the earth's history, ice-derived materials probably constituted a greater proportion of the detritus transported to the oceans than today, but during most of the past they have probably constituted a much smaller fraction than today.

MARINE EROSION FLUX

The erosive energy of waves and currents along coastal shorelines adds material to the oceanic system. Kuenen (1950) calculated that 0.12 km³/year of material is eroded by marine processes, and Barrell (1925) estimated 0.08–0.4 km³. At an average porosity of 15 percent and a solid density of 2.7 g/cm³ for the material eroded, these estimates would amount to $2-9 \times 10^{14}$ g of solids released to the ocean by marine erosion annually.

We can calculate the flux of materials that reach the oceans through marine erosion by using 2.5×10^{14} g/year as the total flux and the composition of the average sedimentary rock as representative of wave-eroded materials (Table 4.11).

ATMOSPHERIC DUST FLUX

Atmospheric transport of materials into the oceanic system is the least important pathway by which the major constituents reach the ocean. This is fortunate because, as will become apparent from the following discussion, the estimates are based on rather tenuous assumptions. There are two chief processes by which eolian materials enter the ocean: (1) fine-grained solids transported to the ocean by wind storms and (2) dust, which has been carried into the upper atmosphere from the continents or from active volcanoes, and falls slowly on the earth's surface.

The importance of tropospheric transport can be estimated by using the area of major occurrences of dry atmospheric haze at sea as a measure of the area of dust fall upon the ocean due directly to wind storms in arid regions on the continents. This area is about 1 percent (36×10^{14} cm^2) of the sea surface. An average dust storm deposits about 0.3 mg/cm^2 of dust. Assuming an average of five storms per year over the sea surface, about 0.54×10^{14} g/year of wind-storm-transported materials falls on the ocean.

Slowly settling dust from the stratosphere also falls into the sea. We assume that the amount of dust in the troposphere is nearly a steady-state concentration. At any moment the air contains about 10^{-12} cm^3 of dust/cm^3 of air. At a density of 2.5 g/cm^3 for atmospheric dust, and an average density of the troposphere of 1.223×10^{-3} g/cm^3 and a mass of 51.3×10^{20} g, the troposphere at any moment contains about 0.105×10^{14} g of dust. The residence time of atmospheric dust is about 1 year; thus 0.074×10^{14} g/year of slowly settling dust falls on the ocean surface. Therefore, the total amount of material from storms and stratospheric dust that reaches the ocean is approximately 0.6×10^{14} g/year. This is 20–30 percent of the rate of sedimentation of fine-grained materials on the abyssal plains of the ocean basins. Assuming that atmospheric dust has the approximate chemical composition of loess (eolian-deposited, fine-grained sediment), the flux of major individual eolian constituents may be calculated and is given in Table 4.11.

Cosmic dust, including iron-nickel spherules and crystalline olivine-pyroxene chondrules, enters the atmosphere as the earth moves through space. This dust is a third, but minor, source of atmospheric materials for the ocean. Wasson et al. (1967) estimate that 10^{-7} g/cm^2 of extraterrestrial dust falls on the earth's surface each year. Therefore about 0.036×10^{14} g of cosmic dust enter the ocean annually. This estimate may be much too high.

Table 4.10
Various Estimates of Excess Volatiles (units of 10^{20} g)

	Goldschmidt (1933)	Rubey (1951)	Nicholls (1965)	Horn (1966)
H_2O	13,500	16,600	16,300	16,700
Total C and CO_2	362	910	2,490	1,110
S	13	22	24	31
N	39	42	44	39
Cl	263	300	335	560
H	7	10	} 13 {	16
B, Br, A, F, etc.		4		

FLUX OF MATERIALS DERIVED FROM THE SUBSURFACE

The flux of materials from the subsurface can be estimated from the quantity of "excess volatiles" that has been delivered to the earth's surface through time. The term *excess volatiles* was defined by Rubey (1951) as those materials that are far too abundant in the atmosphere, hydrosphere, and biosphere and in sediments to be accounted for simply by rock weathering and that are derived from subcrustal sources. Various estimates of the mass of the excess volatiles (water, carbon, sulfur, nitrogen, chlorine, hy-

Table 4.11
Summary Table Showing Estimated Annual Flux of Various Chemical Species into and out of the Ocean Basins (all values in units of 10^{14} g)

Chemical species	Streams		Ground water	Aerosols
	Dissolved	Suspended		
SiO_2	4.4	121.0	0.5	—
Al	—	14.9	—	—
Fe	0.2	12.2	—	—
Ca	5.0	5.5	0.6	—
Mg	1.4	2.8	0.2	0.1
K	0.8	3.9	0.1	0.1
Na	2.1	2.0	0.2	0.7
Cl	2.6	—	0.3	1.4
SO_4	3.8	—	0.4	0.2
HCO_3	19.0	8.3	2.0	—
Total	39.3	170.6	4.3	2.5
Percent of total	16.6	72.0	1.8	1.0
Others (NO_3^-, organic matter, fluorine, sulfide, etc.)				
Estimated grand total				

drogen, boron, bromine, argon, and fluorine) have been made (Table 4.10). Considering the variables included in these estimates, they are in relatively good agreement.

The relative importance of the various pathways by which these materials reach the earth's surface is not very well known. Volcanoes, fumaroles, hot springs, extrusive volcanic lavas, and slow seepage through the earth's crust provide possible mechanisms by which these constituents reach the earth's surface. Rubey concluded that these materials are derived primarily during the crystallization of magmas and that thermal springs may be their most important avenue of escape from the earth's subsurface to its surface. In any case, an estimate of the present flux of subsurface materials that is probably a maximum may be made by assuming that their release has been essentially linear through time, and thus calculable by dividing the amounts of those materials by the age of the earth, 4.5×10^9 years. Figure 4.1 and Table 4.11 give the subsurface flux based on Horn's and Rubey's estimates of total amounts of excess volatiles; the estimate in Figure 4.1 includes H_2O.

The present subsurface flux of materials to the ocean is surely slow compared with fluxes from other sources. For example, from the preceding calculation of the rate of addition of H_2O to the earth's surface, it would take all of geologic time to fill the present ocean basins, whereas streams would take only 42,000 years. Nevertheless, all the water and chlorine in the oceans have presumably been derived by degassing of the earth's interior during and since formation of the primordial crust.

Dust	Marine erosion	Ice	Subsurface	Totals	Percent of total
0.4	1.5	13.0		140.8	59.5
0.1	0.2	1.8		17.0	7.2
	0.1	0.6		13.1	5.5
	0.1	0.6		11.8	5.0
	0.1	0.3		4.9	2.1
	0.1	0.6		5.6	2.4
	—	0.5		5.5	2.3
	—	—	0.1	4.4	1.8
	—	—		4.4	1.8
	0.2	—	0.3	29.8	12.5
0.5	2.3	17.4	0.4	237.3	
0.2	1.0	7.3	0.2	~	100.1
				~ 3–4	
				~ 240×10^{14} g/year	

SUMMARY

Data on the rates of transport of material to the oceans are summarized in Table 4.11. Streams are responsible for most of the materials being deposited in the oceans today. The net addition by all agents is about 250×10^{14} g/year, and the chief constituents of this addition are sodium, potassium, calcium, magnesium, chlorine, silicon, sulfur, and carbon. Particulate material composes about 80 percent of the stream load and dissolved material about 20 percent.

The preceding analyses of the rates at which materials are being added to the ocean basins provide some useful data. Even though the numbers obtained are hardly more than orders of magnitude, because of the uncertainties in and incompleteness of the data, the order of importance of the various agents is clearly established, with streams doing about 85–90 percent of the material transport, ice about 7 percent, ground water and waves about 1–2 percent, and wind and deep-seated sources less than 1 percent each. The only obvious losses from the ocean basins are evaporated water with its contained aerosols, chiefly sodium chloride, which fall on the land and return deviously to the sea. This cycled water is sufficient to cause all of the volume of the ocean to reflux through streams in about 42,000 years, but the contained salts in evaporated water that falls as rain are so low that the renewal time for the NaCl through the atmosphere is of the order of 100 million years.

Thus the ocean basins today are an effective trap for the materials derived from the continents, and the overall process of erosion and deposition is similar to the operation of a giant reflux condenser, or to the continuous addition of tap water to a boiling kettle, while the solution volume in the kettle is held constant. At the present rate, the mass of dissolved and suspended materials carried by streams, and left dissolved in the oceans or sedimented on to their floors, would be equal to the mass of the oceans in about 1 billion years. However, the rate is such that if it continued undiminished (which it presumably could not), the continents would be reduced close to sea level in about 10 million years or so. Yet we are sure that continents have existed for billions of years.

It is as if we had investigated someone's checking account at a bank and discovered that for the current month expenditures were sufficient to exhaust the balance within a year and deposits miniscule compared with expenditures. Our only other piece of information is that the account had been in existence for many years. The continents represent the present account balance, erosion represents the checks drawn, and aerosol recycling plus cosmic dust addition are the deposits to the account. There are obviously many ways to account for the situation at the bank—withdrawals could have been less in the past, deposits could have been greater, the initial balance could have been large; it is even possible that the bank suspended business for a long time interval. But in the absence of any records except those for the current month, can we determine how the many variables fluctuated with time to give the end result?

Let us now look at erosion of the continents in more detail to see what controls present-day "expenditures." If it can be shown that the rate-determining factors are not transitory phenomena, such as particular kinds of vegetation, then we have a hope of deducing the order of magnitude of erosional rates of the past. At this stage of the investigation it seems inescapable from the mere presence of large continental masses that major additions have been made to the continental bank account during the past.

REFERENCES

Alekhin, O. A. and Brazhnikova, L. V., 1963, Removal of solutes from continents by rivers and the relationship of this process to the mechanical erosion of the Earth's surface: in *Chemistry of the Earth's Crust*, A. P. Vinogradov, ed., Israel Program for Scientific Translations (1966), S. Monson, Jerusalem, 291–303.

Barrell, J., 1925, Marine and terrestrial conglomerates: *Bull. Geol. Soc. Am., 36*, 279–342.

Carroll, D., 1962, Rain-water as a chemical agent of geologic processes—a review: *U.S. Geol. Surv. Water Supply Paper, 1535-G.*

Clarke, F. W., 1924, The data of geochemistry: *U.S. Geol. Surv. Bull, 770.*

Conway, E. J., 1942, Mean geochemical data in relation to oceanic evolution: *Roy. Irish Acad. Proc., 48,* sec. B., 119–159.

Edwards, M. L., Kister, L. R., and Scarcia, G., 1956, Water resources of the New Orleans area, Louisiana: *U.S. Geol. Surv. Circ., 374.*

Eriksson, E., 1952, Composition of atmospheric precipitation. II. Sulfur, chloride, iodine compounds: *Tellus, 4,* 280–303.

Gambell, A. W., and Fisher, D. W., 1966, Chemical composition of rainfall of Eastern North Carolina and Southeastern Virginia: *U.S. Geol. Surv. Water Supply Paper, 1535-K.*

Gibbs, R., 1967, The geochemistry of the Amazon River Basin: Part I: The factors that control the salinity and the composition and concentration of suspended solids: *Bull. Geol. Soc. Am., 78,* 1203–1232.

Goldberg, E. D., 1957, Biogeochemistry of trace metals: in Treatise on marine ecology and paleoecology, vol. 1, J. W. Hedgepeth, ed., *Geol. Soc. Am. Mem., 67,* 345–357.

Goldberg, E. D., 1965, Minor elements in sea water: in *Chemical Oceanography,* vol. 1, J. P. Riley and G. Skirrow, eds.: Academic Press, Inc., New York, 163–196.

Goldschmidt, V. M., 1933, Grundlagen der quantitativen Geochemie: *Fortschr. Mineral., 17,* 112–156.

Holeman, J. N., 1968, The sediment yield of major rivers of the world: *Water Resources Res., 4,* 737–747.

Holser, W. T., and Kaplan, I. R., 1966, Isotope geochemistry of sedimentary sulfates: *Chem. Geol., 1,* 93–135.

Horn, M. K., 1966, Written communication.

Hutchinson, G. E., 1967, A treatise on limnology, vol. *1: Geography, Physics, and Chemistry:* John Wiley & Sons, Inc., New York.

Jacobs, W. C., 1937, Preliminary report on the study of atmospheric chlorides: *Monthly Weather Rev., 65,* 147–151.

Judson, S., 1968, Erosion of the land: *Am. Scientist, 56,* 356–374.

Junge, C. E., and Werby, R. T., 1958, The concentration of chloride, sodium, potassium, and sulfate in rain water over the United States: *J. Meteorol., 15,* 417–425.

Kalle, K., 1943, *Der Staffhaushalt des Meeres, Probleme der Kosmichen Physik,* vol. 23: Leipzig Akad. Verlagsegesellschaft, Becker and Erler, Leipzig.

Kennedy, V. C., 1965, Mineralogy and cation-exchange capacity of sediments from selected streams: *U.S. Geol. Surv. Profess. Paper, 433-D.*

Kossinna, E., 1921, *Die Tiefen des Weltmeeres:* Veroffentlichungen des Instituts fur Meereskunde, Univ. Berlin, Neue Folge, A, 9.

Kuenen, Ph. H., 1950, *Marine Geology:* John Wiley & Sons, Inc., New York.

Livingstone, D. A., 1963a, Chemical composition of rivers and lakes: in Data of geochemistry, 6th ed., M. Fleischer, ed., *U.S. Geol. Surv. Profess. Paper, 440-G.*

Livingstone, D. A., 1963b, The sodium cycle and the age of the ocean: *Geochim. Cosmochim. Acta, 27,* 1055–1069.

Meinardus, W., 1928, *Der Kreislauf des Wassers:* Festrede Georg-August Univ., June 1, 1927, Gottingen.

Mordy, W. A., 1953. A note on the chemical composition of rainwater: *Tellus, 5,* 470–474.

Murray, Sir John, 1887, On the total annual rainfall on the land of the globe, and the relation of rainfall to the annual discharge of rivers: *Scottish Geograph. Mag., 3,* 65–77.

Nicholls, G. D., 1965, The geochemical history of the oceans: in *Chemical Oceanography,* vol. 2, J. P. Riley and G. Skirrow, eds.: Academic Press, Inc., New York, 277–294.

Rubey, W. W., 1951, Geologic history of sea water: *Bull. Geol. Soc. Am., 62,* 1111–1147.

Sackett, W. M., and Arrhenius, G. O. S., 1959, Aluminum content of ocean and natural waters: *Preprints Intern. Oceanog. Congr.,* M. Sears, ed.: A.A.A.S., Washington, D.C., 824.

Strakhov, N. M., 1967, *Principles of Lithogenesis,* vol. 1: Consultants Bureau, New York.

Wasson, J. T., Adler, B., and Oeschger, H., 1967, Aluminum-26 in Pacific Sediment: Implications: *Science, 155,* 446–448.

Whitehead, H. C., and Feth, J. H., 1964, Chemical composition of rain, dry fallout, and bulk precipitation at Menlo Park, California, 1957–1959: *J. Geophys. Res., 69,* 3319–3333.

Williams, P., 1966, Oral communication.

Yevteyev, S. A., 1959, Determination of the amount of morainal material carried by glaciers to the east Antarctic coast: *Inform. Bull. Soviet Antarctic Expedition, 11,* 14–16.

5 | # Rates and Controls of Continental Erosion

Because streams are the dominant purveyors of material to the oceans, and probably have been so in the past, we now look at their work in terms of the similarities and differences that are observed from continent to continent. If present-day rates are to be a guide to those of the past, we must be able to assess at least the effects of variations of latitude (and hence climate and vegetation), average elevation, relief, and rock type exposed to erosion. We know that each of these factors can cause extreme variations in rates of material transport when small areas are considered. The only hope is that their effects tend to counterbalance when regions as large as continents are considered and that differences between continents are not so great that we are prevented from predicting the effects of different configurations of the continents in the past.

In this chapter, chemical denudation will be considered in greater detail than erosion by removal of particulate debris, largely because of the relative abundance of data.

RUNOFF FROM THE CONTINENTS

The total runoff from the continents, as shown in Chapter 4, is about 0.32×10^{20} g/year, or 36.5×10^6 ft^3/sec. If we make a cumulative plot of the discharge of the individual continents versus their areas (Figure 5.1), it emerges that although some continents are wetter and others drier, their deviations from the world average are not extreme. The runoff per unit area from Asia and Europe is almost exactly equal to the world average; Africa and North America are a little lower; South America is considerably wetter. Africa, South America, and North America, considered as a unit, have almost exactly the same discharge per unit area as Asia. Antarctica is frozen, and Australia is arid; as a consequence they contribute little runoff, but their areas are not sufficient to influence the world average very much.

What seems to be emerging is that if the total area of land has not changed markedly throughout the past, total stream discharge may well have always been of the same order of magnitude. The deviations of the individual continents today, because of their differences of latitude and elevation, are not extreme. If all the continents could somehow be situated like South America, runoff would almost be doubled, and if like Africa, runoff would be diminished about 20 percent. Today average rainfall on the land is about the same as that upon the seas. It does not seem likely that shifting the various continents relative to each other would have drastic effects on total stream discharge.

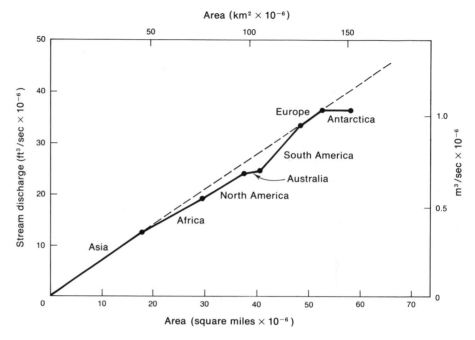

Figure 5.1 Cumulative plot of river discharge versus continental area. Note that Asia and Europe are representative of the world average, Australia and Africa are dry, and South America is very wet (data from Livingstone, 1963).

There are several variables we have not been able to assess; what, for example, would be the effects on total precipitation of an increase in the average annual temperature of the earth's surface environment? What would be the effects on runoff if all land vegetation were eliminated?

Table 5.1
Chemical and Mechanical Denudation of the Continents

Continent	Annual chemical load delivered to oceans (units of 10^{14} g)	Annual chemical denudation (metric tons / km²)
North America	7.0	33
South America	5.5	28
Asia	14.9	32
Africa	7.1	24
Europe	4.6	42
Australia	0.2	2
Total	39.3	

CHEMICAL VERSUS MECHANICAL DENUDATION

Although the amount of water that leaches the continents and sweeps debris to the oceans each year is a major factor in the rate at which transport to the oceans is accomplished, it does not necessarily hold that there is a one-to-one correspondence between water volume and material moved. Table 5.1 gives the amount of material removed from each continent as dissolved or suspended load, as well as the rate of removal per square kilometer of the total continental area. Specific discharge (discharge per unit area) seems to be uncorrelated with either the rate of chemical or mechanical denudation. South America, with the highest specific discharge, has the second lowest rate of chemical denudation; Europe, with an intermediate rate of specific discharge, has the highest rate of chemical denudation. On the other hand, Asian streams, with an intermediate specific discharge, are transporting about 20 times as much solid material to the sea from each square kilometer as are the streams of Africa.

Consequently, although the continents tend to be fairly uniformly watered, they show quite different resistances to being worn away. Furthermore, their chemical resistance apparently has primary controls that are different from their mechanical resistance, as shown by lack of correlation between chemical and mechanical denudation rates. A graph showing average continental elevation as related to the two types of denudation (Figure 5.2) provides some insight into the fundamental differences between the two types of erosion. Chemical denudation rates are apparently independent of continental elevation; mechanical rates show a tendency toward exponential increase with increasing elevation. Analysis of data for individual drainage basins usually shows an increase in chemical denudation rate with increasing elevation, but on a continental scale this effect seems to be overwhelmed by rock type.

Annual mechanical load delivered to oceans (units of 10^{14} g)	Annual mechanical denudation (metric tons/km²)	Ratio mechanical to chemical
17.8	86	2.6
11.0	56	2.0
145.0	310	9.7
4.9	17	0.7
2.5	27	0.65
2.1	27	>10.0
183.3		Overall 4.7

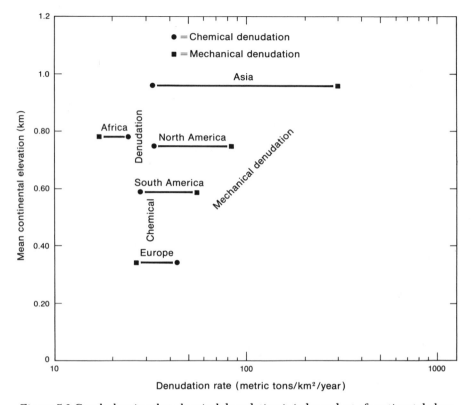

Figure 5.2 Graph showing that chemical denudation is independent of continental elevation, whereas mechanical denudation generally increases with increasing elevation. Africa shows "abnormally" low values for both types of denudation (data from Livingstone, 1963; Holeman, 1968).

Apparently the two phenomena can be considered separately for the time being, even though we are aware that they are intimately interrelated in some important instances. Granites are slow to erode mechanically before the grains are loosened chemically. On the other hand, the erosion of shales and sandstones may be overwhelmingly controlled by purely physical factors such as water velocity and steepness of slope.

CHEMICAL DENUDATION

In our attempt to elicit the primary controls of chemical denudation, the basic data to be explained are the chemical compositions of the river discharges of the individual continents. These are shown in Figure 5.3. Some of the important relations illustrated by Figure 5.3 are that the waters are dominated by Ca^{2+} and HCO_3^-, and it is these two species that account for most of the difference in total dissolved load between the dilute waters from South America and the more concentrated ones from Europe. Silica and Cl⁻ are the only constituents that show a clear-cut inverse relation with total dissolved solids. Data for Na and K are not complete, but K tends to be low and nearly constant, while Na is irregularly variable. If Asia is excluded, the waters from the other continents have nearly equal concentrations of Mg^{2+} and SO_4^{2-}. Also, it is apparent that the waters of Africa and South America differ by a nearly constant concentration factor for all species; the composition of African waters would be approximated by evaporating South American waters to about one half their present volume.

The reciprocal relation between Ca^{2+} and SiO_2 concentrations is suggestive of control by rock type. Calcium can be expected to be derived chiefly from limestones and evaporites, and silica from the weathering of silicate minerals. Also, the concentrations of HCO_3^- in the rivers discharg-

Figure 5.3 Concentrations of dissolved species in continental drainage. The major differences from continent to continent are in the contents of Ca^{2+} and HCO_3^- (data from Livingstone, 1963).

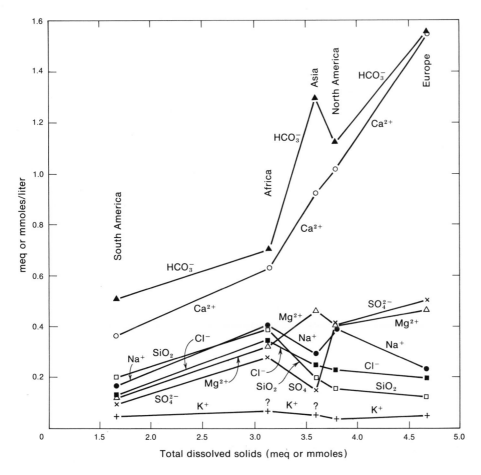

ing from the continents are a measure of the "reactivity" of the continents; they reflect the CO_2 neutralized by the rocks that have been encountered by the water during its residence on the land. A "limy" continent should be more reactive than one made up of silicate minerals.

A fair mass balance can be achieved if it is assumed that, on the average, silicate minerals produce one HCO_3^- from each CO_2 molecule while releasing $2SiO_2$ (silicate $+ CO_2 = HCO_3^- + 2SiO_2$), and that all the Ca^{2+} and Mg^{2+} comes from carbonate minerals, except for enough to balance SO_4^{2-}. Table 5.2 shows that the HCO_3^- content of streams can be divided into that formed by reaction of CO_2 with limestones and that resulting from reaction with silicate minerals. Each molecule of CO_2 neutralized by a carbonate mineral yields $2HCO_3^-$; each one neutralized by a silicate mineral produces a single HCO_3^-.

Only Africa is strongly anomalous in terms of calculated versus observed HCO_3^-, indicating perhaps that the silicate reactions there give a higher than 2:1 silica yield per molecule of CO_2 consumed. On the other hand, South America is a little deficient in its balance, indicating a lower yield. The bicarbonate values for Africa, because of the high silica content of its waters, probably should have the value for limestone bicarbonate corrected for Ca^{2+} and Mg^{2+} released from the silicates.

A comparison of the percentage of CO_2 neutralized by silicates with the total dissolved load in the streams shows that if a high percentage of the load is won from silicate minerals, the total load is low. By plotting rates of chemical denudation as a function of the ratio $(SiO_2 + HCO_3^-)/SiO_2$, as in Figure 5.4, a roughly linear relationship is obtained. The plot has the advantage of permitting a short extrapolation to a value of 1.5 on the abscissa, which would be the limiting rate of chemical denudation of a continent from which all bicarbonate was produced by reaction with silicates. Even such a short extrapolation is open to question, but it looks as if a denudation rate of about 20 tons/km^2/year for a continent composed entirely of crystalline or argillaceous sediments is a reasonable minimum.

Table 5.2

Continent	Limestone bicarbonate (meq Ca^{2+} + meq Mg^{2+} − meq SO_4^{2-})		Silica bicarbonate (½ mmoles SiO_2)	
	HCO_3^-	CO_2 neutralized	HCO_3^-	CO_2 neutralized
Europe	1.49	0.75	0.06	0.06
Asia	1.10	0.55	0.10	0.10
North America	1.06	0.53	0.07	0.07
Africa	0.65	0.33	0.20	0.20
South America	0.38	0.19	0.10	0.10
World average	0.85	0.42	0.11	0.11

Rain waters have a short residence time on the land; to what extent do streams saturate with respect to the materials they traverse? Because CO_2 is the chief aggressive constituent, let us first investigate the CO_2 available for reaction in streams. In Figure 5.5 the internal CO_2 pressure of a number of streams has been plotted as a function of total dissolved solids. The apparent pressure of CO_2 was obtained from the HCO_3^- content and the pH, utilizing the relation $P_{CO_2} = (H^+) (HCO_3^-) (10^{7.8})$, which is valid at 25°C. No attempt was made to correct for temperature or to convert concentrations of HCO_3^- to activities, so the individual values shown are only approximate. However, several major conclusions can be drawn; the CO_2 pressure is independent of total dissolved solids, and it shows no obvious correlation with the climatic zone in which the stream flows. Also, the average pressure is about $10^{-2.5}$ atm of CO_2—about 10 times that of the earth's atmosphere. A pressure of $10^{-2.5}$ atm for stream waters is just about that estimated as the average for waters from soils. Perhaps the CO_2 pressure of streams reflects a steady-state condition resulting from production of CO_2 by oxidation of organic matter in the water, resulting in continuous loss to the atmosphere by diffusion from the stream surface.

If streams obtain enough Ca from the materials over which they flow to equilibrate with calcite, their internal CO_2 pressure should be calculable from the relation

$$H_2O + CO_{2(gas)} + CaCO_{3(solid)} = Ca^{2+} + 2HCO_3^-,$$

using the observed concentrations of Ca^{2+} and HCO_3^- to obtain a value for P_{CO_2}. The results of this calculation are shown in Figure 5.6, in which P_{CO_2} values that would result if equilibrium with calcite had been obtained are plotted against the ratio $(SiO_2 + HCO_3^-)/SiO_2$, which has been shown to be an index of the "argillaceous" or "limy" nature of the terrain being eroded. Only the waters from Europe approximate saturation with $CaCO_3$ at the P_{CO_2} of the stream waters. Waters from the other continents are markedly undersaturated, the degree of undersaturation increasing as the areas

Total A + B (meq)	Observed bicarbonate (meq)	Percent CO_2 neutralized by silicates
1.54	1.54	7
1.20	1.20	15
1.13	1.11	12
0.85	0.72	38
0.48	0.51	35
0.96	0.96	19

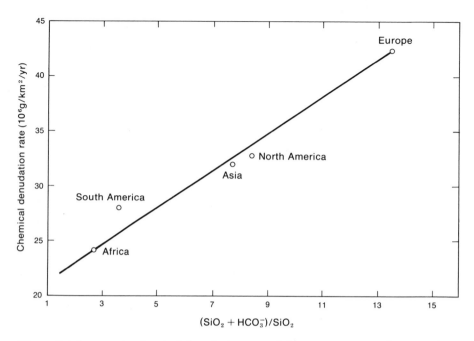

Figure 5.4 Increase in chemical denudation rate of the continents as a function of the silica and bicarbonate content of the stream waters, indicating that denudation rate increases as the proportion of carbonate rocks eroded increases (data from Livingstone, 1963).

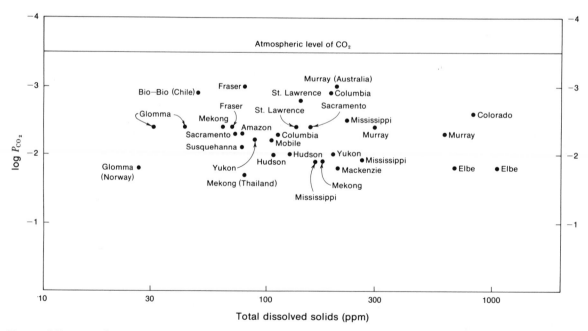

Figure 5.5 Internal CO_2 pressure of some major rivers as a function of total dissolved solids. The CO_2 pressure was estimated from the pH and HCO_3^- content of the streams. Apparently CO_2 content is nearly independent of total dissolved solids, river size, rock type drained, or latitude. Values for several streams have been plotted at times of marked differences in discharge and hence total dissolved load. The average CO_2 pressure calculated is about 10 times that of the atmosphere (data from Durum et al., 1960).

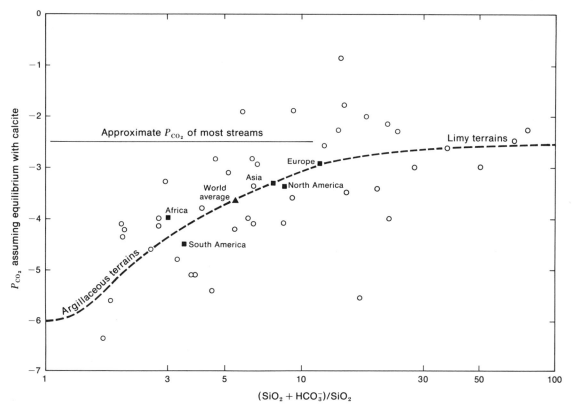

Figure 5.6 Demonstration that the CO_2 pressure at which stream waters would be approximately saturated with calcite for their contents of Ca and HCO_3^- is related to the silica content of the waters and hence to the type of terrain being denuded. Circles represent individual rivers.

drained expose increasing proportions of silicate-bearing materials to the waters moving across or through them. It is apparently a coincidence that the average river water of the world is saturated with calcium carbonate at the P_{CO_2} of the earth's atmosphere.

RATE OF ACID CONSUMPTION BY THE CONTINENTS

The continents are continuously using CO_2 from the atmosphere as their rocks react. Each liter of water contains about 0.5 meq of bicarbonate that comes from neutralization of CO_2 (Table 5.2). As we shall see later (Chapter 11), this CO_2 must eventually be restored to the atmospheric system by reactions in the oceans or in deposited sediments. However, at this juncture we can ask how long it would take to deplete the CO_2 of the atmosphere if there were no restorative process.

The mass of CO_2 in the atmosphere is about 2.3×10^{18} g and that neutralized by the continents is 7.0×10^{14} g/year. Therefore, removal of all CO_2 at

Table 5.3

Continent	Millimoles of CO_2 neutralized per liter of river discharge	Ratio total load to millimoles CO_2 neutralized
Europe	0.81	5.7
Asia	0.65	5.5
North America	0.60	6.3
Africa	0.53	5.9
South America	0.29	5.7

the present rate of consumption would require only $(2.3 \times 10^{18})/(7.0 \times 10^{14}) \simeq 3300$ years. This number gives us a feel for the rate required for restorative processes if the CO_2 content of the atmosphere is to be kept at the present level, and shows the approximate time scale on which the atmosphere might change if they failed to operate. The relative effectiveness of the continents is shown in Table 5.3. Europe can thus be considered the most basic of continents and South America the least. The table also demonstrates that the total dissolved load is largely dependent on the reaction with CO_2 and that solution of minerals that do not require CO_2 does not obscure the relation.

SUMMARY OF CONTROLS OF CHEMICAL DENUDATION

What seems to emerge from the preceding analysis of the composition of the waters draining from the continents is a dominant control by rock type.

Table 5.4
Water Types[a]

	Limy		Argillaceous	
	Miocene limestone	Danube River	Wissahickon schist	Amazon River at Obidos
SiO_2	8.9	5.6	14.0	10.6
Ca^{2+}	48.0	43.9	3.1	5.4
Mg^{2+}	5.8	9.9	1.2	0.5
Na^+	4.0	2.8	3.3	1.6
K^+	0.7	1.6	0.8	1.8
HCO_3^-	168.0	167.0	21.0	17.9
SO_4^{2-}	6.4	14.7	1.2	0.8
Cl^-	4.8	2.4	2.4	2.6
Total	246.6	247.9	47.0	41.2

[a] Analyses of ground waters from White et al. (1963). Analyses of river waters from Livingstone (1963). Analyses in ppm.

The concentrations of the various chemical species that are dissolved by reaction with CO_2 reach levels that may be almost independent of the time interval of contact with the rocks. This contention receives some support from the chemical analyses presented in Table 5.4, in which selected analyses of underground waters from various rock types have been matched with analyses of rivers whose drainages are dominated by the same kinds of rocks. Water from limestone is very like that of the Danube, and water from the Wissahickon schist is almost identical to that of the Amazon. Such relations tend to relegate climatic or vegetational factors to secondary roles in chemical denudation. The amount of material that is won from a continent is dictated by its total area, modified by a rock-type factor. The rates of chemical lowering of the present continents differ only by about a factor of 2, and variations expected from continents made up respectively of the most reactive (limestones) and the least reactive (argillaceous materials) rocks probably would be only a factor of 3.

MECHANICAL DENUDATION

The controls of mechanical denudation of the continents are so complex and the data so sparse that we shall make little attempt here to assess the relative roles of the variables even semiquantitatively. As pointed out in Chapter 4, about 80 percent of the particulate material carried into the oceans by streams comes from Southeast Asia. Thus at a given moment in time a small fraction of the total land area can dominate the rest; this information alone almost precludes reconstruction of mechanical denudation rates for some given instant of the geologic past. There does, however, as

Crystalline		Salty	
Granite, South Carolina	Nile River	Ecca shale, South Carolina	Rio Grande River at Laredo, Texas
35.0	20.1	32	30.0
13.0	15.8	62	109.0
4.3	8.8	64	24.0
8.4	15.6 ⎱	92	⎰ 117.0
3.5	3.9 ⎰		⎱ 6.7
72.0	85.8	362	183.0
6.9	4.7	106	238.0
3.8	3.4	140	171.0
146.9	158.1	858	878.7

shown in Figure 5.2, seem to be a gross correlation of denudation rate with continental elevation. The rate appears to increase exponentially with increasing elevation, suggesting that if in the geologic past the continents were uniformly high, the amount of debris shed to the ocean basins per year would be many, many times what would be shed if they were nearly awash.

Some striking studies have been made of the effects of effective precipitation (that which runs off) in some small basins on sediment yield, as well as on the intercorrelation of precipitation and vegetation type. As effective precipitation increases from zero, the amount of sediment eroded rises steeply. A small amount of runoff does a lot of erosion in desert areas, where there is little vegetation cover. With increase of precipitation grasses cover the landscape and erosion rate diminishes. Further increase results in forests and still further reduction of sediment washed down the streams. Although these relations are oversimplified in terms of the behavior of many drainage basins, they do emphasize the important interplay between runoff and vegetation in relation to mechanical denudation rates. The Colorado River, which drains a relatively high and semiarid region, is mechanically denuding its drainage area more than 4 times as fast as the Mississippi.

Windblown material, or *loess,* is particularly susceptible to erosion when its vegetative cover is breached. The silt-laden rivers of China owe most of their load to the results of human occupation of extensive areas of loess. A 10-fold increase in erosional rate as a result of a change of land area from pasture to cropland is not at all uncommon.

It may not be possible to get a reasonable estimate of the mass or chemical composition of the mechanical load of the world's rivers as it was prior to man's significant alteration of the landscape. About all we can do at the moment is to estimate the extreme variations we might expect from the present-day average rate of 120 metric tons/km²/year. Perhaps these variations are indicated by the range in ratios of suspended to dissolved load of various major rivers (see Table 4.5, Chapter 4) from about 0.3:1 for the Congo to 48:1 for the Rio Grande, a factor of about 150.

SUMMARY OF CONTINENTAL DENUDATION

The picture that emerges from the brief analysis in this chapter is that chemical denudation rates are dominated by the kinds of rocks that are being eroded and range from continent to continent only by a factor of 3 or 4. This implies a relative constancy through time, controlled largely by land area subject to attack.

On the other hand, mechanical erosion rates of the continents range greatly today, by a factor of 25 or more, and may well have ranged as much or more in the past.

It may be that chemical erosion can be regarded as a continuous and relatively constant function of time, whereas mechanical erosion occurs in great pulses. As the elevation of the lands above sea level increases, many factors come into play that tend to maximize erosion. Climates change and become more highly differentiated; vegetation types respond to these changes. Many areas in which particulate material accumulates on low slopes or in floodplains as a result of continuous chemical leaching become unstable and the debris is washed off to the oceans. Water continuously denudes and alters minerals chemically, but mechanical transport has threshold conditions that must be reached before transport takes place. Perhaps chemical and mechanical erosion are related like two ends of a steel spring being dragged by one end. The end being pulled has a constant velocity, but the other end moves by a series of leaps. Each leap must await a tension sufficient to overcome friction.

Consequently we can only guess at this time about the average ratio of suspended to dissolved load through all of time, but it is possible that a fairly constant ratio would be obtained if time intervals of tens or hundreds of millions of years are integrated.

REFERENCES

Durum, W. H., Heidel, S. G., and Tison, L. J., 1960, World-wide runoff of dissolved solids: *I.A.S.H. Publication No. 51*, 618–628.

Holeman, J. N., 1968, The sediment yield of major rivers of the world: *Water Resources Res. 4*, 737–747.

Livingstone, D. A., 1963, Chemical composition of rivers and lakes: Data of geochemistry, 6th ed., *U.S. Geol. Surv. Profess. Paper*, 440-G.

White, D. E., Hem, J. D., and Waring, G. A., 1963, Chemical composition of subsurface waters: Data of geochemistry, 6th ed., *U.S. Geol. Surv. Profess. Paper*, 440-F.

6 | # Weathering Processes

Now that we have examined the rates at which the land areas are being destroyed and have an idea of the general chemical composition of the dissolved and particulate loads of streams, we can look in more detail at the weathering processes that make it possible for streams to do their work. We shall investigate the compositions of rains and then those of the soil waters that do the chemical alteration of surface rocks. The reactions between soil waters and the common rock minerals will be presented in some detail, so that the controls of river composition and thus the dissolved materials carried to the oceans are delineated. Finally, the bulk chemical changes involved in soil formation will be presented, and the compositions of soils, which eventually become the particulate materials carried by streams, will be related to rock type and climate. Fundamentals of mineral chemistry used in this chapter are discussed in Appendix B.

Hopefully this investigation will give us a basis for the interpretation of the mineral assemblages we see in the sedimentary rocks that constitute our chief record of the geologic past. Our ability to decipher environmental conditions in bygone days depends, first, on our competence to relate the characteristics of present-day deposits to their genesis, and, second, on our ability to distinguish between postdepositional changes and primary characteristics. If the primary characteristics of ancient sediments can be reconstructed, it follows that postdepositional history must also be in part revealed in the process.

If, by examining a sedimentary rock, we could determine the nature of the land area from which it was initially derived in terms of climatic conditions and then the rock types present in the area (as well as their ages and relative proportions), the nature of the ancient topography with respect to relief and elevation, and the flora and fauna that inhabited the region, we would have answered most of the basic questions concerning the "provenance" of the rock. But even this is not enough; we also want to read from the rock the conditions of the depositional site. Was it deposited in fresh water or sea water? If in sea water, were the sea water composition and temperature the same as those of today? Was the sediment deposited in deep or shallow water? Was it deposited near or far from land? How much time is represented per unit of thickness? What creatures were living in the sea?

We cannot hope to answer all of these questions, of course, but we must examine, one by one, the parts of the sedimentary rock cycle in order to bring to bear every possible bit of information helpful to our deductions.

WEATHERING OF MINERALS AND ROCKS

Rocks exposed at the earth's surface are subjected to continuous alteration. Although chemical, biological, and physical processes are all involved in rock weathering, by far the most important of these is chemical processes. A discussion of weathering in a purely chemical framework is almost sufficient to cover the major aspects of the problem.

Chemical reactions between rock minerals and soil waters produce dissolved constituents and solid residues. The dissolved constituents enter the ground-water system and eventually move into streams. Many ground waters have compositions near those predicted entirely from rock mineral-soil water reactions. The solid residues, from the moment of their formation, are subject to dispersal by the agents of erosion, primarily running water.

Physical and Biological Weathering

Perhaps a brief discussion of some aspects of physical and biological weathering is sufficient to demonstrate the kinds of processes involved and to give an idea of their supplementary roles in the overall chemically dominated phenomenon.

An example of purely physical weathering is perhaps that of glacial plucking. As a glacier moves across an area, it tears rock material from the land surface, and this detritus is incorporated into the ice. The rock detritus is essentially unaltered and has the mineralogical and chemical composition of the original rock. Glacial erosion is not very significant in the broad picture of weathering because continental and valley glaciers cover less than 12 percent of the present land area. Also, as shown in Chapter 4, the mass of glacially derived material delivered to the oceans is only about 10 percent of the amount of sediment carried by streams to the oceans. During most of the geologic past glaciers probably have been less important than they are today. Glaciation apparently has been extensive only 6 times or so during the earth's history; each episode lasted perhaps a few million years.

Another weathering mechanism that is often cited as an example of physical weathering is frost heaving. Water in small cracks and crevices in rocks may freeze at night and thaw during the heat of day. Changes in the specific volume of H_2O as it alternately freezes and thaws may force the cracks to enlarge, and eventually the rock is weakened and broken apart. This mechanism is operative particularly in semiarid regions or mountainous areas where temperature extremes occur frequently. Even freezing and thawing cannot break rock without earlier weakening of the bonds between the grains by chemical reactions between water in the cracks and crevices and the rock minerals.

Weathering of rock materials in desert regions takes place largely by physical mechanisms. In such an environment water plays a minor role, and wind is an important agent of both weathering and erosion. During wind storms transported rock fragments bombard outcropping rocks, abrading them and slowly breaking the rock into smaller pieces. Acids generated around the root hairs of living plants and by the bacterial decomposition of plants interact with rock minerals and weaken the rock surface, thus making the surface more susceptible to physical weathering and erosion by wind.

Wind erosion and frost heaving both illustrate the importance of chemical "preparation" of a rock prior to effective physical destruction.

Chemical Weathering

Rain

The chief medium of transfer of water from the continents to the oceans is, of course, the rivers. Rain derived from the oceans falling on the continents keeps the system in balance. Total water at any one time in the atmosphere as water vapor is approximately 0.13×10^{20} g, and this water turns over approximately every 3 or 4 weeks. The average rainfall on the land areas of the earth is about 66 cm/year. The variability is extreme, ranging from places in the Atacama Desert of Chile, where rainfall has never been recorded, to a record of 2647 cm/year at Cherrapunji, India. Not only is there a tremendous range of yearly precipitation but the percentage of precipitation that re-evaporates, percolates into the ground, or runs out over the surface also is subject to extreme variation. In some areas of high temperature and low humidity almost all the rain that falls is returned quickly to the atmosphere. Depending on specific circumstances, any one of the three processes of percolation, runoff, or evaporation may account for nearly 100 percent of the precipitation recorded. Figure 6.1 shows the manner in which rainfall is distributed among these processes under an "average" set of conditions. Generally, a high percentage of evaporation accompanies high temperatures, low humidities, and a slow rate of precipitation. High percolation generally requires a nearly flat topography, permeable soils, good drainage, and slow rate of precipitation; for high runoff these conditions are reversed.

On a global basis, about 35 percent of the water that falls on the continents runs off in streams or through the ground-water system into the oceans. Thus the "effective precipitation" is about 24 cm/year. This number, coupled with the total land area of 149×10^{16} cm^2, yields the figure previously cited of 0.36×10^{20} g for the water flux from continents to the oceans (Chapter 4).

As more and more careful chemical analyses are made of rain it becomes obvious that the compositions of individual rainfalls are variable and that rain deviates markedly from the composition H_2O (See Table 4.7). Figure 6.2 shows some typical maps of the chloride and sodium content of

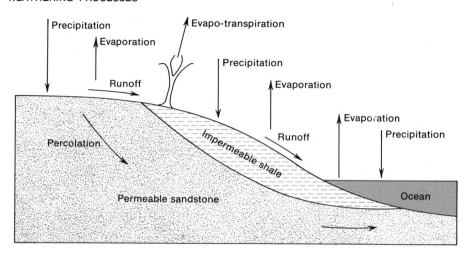

Figure 6.1 Distribution of rainfall in a temperate climate. Low slope and permeable rocks, as at the left, lead to a high percentage of percolation, whereas steep slopes and impermeable rocks maximize runoff. The worldwide average for runoff plus percolation is about 35 percent of the total precipitation.

rain. The chloride and sodium contours in Figure 6.2 are subparallel to the coast lines; and the analyses of rain from some areas such as Bermuda show that the ratios of cations and anions are fairly close to those of sea water.

The discovery of the high salt content of rain near the oceans was a little surprising because sea salts are not volatile, and one might expect that evaporation of water from the oceans would leave the salts behind. Such analyses led to detailed studies of the origin of the salts in rain. A large percentage of the salts apparently is derived by the bursting of small bubbles on the sea surface owing to capping of waves or the impact of raindrops. Droplets of sea water are ejected into the atmosphere; these droplets evaporate with resultant precipitation of the solids as tiny particles that are subsequently carried high into the atmosphere by turbulent winds. These particles are transported over land and brought down by rain.

In many instances the deviations of the ratios of the ions in rain water from those of sea water are marked. The processes that put sea salts *selectively* into the atmosphere are little understood. Among the suggested mechanisms for fractionation are escape of Cl as gaseous HCl from sea-salt aerosol with consequent relative enrichment in Na, and bubbling and thermal diffusion. In general, the salt content of rain diminishes toward continental interiors if the rain is not contaminated by materials from industrial areas.

The map of Figure 6.3 shows the high sulfate content of rain in the southwestern United States, which is due to sulfate-bearing particles from the desert soils that have been blown into the atmosphere. Rain near industrial areas commonly exhibits high contents of sulfate and CO_2 largely

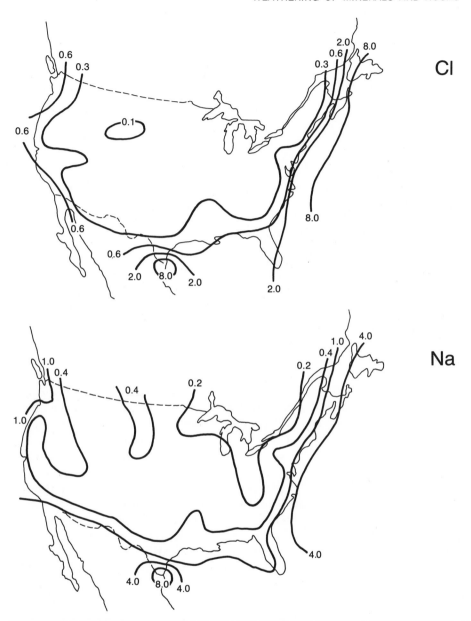

Figure 6.2 Chloride and sodium content of rain over the continental United States. Contoured in parts per million (after Junge and Werby, 1958).

derived from the burning of coal and oil. SO_2 derived from man's activities or natural sources reacts with H_2O and O_2 to form H_2SO_4 in rain. The reaction is

$$2SO_2 + O_2 + 2H_2O = 4H^+ + 2SO_4^{2-}.$$

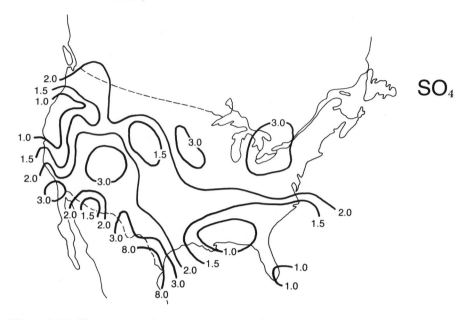

Figure 6.3 Sulfate content of rain over the continental United States. Contoured in parts per million (after Junge and Werby, 1958).

The CO_2 content of rain is of particular importance in weathering processes. If rain were initially pure H_2O that had equilibrated with the atmosphere, rain should have a pH of 5.7 and should contain about 10^{-5} moles (about 1 ppm) of dissolved CO_2. Rain commonly has a pH less than the idealized 5.7. The lowering usually is attributed to industrial contamination, especially by sulfur gases released from fuels. Also, there is commonly a remarkable change in the pH of rain from the beginning of a shower until the end, with the first drops showing an abnormally low pH which rises as acidic particles are literally rained out of the atmosphere. Rain-water pH values higher than 5.7 occur near coasts where there is addition of sea aerosol. Land-derived, fine-grained calcium carbonate dust carried into the atmosphere by wind may also increase the pH by reacting with the rain water. As a chemical reagent the effectiveness of rain is largely due to its acid content, which amounts to 10^{-4}–10^{-3} moles of available H^+ per liter.

Soil Water

After rain water has entered the soil, its characteristics are changed markedly. Table 6.1 lists some analyses of waters from various soil types and shows how the usual few parts per million of rain water increase markedly as the water reacts. Also, the chemical contents of the waters are quite variable, although only a few types of occurrences are represented. In fact, it is difficult to characterize soil waters chemically because their composi-

Table 6.1
Chemical Composition of Some Soil Waters

Locality	Concentration (ppm)											Reference
	SiO_2	Al	Fe	Ca^{2+}	Mg^{2+}	Na^+	K^+	HCO_3^-	SO_4^{2-}	Cl^-	pH	
Mor humus layer, English Lake District	—	—	—	1.1	0.9	9.2	5.8	—	17.5	11.5	4.2	Gorham (1961)
Mull humus layer, English Lake District	—	—	—	10.0	1.6	8.7	4.4	8.1	9.6	7.6	6.2	Gorham (1961)
Soil developed on Wissahickon schist, Piedmont Province, Maryland	5.45	Trace	Trace	4.15	3.29	5.45	4.86	—	—	—	5.45	Bricker (1968)
Weathered zone, Woodstock granite, near Baltimore, Maryland	23	0.1	0.00	5.8	15	4.9	0.9	80	11.6	0.8	8.4	Wolff (1967)
Soil developed on Sierra Nevada granitic terrain, California	23	0.00	0.12	7.0	0.3	3.9	1.0	34	0.0	0.4	6.9	Feth et al. (1964)
Soil developed on basalt, Kauai, Hawaii	1.2	0.00	0.18	0.9	1.9	6.3	0.6	6	4.0	9.1	5.1	Patterson and Roberson (1961)

tions change during percolation, showing that they react rapidly with the solid constituents of the soil.

The upper part of the soil is a zone of intense biochemical activity. The bacterial population near the surface is large but decreases downward with a steep gradient. One of the major chemical processes is the oxidation of organic material, which releases CO_2. Samples of soil gases obtained above the zone of water saturation commonly show CO_2 contents 10–40 times that of the free atmosphere and in some instances CO_2 makes up as much as 30 percent of the soil gases as opposed to 0.03 percent of the free atmosphere. In addition to the acid effects of CO_2 produced by oxidation, a highly acid microenvironment is created by the roots of living plants. Values of pH as low as 2 have been measured immediately adjacent to root hairs. A single plant may have several miles of rootlets so that the chemical effects are formidable.

In the humid tropics rainfall and temperatures are high and bacterial oxidation is so rapid that little organic material has an opportunity to become incorporated in the soil. Oxidation takes place at the surface and the CO_2 is lost to the atmosphere rather than becoming dissolved in the percolating waters. Consequently, many waters from tropical soils have CO_2 contents near that of the free atmosphere, and pH values around 5, and are thus low in reactivity. In waterlogged soils all oxygen may disappear, and methane and hydrogen are produced. This change to strongly reducing conditions results from depletion of the oxygen by reaction with organic materials and also from oxidation of inorganic species, such as ferrous and manganous ions, to higher valences. Although the reactions and the nature of the compounds have not been worked out in detail, the organic materials in soil waters form strong complexes with many elements and promote their solution and transport.

Reaction of Rock Minerals with Soil Waters

Chemical weathering of rocks is the integration of the reactions of soil waters with the individual minerals composing rocks. As a result of these reactions, dissolved constituents are produced and added to the reacting aqueous solutions. Also, new minerals are formed and the original rock minerals are partially or completely destroyed. These solid inorganic products of weathering, along with organic substances derived from biochemical processes in the soil zone, are transported from their site of origin by erosional agents or remain in place as residual materials, i.e., soils.

We can now look at the chemical weathering of rocks by examining the chemistry of individual mineral-water reactions.

CONGRUENT SOLUTION—NaCl. A number of highly soluble constituents of rocks that are being weathered appear in solution in the soil as ground waters as a result of a single leaching process. These soluble constituents are present in minor amounts in most rocks; in many instances we do not even know specifically the form in which they are present. In some cases they may occur as films of concentrated brines on the surfaces of other minerals, left over from the expulsion of pore waters during compaction of the rock. In other instances they may have been released from minute inclusions of fluid trapped within individual rock minerals. Such inclusions, although they make up a minuscule percentage of the total rock, are usually highly concentrated salt solutions. Third, these soluble salts may be present as scattered mineral grains.

Ordinarily they do not contribute a significant percentage of the major dissolved species of soil waters; the notable exceptions are contributions that appear in solution as Na^+, Ca^{2+}, Cl^-, and SO_4^{2-} . Much of the Na^+ and

Cl^- apparently are derived from mineral NaCl (halite), whereas the Ca^{2+} and SO_4^{2-} come from the minerals gypsum ($CaSO_4 \cdot 2H_2O$) and anhydrite ($CaSO_4$). Let us look at some of the controls of their weathering.

Much halite occurs in bedded salt deposits or is dispersed in other rocks. The solubility of NaCl in pure water is about 6 moles/liter, or about 350 g/1000 g of H_2O at 25°C. Thus, water percolating through a salt bed or a rock containing NaCl can dissolve a great deal of NaCl before the solution becomes saturated, and beds of salt are rapidly dissolved away. Any relatively insoluble constituents in the salt bed such as quartz and clay minerals remain at the source as a residue.

Table 6.2 shows the Cl content of some typical rock types. The relative uniformity of the concentration of Cl in all these different rock types is hard to explain. Presumably the NaCl occurs in fluid inclusions in igneous and metamorphic rocks, as fluid inclusions in mineral grains, and as relict brines in shales and limestones. The high concentrations in the syenites and nepheline syenites can be correlated with the presence of primary igneous minerals that have Cl as an essential constituent in these relatively rare alkali-rich igneous rocks. The two discrepant values for shales apparently represent sampling differences. The low values are from shales ex-

Table 6.2
Chlorine Content of Terrestrial Rocks [a]

Rock type	Cl (ppm)
Igneous rocks	185
Ultramafic rocks	100
Mafic rocks	160
Intermediate rocks	180
Acidic rocks	200
Syenites	430
Nepheline syenites	2,170
Sedimentary rocks	105
Shales and clays	100 (1466[b])
Sandstones	20
Limestones	130
Dolomites	660
Natural halite rock	605,800[c]
Schists	354
Gneisses	207
Amphibolites	300
Earth's crust	180
Earth's upper mantle	100

[a] After Johns and Huang (1967).
[b] Billings and Williams (1967).
[c] Clarke (1924).

posed at the surface and probably partly weathered before analysis; the higher value of about 1500 ppm is likely to be more representative of unweathered material, because the samples were obtained from boreholes. Sandstones are low, even before weathering at the surface, because they are accumulations of chemically inert materials in the first place and are subject to subsurface leaching because of their high permeability and consequent performance as aquifers.

In Chapter 4 we estimated that 55 percent of the Cl^- in rivers is derived from rocks and 45 percent from cycling through the atmosphere and that chloride is about 1.2 percent of the total load of suspended plus dissolved material carried by rivers each year. If so, the average content of Cl in the rocks weathered to produce the load should be 0.6 percent, or 600 ppm. This concentration is only about 3 times the average crustal content of 180 ppm and is in reasonable accord with what we might expect if the Cl leached from rocks today is controlled chiefly by weathering of shales with Cl contents of more than 1000 ppm.

CONGRUENT SOLUTIONS—GYPSUM AND ANHYDRITE. Gypsum ($CaSO_4 \cdot 2H_2O$) and anhydrite ($CaSO_4$) dissolve in soil waters to produce Ca^{2+} and SO_4^{2-}. If the mineral of the rock being weathered is anhydrite, the first step of the weathering process may be hydration to gypsum, followed by congruent solution of the gypsum. The reactions are

$$CaSO_4 = Ca^{2+} + SO_4^{2-} \text{ (direct solution of anhydrite)},$$

$$CaSO_4 + 2H_2O = CaSO_4 \cdot 2H_2O \text{ (hydration of anhydrite to gypsum)},$$

and

$$CaSO_4 \cdot 2H_2O = Ca^{2+} + SO_4^{2-} + 2H_2O \text{ (direct solution of gypsum)}.$$

The hydration of anhydrite to gypsum involves a 26 percent increase in mineral volume.

The solubility curves for gypsum and anhydrite in pure water as a function of temperature are shown in Figure 6.4. The temperature at which gypsum and anhydrite are in equilibrium in the system $CaSO_4 - H_2O$ has been a subject of debate for many years, but it is probably close to 57°C. For the reaction

$$CaSO_4 \cdot 2H_2O = CaSO_4 + 2H_2O,$$

the equilibrium constant is

$$K_{(T,P)} = a^2_{H_2O}.$$

Thus equilibrium between gypsum and anhydrite at a given T and P is at a fixed value of the activity of H_2O.

At room temperature, the equilibrium activity of water is considerably less than that of pure water; that is, dilute soil solutions tend to convert anhydrite to gypsum. On the other hand, if gypsum were placed in a saturated NaCl solution, it would tend to dehydrate. Gypsum behaves pretty much as expected experimentally, but anhydrite is a "difficult customer"; it

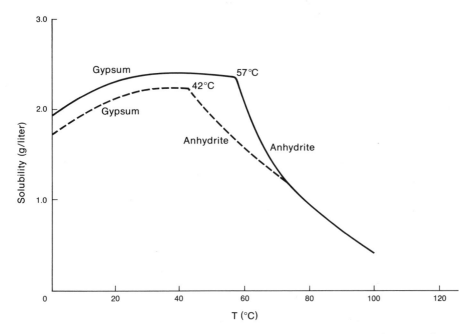

Figure 6.4 Solubility curves for gypsum and anhydrite in pure water as a function of temperature and at ordinary pressure. Solid and dashed lines show the solubilities based on two different sets of data. Note the difference in the equilibrium temperature (after Kinsman, 1966).

dissolves slowly, refuses to crystallize when it should, and in general is kinetically recalcitrant. The rate of solution of the calcium sulfate minerals is much lower than that of halite. A pinch of salt will dissolve in a bucket of water in a few minutes; several hours are required for a similar sample of gypsum, and perhaps weeks for anhydrite. Also, the solubility of gypsum is only about one fortieth that of halite, i.e., about 0.015 moles/liter or about 2000 ppm in water at room temperature.

CONGRUENT SOLUTION—CARBONATE MINERALS. The carbonate minerals calcite and dolomite are the major minerals found in carbonate rocks, whereas quartz, clay minerals, feldspars, and iron oxides are in small percentages. Minor amounts of aragonite, the dimorph of calcite, and magnesian calcites are found in relatively young carbonate rocks.

When an excess of $CaCO_3$ is placed in a beaker of pure water, the Ca^{2+} and CO_3^{2-} are released into the aqueous solution. Ca^{2+} does not hydrolyze, whereas CO_3^{2-} reacts appreciably with the water in a succession of reactions

$$CO_3^{2-} + H^+ \rightarrow HCO_3^-, \qquad (1)$$

$$HCO_3^- + H^+ \rightarrow CO_{2aq} + H_2O \qquad (2)$$

and if there is an air space above the beaker for a gas phase to occupy,

$$CO_{2aq} \rightarrow CO_{2g}. \qquad (3)$$

When equilibrium is achieved, these various species are distributed in accord with the equilibrium constants for the reactions.

The H^+ in reactions (1) and (2) is derived from the ionization of H_2O. The tendency of CO_3^{2-} to react with pure water to form the stable HCO_3^- complex, which in itself is a very weak acid, leads to the solution of carbonate minerals.

If we add a concentrated strong acid to a beaker containing $CaCO_3$, the $CaCO_3$ reacts rapidly with the acid and evolves a whole series of carbon-bearing species according to the successive reactions (1), (2), and (3). A Ca^{2+} enters the solution for each CO_3^{2-}, HCO_3^-, $CO_{2\,aq}$, and $CO_{2\,g}$ formed, and thus the $CaCO_3$ dissolves. In a dilute acid, little CO_2 gas is evolved because reactions (1)–(3) are not driven far to the right before the acid is used up, and the dominant carbonate species in solution is HCO_3^-. The other common carbonate minerals react to acid in much the same way as pure $CaCO_3$, although their rates of solution are far less. The solubilities of the common carbonate minerals, in order from most to least soluble, are magnesite, aragonite, calcite, dolomite, siderite, and rhodochrosite. The solubilities of the first four are nearly the same under weathering conditions; siderite and rhodochrosite are much less soluble. Magnesite and dolomite dissolve much more slowly than calcite and aragonite.

In nature the reaction of dissolved CO_2 with carbonate minerals to produce bicarbonate ion and cations is responsible for the solution of carbonate rocks and minerals. The $CO_{2\,aq}$ in the soil water is derived from CO_2 produced by the oxidation of organic materials by bacteria in the soil. Reactions (3) and (2) are driven to the left.

Using $CaCO_3$ as a representative carbonate mineral, the solution of carbonate minerals and rocks may be generalized by the reaction

$$CaCO_3 + CO_{2g} + H_2O = Ca^{2+} + 2HCO_3^-. \qquad (4a)$$

In this reaction, one of the HCO_3^- ions comes from the CO_{2g} and the other from $CaCO_3$. The equilibrium constant for this reaction at room temperature is

$$K = \frac{a_{Ca^{2+}}\, a_{HCO_3^-}^2}{a_{CO_{2g}}} = 10^{-5.8} \qquad (4b)$$

The amount of $CaCO_3$ dissolved according to reaction (4a) depends on the temperature, pressure, original amount of HCO_3^- in the solution, and the activity or partial pressure of CO_2. In the weathering of carbonate minerals and rocks, the partial pressure of CO_2 plays the dominant role in the amount of carbonate dissolved per unit of reactive soil water.

From equation (4a) it is apparent that as CO_2 pressure is increased, the amount of $CaCO_3$ dissolved also increases. The higher the initial CO_2 pressure and hence dissolved CO_2 content of the waters, the more aggressive the water.

How effective is rain as a chemical reagent versus CO_2-charged soil water in dissolving carbonate minerals? We can answer this question by considering the basic carbonate dissolution equations (4a) and (4b).

Each molecule of CO_2 in rain or soil water produces one Ca^{2+} ion and two HCO_3^- ions from a molecule of $CaCO_3$. Consequently, we can let the concentration of Ca^{2+} be x and that of HCO_3^- be $2x$, postulate a CO_2 pressure, and solve for Ca^{2+} activity. For example, for a CO_2 pressure of $10^{-3.5}$ atm,

$$\frac{(x)(2x)^2}{10^{-3.5}} = 10^{-5.8} \qquad (4b)$$

$$x^3 = \frac{10^{-9.3}}{4} = 10^{-9.9}$$

$$x = 10^{-3.3} = a_{Ca^{2+}}$$

In dilute solution, the activity of Ca^{2+} is very nearly its concentration in moles per 1000 grams of water, so that there should be $10^{-3.3}$ moles of Ca^{2+} or about 20 ppm. In a typical soil water, with P_{CO_2} of 10^{-2} atm, Ca^{2+} would be 65 ppm; in an atmosphere of nearly pure CO_2 it would be about 300 ppm.

The chemical weathering of carbonate rocks produces Mg^{2+} in solution, as well as Ca^{2+} and HCO_3^- if the mineral dolomite is present. Also, an insoluble residue of quartz, clay minerals, and iron oxides is often found in soils developed from carbonate rock weathering. Large solution cavities filled with deep-red soil produced as a result of solution of limestone or dolomite and accumulation of solid weathering products are commonly found developing on carbonate rocks in temperate and tropical climatic regions. Solution of carbonate rocks may lead to the development of caves with complex channel systems through which water flows underground. Regions of intense carbonate solution, with resultant cavern collapse and underground drainage, are characterized by a hummocky, rolling surface topography pocked with sinkholes, called Karst. Some of the better-known areas of Karst are found in Kentucky, on the north coast of Puerto Rico, and in the place of name-origin, the Karst of Yugoslavia.

Chemical weathering of pure calcite produces Ca^{2+} and HCO_3^- in the ratio of 1:2, whereas the solution of pure dolomite produces a water containing Ca^{2+}, Mg^{2+}, and HCO_3^- in the ratios 1:1:4 according to the reaction

$$CaMg(CO_3)_2 + 2CO_2 + 2H_2O = Ca^{2+} + Mg^{2+} + 4HCO_3^-.$$

The results of these idealized reactions are expressible as bar graphs with water compositions shown as mole ratios of dissolved species to HCO_3^-, assigning a value of unity to the HCO_3^- content of the water (Figure 6.5). Such graphs are useful in determining the mineral source of the dissolved species in ground waters. Some natural waters derived from carbonate rocks are also plotted in Figure 6.5. Notice that the natural ratios of $Ca^{2+}/Mg^{2+}/HCO_3^-$ are nearly those calculated.

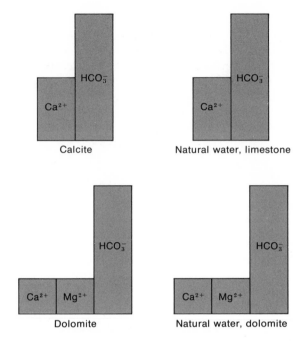

Figure 6.5 Bar graphs of water compositions calculated from the reaction of CO_2-charged H_2O with $CaCO_3$ and $CaMg(CO_3)_2$, compared with bar graphs of ground waters derived from a limestone and dolomite. Water compositions are shown as mole ratios of dissolved species to HCO_3^-, in which the HCO_3^- content is set at unity.

CONGRUENT SOLUTION—SILICA. Silica in one form or another—quartz, opal, chalcedony, tridymite, cristobalite, etc.—is nearly ubiquitous in rocks. The chemical weathering of silica occurs by congruent solution and hydration according to the reaction

$$SiO_{2(s)} + H_2O = H_4SiO_4.$$

Quartz has a solubility at earth-surface temperatures of about 6.5 ppm of dissolved silica,[1] whereas the other forms of silica have higher solubilities. Fresh silica gel precipitated in the laboratory has a solubility of 115 ppm of dissolved silica at 25°C. The first dissociation constant of silicic acid is $10^{-9.8}$; at pH values above 9, silicic acid dissociates significantly according to the reaction

$$H_4SiO_4 = H^+ + H_3SiO_4^-,$$

and the solubility of silica increases if a source of SiO_2 is present. Figure 6.6 shows the solubility of silica as a function of pH and temperature.

The solubility of quartz has been difficult to determine in the laboratory because it is slow to react. The average dissolved silica concentration of the world's rivers is 13 ppm, 2 times the solubility of quartz at earth-sur-

[1] Silica solubilities are most commonly reported as SiO_2, without regard to the actual nature of the dissolved species.

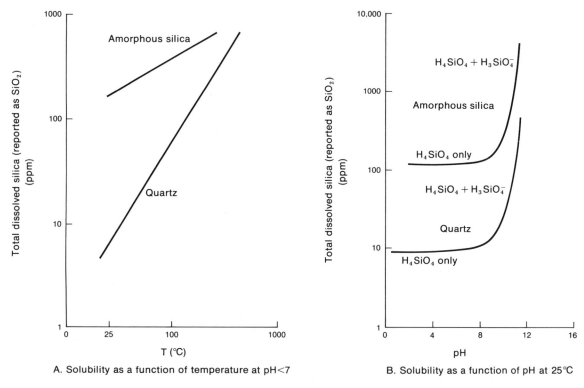

Figure 6.6 The solubility of silica as a function of pH and temperature (after Krauskopf, 1967).

face temperatures and only about one tenth the solubility of amorphous silica. The slow reactivity of quartz in the laboratory, coupled with the usual supersaturation of stream waters with respect to quartz, suggest that it may be nearly chemically inert at the earth's surface. We must look to silicates other than quartz to account for the dissolved silica in streams.

Table 6.3 shows a composite analysis of waters draining a nearly pure quartz rock. Although the total dissolved solids are low and the dissolved silica concentration is less than quartz solubility, the dissolved silica probably came from trace amounts of feldspar rather than from solution of the abundant quartz. Otherwise there would be no adequate source of the K^+ and Na^+.

INCONGRUENT SOLUTION—IRON MINERALS. The chemical weathering of iron-bearing minerals and rocks is somewhat more complex than the weathering of halite, calcium sulfate, or carbonate minerals in that most iron mineral-soil water interactions involve oxidation-reduction reactions and may also involve the formation of several residual solid phases. Most of the major iron minerals, e.g., magnetite (Fe_3O_4), siderite ($FeCO_3$), pyrite (FeS_2), pyrrhotite (FeS), and iron silicates of various compositions, weathering at the earth's surface, contain ferrous iron in their structure. These minerals

Table 6.3
Composite Analysis of Waters Draining a Quartzite, Sangre de Cristo Range, New Mexico[a]

Constituent	Concentration (ppm)
HCO_3^-	5.7
SiO_2	3.6
Ca^{2+}	1.9
SO_4^{2-}	1.8
Na^+	0.5
K^+	0.3
NO_3^-	0.1
Mg^{2+}	ca. 0.06
Cl^-	ca. 0.06

[a]Data from Miller (1961).

usually occur dispersed throughout the rock but may be in discrete bands or segregations. Ferrous and ferric iron also are found in rock minerals, other than those in which iron is a major element, substituting for ions of similar size and charge, e.g., Fe^{2+} for Mg^{2+} in dolomite and Fe^{3+} for Al^{3+} in montmorillonite.

During the course of weathering, ferrous iron is released from the minerals and oxidized by O_2 in the soil water to ferric ion which hydrolyzes to form ferric oxides. The minerals include the hydrated ferric oxide dimorphs goethite ($\alpha FeO \cdot OH$) and lepidocrocite ($\gamma FeO \cdot OH$) and the anhydrous dimorphs hematite (αFe_2O_3) and maghemite (γFe_2O_3, magnetic hematite). Because these alteration products are yellow and orange-brown to red in color, the weathering of iron-bearing rocks usually results in vividly colored soil zones that contrast sharply with the drab fresh rock from which the soils are derived. A few percent total iron in a rock is all that is necessary to impart strong reddish coloration to a soil.

The marked tendency of iron to form oxide minerals in the weathering zone and the remarkable persistence of these minerals in soils can be explained in terms of their vanishingly low solubilities under oxygenated conditions. There has been much confusion concerning the relative stabilities of the hydrous and anhydrous iron oxides in the soil environment. It now appears that hematite is the stable species, even in the presence of highly dilute soil solutions in which conditions would be most favorable for formation of a hydrate. However, all four of the oxides listed above are so extremely insoluble, and their solubility differences so small, that any one may form and persist for very long times, even from a geologic standpoint. Broadly speaking, there is a tendency for hematite to develop in the soils of warm, well-watered regions and for goethite to form elsewhere. The great range of conditions under which the iron oxides are extremely insoluble is illustrated diagramatically in Figures 6.7 and 6.8 using partial pressures of gases or Eh and pH as descriptive variables. (See Appendix B

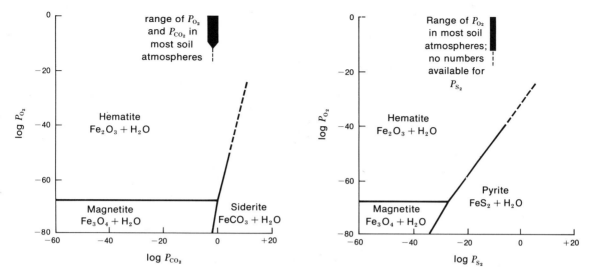

Figure 6.7 Partial pressure diagrams showing the stability of some iron minerals as functions of the partial pressure of O_2, CO_2, and S_2 (after Garrels and Christ, 1965).

for discussion of such diagrams.) On these diagrams hematite is used to represent the whole gamut of ferric oxides and iron metasilicate to represent that of the iron silicates. The differences in stabilities of the various species are so small that the relations shown are not affected by these generalizations.

Figure 6.7 shows that ferric oxide should form at the expense of magnetite, siderite, or pyrite in most soil environments, which range approximately from 10^{-1} to $10^{-3.5}$ atm of CO_2, $10^{-0.7}$ to 10^{-10} atm of O_2, and vanishingly small pressures of S_2.

Figure 6.8 shows the stability relations of a group of iron compounds as a function of oxidation potential (Eh) and pH. A number of specifications must be made about the chemical composition of the system (see the caption) in order to delineate the exact boundaries of the mineral fields, but the diagram demonstrates the general pattern of iron mineral behavior. The stippled area indicates the range of conditions found in soils; in general, the upper part of the stippled area can be considered to represent the upper, moist, aerated zone of the soil and the lower portion corresponds to conditions in the lower part of the soil zone, where the soil waters are reacting with the mineral species containing reduced iron.

Ferrous and ferric iron are the two important species of dissolved iron encountered in most natural solutions. Fe^{3+} is important only in highly oxidizing and strongly acidic environments. Fe^{2+} is present in significant concentrations ($>10^{-6}$ m) over a range of oxidizing and acidic to slightly alkaline conditions. The great stability of ferric oxide is evident from these Eh-pH diagrams, as is the relation that the numerous iron-bearing minerals containing ferrous iron are stable only under reducing, moderately acidic

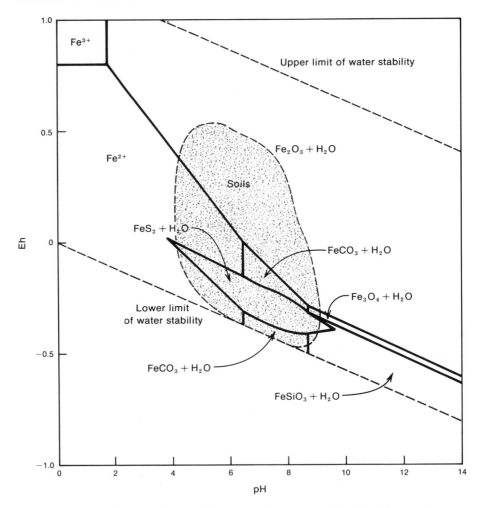

Figure 6.8 Eh-pH diagram showing the areas of dominance of dissolved iron species and stability fields of some iron minerals at 25°C and 1 atm total pressure. The fields of the dissolved species are shown where the total activity of the ions is greater than 10^{-6}. The environmental conditions are total dissolved $CO_2 = 10^0$ moles, total sulfur $= 10^{-6}$ moles, and amorphous silica present. The upper and lower dashed lines are the stability boundaries of H_2O (after Garrels and Christ, 1965).

to highly alkaline conditions. As ground waters become more concentrated in dissolved carbonate, sulfur, or silica, the stability fields of the iron minerals containing these species increase in area. In most soil-water environments, dissolved oxygen is too high, and carbonate, sulfur, or silica are too low in concentration to stabilize the primary iron-bearing minerals. Some conclusions concerning the chemical weathering of iron-rich minerals and rocks can be reached by consideration of these diagrams and some simple chemical reactions.

The production of iron oxides can be looked upon as a two-stage process. First, ferrous iron is released from ferrous-containing minerals by acid attack by CO_2, and then the released ferrous iron is oxidized by dissolved oxygen.[2] The Fe-silicates and carbonates react with the carbonic acid in the waters to yield Fe^{2+}:

$$FeSiO_3 + 2CO_2 + 3H_2O = Fe^{2+} + H_4SiO_4 + 2HCO_3^-,$$

and

$$FeCO_3 + CO_2 + H_2O = Fe^{2+} + 2HCO_3^-.$$

When the ferrous iron derived from the solution of these minerals comes in contact with dissolved oxygen, the Fe^{2+} is oxidized and precipitates as ferric oxide:

$$2Fe^{2+} + 4HCO_3^- + H_2O + 1/2O_2 = Fe_2O_3 + 4CO_2 + 3H_2O.$$

The weathering of iron sulfides involves the oxidation of sulfur as well as iron. Solutions derived from weathering of pyrite are very acidic due to the production of sulfuric acid during the formation of ferric oxide. Under humid weathering conditions, the reaction is

$$2FeS_2 + 15/2O_2 + 4H_2O = Fe_2O_3 + 4SO_4^2 + 8H^+.$$

In this reaction iron is oxidized from $+2$ to $+3$ and sulfur from -1 to $+6$. The oxidant is oxygen dissolved in the soil waters. In arid regions under highly oxidizing and acidic conditions, ferric sulfate minerals may form in addition to or in place of ferric oxides as the result of pyrite oxidation. Eventually these sulfates convert to ferric oxides.

The weathering of Fe_3O_4 (magnetite) is complicated, and the processes are not completely understood. The weathering reactions leading to the alteration of magnetite must be exceedingly slow because magnetite is one of the minerals that remains unaltered until the last stages of soil formation and is often found as a heavy mineral component of sedimentary rocks. Apparently magnetite grains become armored by a durable impermeable film of Fe_2O_3.

The above discussion and examples emphasize the point that the chemical weathering of iron-bearing minerals and rocks results in the formation of ferric oxides because of their wide range of environmental stability. Ferrous iron is the primary iron species produced in solution as a result of weathering, but it is found only at the lower part of the soil zone and is oxidized when the soil waters eventually get back to the surface. Consequently, the iron initially dissolved in slightly acid ground waters is transported in streams as discrete ferric oxide particles or as ferric oxide coatings on mineral grains rather than in solution.

[2] Conversion of ferrous compounds to ferric oxides does not require preliminary acid attack and can be accomplished by alkaline oxygenated waters. However, the usual process involves both acid attack by CO_2 and oxidation of ferrous iron by dissolved oxygen.

INCONGRUENT SOLUTION—SILICATE MINERALS. Silicate minerals compose more than 75 percent of the rocks with which soil and ground waters come in contact. These minerals include chiefly quartz, olivine, pyroxenes, amphiboles, feldspars, micas, and clay minerals (see Appendix A). Because of the great abundance and varied composition of silicates, the chemical weathering of silicate minerals, particularly the aluminosilicates, plays a major role in supplying dissolved species to natural waters. Indeed, the concentrations of many constituents in natural waters are controlled by equilibria involving one or more silicate minerals.

It has long been recognized that waters derived from feldspathic igneous rocks contain dominantly Na^+, Ca^{2+}, and HCO_3^- and are high in dissolved silica (Headden, 1903). The obvious sources of these constituents are the silicate minerals that make up the igneous rocks and that are chemically altered by waters percolating through them. However, surprisingly so, it is only recently that the silicate minerals have been recognized as rapidly reacting phases that exert a control on the compositions of natural waters.

Silicates. The chemical weathering of silicates is accomplished in a manner analogous to that of the carbonate minerals, that is, by hydrolysis and reaction with CO_2. For a nonaluminous silicate (using Mg-olivine as an idealized silicate), we may write

$$Mg_2SiO_4 + 4CO_2 + 4H_2O = 2Mg^{2+} + 4HCO_3^- + H_4SiO_4 \qquad (5)$$

H^+ ions derived from the H_2O and CO_2 unite with the silicate group to form the silicic acid, H_4SiO_4. Notice that in reaction (5) acid is consumed, and any unit of soil water reacting with a silicate mineral becomes more alkaline as the reaction proceeds.

In nature the cation composition of silicates is more complex than the idealized Mg-silicate used in reaction (5). The cations go into solution at different rates and the surface of each silicate grain in contact with the reactive water is armored by a coating of somewhat different composition than the original mineral composition. Once this coating is built up, alteration of the silicate mineral is slow because solution must now take place primarily by diffusion of cations through a nearly inert surface layer. This process of incongruent solution, whereby an original mineral dissolves to form a new mineral plus dissolved constituents, is most important in the chemical weathering of aluminosilicates.

Aluminosilicates. Aluminum is nearly immobile as a dissolved constituent in most natural waters; only about 0.1 ppm is found as a dissolved constituent in waters, and most of the aluminum under earth-surface conditions is transported in solids. Because Al is so immobile, aluminosilicates weather to more aluminous solid phases by incongruent solution. CO_2 charged soil waters attack the primary aluminosilicates and constituents such as Na^+, Ca^{2+}, K^+, Mg^{2+}, and dissolved silica are released into solution. The chief alteration product of this aggressive attack, and usually the first to form, is kaolinite. However, other alteration products may form depending on the composition of the primary minerals being weathered and the aggressiveness (initial CO_2 content) and residence time of the attacking waters. These products include montmorillonite, chlorite, illite, various iron and aluminum oxide hydrates, X-ray amorphous or ill-defined aluminum oxide and aluminosilicate hydrates, and, more rarely, zeolites.

As an example of the chemical weathering of aluminosilicates, we may use the reaction of albite to kaolinite plus solution:

$$2NaAlSi_3O_8 + 2CO_2 + 11H_2O = Al_2Si_2O_5(OH)_4 + 2Na^+ + 2HCO_3^-$$

$$+ 4H_4SiO_4. \tag{6}$$

Notice that the albite alters to a new mineral—kaolinite in this case—plus solution and that acid is consumed. Thus any unit of reacting soil water would become more alkaline as the reaction proceeded and CO_2 was consumed. Subsequent reaction would proceed slowly by simple hydrolysis. The total dissolved material per liter is only about 60 ppm for a solution initially containing a typical soil CO_2 content, so that fresh CO_2-charged soil water must be added frequently in order for an albite grain to be completely destroyed. Also, because the initial reaction of water with albite involves the surfaces of the mineral grains and formation of kaolinite on these surfaces, further reaction is slow because ions must diffuse through this surface "armor."

The water resulting from reaction (6) would contain HCO_3^-, Na^+, and H_4SiO_4 in the ratios 1:1:2. Similar reactions to kaolinite for other common primary silicates can be written and the resultant waters expressed as the bar graphs, shown in Figure 6.9. The compositions of biotite, hornblende, and pyroxene represent compositions typical of igneous rocks. It has been demonstrated (Garrels, 1967) that the chemical analyses of natural waters from igneous rocks can be stated from knowledge of the approximate mineralogy of a particular igneous rock and the mineral constituent bar graphs shown in Figure 6.9. "Reconstructed" waters from a rhyolite and a basalt, expressed as mole ratios of the constituents to HCO_3^-, are shown in Figure 6.10. These reconstructed water compositions, based on appropriate mineral compositions of a rhyolite and basalt and on the assumption that these minerals weather to kaolinite plus solution, closely approximate the compositions of the natural waters. From the bar graphs for rhyolite, one can determine quickly that most of the silica came from Na-feldspar and a little from K-feldspar and hornblende. Also, the dominating influence of the feldspars on the overall water composition is evident. Such reconstructions work fairly well when uncontaminated waters from igneous rocks with a limited number of source minerals are considered. However, waters derived from rocks containing a complex mineralogy, or waters containing high concentrations of anions other than HCO_3^-, such as Cl^- and SO_4^{2-}, are less susceptible to this type of analysis.

Material Balance. Waters derived from the chemical weathering of igneous rocks provide us with a measure of the weathering reactions involv-

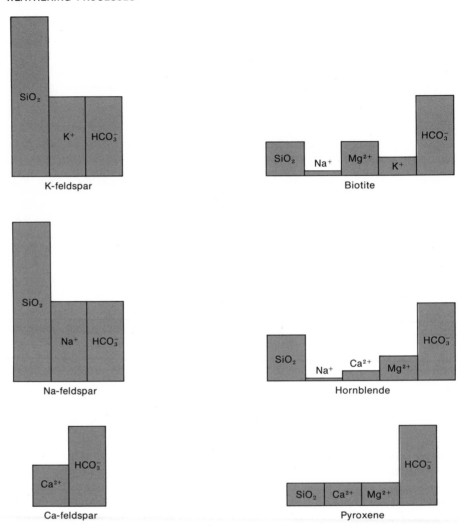

Figure 6.9 Bar graphs of water compositions resulting from the attack of CO_2-charged H_2O on primary silicates to form kaolinite and release dissolved constituents. Water compositions are shown as mole ratios of dissolved species to HCO_3^-, which is set at unity.

ing silicates. One approach to silicate weathering and the genesis of waters derived from weathering processes is to add the dissolved constituents in a water to the known alteration products to see if the original silicate minerals can be produced. Ephemeral spring waters derived from the chemical weathering of granitic rocks in the Sierra Nevada of California and Nevada provide an example of such a calculation.

The ephemeral spring waters in this region of the Sierra Nevada are primarily $Na^+ - Ca^{2+} - HCO_3^-$ waters containing dissolved silica. The average pH of these waters is 6.2. The primary minerals found in the igneous

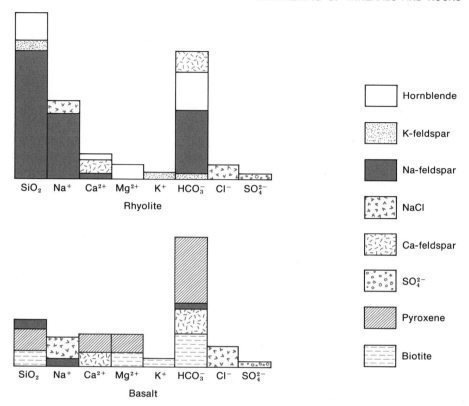

Figure 6.10 "Reconstructed" waters from rhyolite and basalt. The histograms show the mineral sources of the various chemical species in the water. A mass balance is achieved while maintaining the ratios of the constituents required for the individual minerals. Water compositions are shown as mole ratios of dissolved species to HCO_3^- (after Garrels, 1967).

rocks are quartz, plagioclase, K-feldspar, and accessory biotite or hornblende. The chief alteration product of these primary minerals is kaolinite. Table 6.4 shows the results of calculations that involve reacting kaolinite with the dissolved constituents in the Sierra Nevada ephemeral spring waters to produce the original source minerals of the granitic rock. First, the concentrations of the dissolved constituents in snow water were subtracted from the average analysis of the ephemeral spring waters to arrive at the constituents in the spring waters that were derived from the rocks. The concentrations of dissolved constituents in the snow reflect mainly atmospherically derived materials. Second, enough Na^+, Ca^{2+}, HCO_3^-, and dissolved silica were reacted with kaolinite to form plagioclase feldspar; all of the Na^+ and Ca^{2+} was used up in this reaction. The plagioclase feldspar formed has the composition of that found in the rocks. The third step involves the reaction of all of the Mg^{2+}, and enough K^+, HCO_3^-, and dissolved silica with kaolinite to make biotite. Finally the remaining K^+, and enough HCO_3^-, and dissolved silica are reacted with kaolinite to form potash feldspar. About 4 percent of the original dissolved silica is left over; this percentage is within the error of the original water analyses. This example of a material balance works out rather well and points out several aspects of the chemical weathering of silicate minerals and rocks that we have touched upon in previous sections.

The chemical alteration of igneous rocks occurs in a closed chemical system; that is, there is little loss or gain of CO_2 and H_2O during the attack of aggressive soil water on the primary silicates. If this were not so, the balance in Table 6.4 would not work out. Also, although the rocks from

Table 6.4
Calculation of Source Minerals for Sierra Nevada Springs (ephemeral springs, concentrations in moles per liter $\times 10^4$)[a]

Reaction (coefficients $\times 10^4$)	*Na^+*	*Ca^{2+}*	*Mg^{2+}*
Initial concentrations in spring water	1.34	0.78	0.29
Step I Minus concentrations in snow water	1.10	0.68	0.22

Step II
   Change kaolinite back into plagioclase
$1.23Al_2Si_2O_5(OH)_4 + 1.10Na^+ + 0.68Ca^{2+}$
   Kaolinite
   $+ 2.44HCO_3^- + 2.20SiO_2 =$
$1.77Na_{0.62}Ca_{0.38}Al_{1.38}Si_{2.62}O_8 + 2.44CO_2$
   Plagioclase
   $+ 3.67H_2O$
	0.00	0.00	0.22

Step III
   Change kaolinite back into biotite
$0.037Al_2Si_2O_5(OH)_4 + 0.073K^+ + 0.22Mg^{2+} +$
   Kaolinite
   $0.15SiO_2 + 0.51HCO_3^- =$
$0.073KMg_3AlSi_3O_{10}(OH)_2 + 0.51CO_2 + 0.26H_2O$
   Biotite
	0.00	0.00	0.00

Step IV
   Change kaolinite back into K-feldspar
$0.065Al_2Si_2O_5(OH)_4 + 0.13K^+ + 0.13HCO_3^-$
   Kaolinite
   $+ 0.26SiO_2 =$
$0.13KAlSi_3O_8 + 0.13CO_2 + 0.195H_2O$
   K-feldspar
	0.00	0.00	0.00

[a]Adapted from Garrels and Mackenzie (1967).

which the Sierra Nevada spring waters were derived contain an abundance of quartz, virtually all the dissolved silica comes from the alteration of plagioclase feldspar and almost none from quartz. Little dissolved material is derived from the alteration of K-feldspar, and most of the K^+ (and Mg^{2+}) comes from the dark silicate minerals, e.g., biotite or hornblende. Thus K-feldspar and quartz appear as relatively chemically inert minerals under the conditions of weathering. Also, although the Sierra rocks do not contain K-mica, it also tends to be inert. Of course, HCO_3^- is the main anion produced when igneous minerals and rocks are weathered; the Cl^- and

K^+	HCO_3^-	SO_4^{2-}	Cl^-	SiO_2	Products (moles/liter $\times 10^4$)
0.28	3.28	0.10	0.14	2.73	
0.20	3.10	—	—	2.70	Derived from rock
Minus plagioclase					
0.20	0.64	0.00	0.00	0.50	$1.77 Na_{0.62} Ca_{0.38}$ Feldspar
Minus biotite					
0.13	0.13	0.00	0.00	0.35	0.073 biotite
Minus K-feldspar					
0.00	0.00	0.00	0.00	0.12	0.13 K-feldspar

SO_4^{2-} in waters draining igneous rocks come primarily from rain water or dry fallout. Some Cl^- is derived from fluid inclusions in the silicate minerals, whereas some SO_4^{2-} may be derived from solution of calcium sulfate minerals or from oxidation of pyrite in the igneous rocks.

Activity-Activity Diagrams. Perhaps one of the most easily visualized approaches to the chemical weathering of silicate minerals and the waters derived from these weathering processes is the use of activity-activity diagrams (see Appendix B) to quantify the processes leading to the alteration of silicates during weathering. For example, Figure 6.11 is an activity-activity diagram showing the stability relations of some silicate minerals and of gibbsite as functions of the activities of Na^+, H^+, and dissolved silica. Analyses of waters from various rock types also are plotted in the figure. Waters falling in a particular stability field may be in equilibrium with the phase occupying that field and definitely are not in equilibrium with the other phases depicted. Notice that most waters contain between about 2 and 60 ppm of dissolved silica and that there are fairly sharp cutoffs at both values. No waters plotted contain enough silica to be saturated with amorphous silica. Thus the upper silica cutoff at 60 ppm may reasonably represent the equilibrium boundary between kaolinite and montmorillonite. Waters with somewhat higher a_{Na^+}/a_{H^+} ratios and 60 ppm of dissolved silica may be in equilibrium with montmorillonite. Most waters plot in the stability field of kaolinite, which gives support to the generalization that the primary alteration product of the chemical weathering of silicates is kaolinite.

The lower silica cutoff at 2 ppm may be due to a dissolved silica control by the two-phase equilibrium gibbsite-kaolinite. In actual fact, waters draining bauxitic soils in Jamaica and soils developed on basalts in Hawaii —both soils are composed primarily of gibbsite and kaolinite—contain low silica concentrations of from 1 to 6 ppm dissolved silica. Silicate minerals emerge as the important contributor of dissolved silica to natural waters.

Another application of activity-activity diagrams to silicate weathering is shown in Figure 6.12. This diagram shows the fields of stability of some aluminosilicates plus gibbsite as functions of the activities of Ca^{2+}, H^+, and dissolved silica. The compositions of waters from igneous rocks also are plotted on the diagram. The two solid-line arrows show the calculated compositional changes with time as CO_2-bearing waters alter plagioclase feldspar to more aluminous phases. Notice that the calculation for a closed system with an initial dissolved CO_2 of 0.001 moles/liter best fits the observed water analyses. Also notice that the progressive alteration of plagioclase feldspar during the chemical weathering of igneous rocks is quantitatively documented in this diagram. We may use this diagram as a pictorial summary of the processes involved in the chemical weathering of silicate minerals and rocks.

Summary. The chemical weathering of silicate minerals and rocks is accomplished by the aggressive attack of CO_2-bearing soil and ground waters on primary silicate minerals. Quartz, K-feldspar, and K-mica are relatively

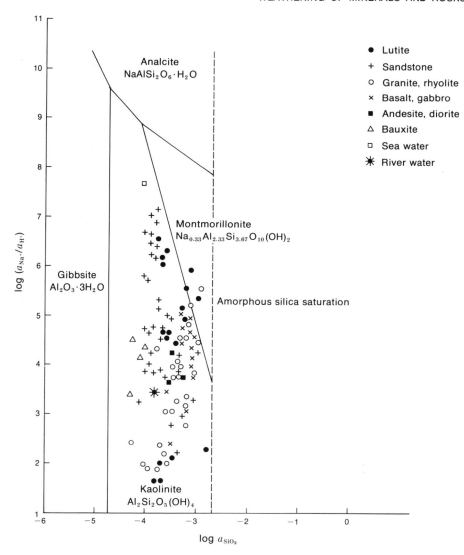

Figure 6.11 Activity-activity diagram for a portion of the system Na_2O-Al_2O_3-SiO_2-H_2O at 25°C and 1 atm total pressure. The stability fields of minerals are shown as a function of the activities of Na^+, H^+, and dissolved silica. Analyses of waters from various rock types are also shown. Notice that most of the water analyses fall within the kaolinite field (after Bricker and Garrels, 1965).

chemically inert, whereas plagioclase feldspars and dark minerals are much more reactive. The primary minerals react and release constituents to the waters. As the concentrations of these constituents rise, kaolinite is formed and dissolved silica and aluminum are abstracted from the waters. Gibbsite, which cannot tolerate much silica in solution without converting to an aluminosilicate, should form only in soils that are continually drained

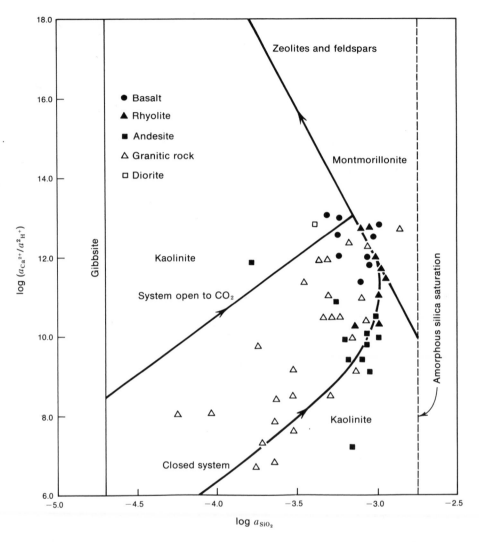

Figure 6.12 Activity-activity diagram for a portion of the system CaO-Al_2O_3-SiO_2-H_2O at 25°C and 1 atm total pressure. The stability fields of the minerals are shown as functions of Ca^{2+}, H^+, and dissolved silica. Analyses of waters derived from igneous rocks are plotted on the diagram. The two solid-line arrows show the compositional changes with time as CO_2-charged water reacts with plagioclase feldspar. Notice that the calculation for a closed system best fits the observed water analyses (after Garrels, 1967).

by nearly pure water. As dissolved CO_2 in the reacting waters is consumed, cations from the dissolution of the primary silicates rise in concentration, as do HCO_3^-, pH, and dissolved silica. Eventually the concentrations of constituents become high enough to form montmorillonite or another clay mineral. Also, with enough initial CO_2, Ca^{2+} and HCO_3^- increase until calcite precipitates. At this point, most of the initial dissolved

CO_2 has been consumed and further reactions proceed very slowly by hydrolysis. Eventually zeolites might form.

RESIDUES FROM ROCK ALTERATION

Compositional Changes

Now that we have looked at the individual chemical reactions involving rock minerals and soil waters, with emphasis on the compositions of the waters, we may gain further insight into the weathering of rocks by considering the compositional changes of the solid residues that take place as a fresh rock is altered. Tables 6.5 and 6.6 show the progressive changes in the chemical and mineralogical composition of a basalt and a granite gneiss as these initially fresh rocks decay. The chemical analyses show only relative changes; the determination of absolute amounts of the various components lost or gained is difficult.

The best way to describe the weathering process would be to refer losses or gains of constituents to an original rock volume, but this procedure cannot be applied very often because the continuous downward collapse of the residual material makes it impossible to determine initial conditions. Notice that as the primary rock silicates are altered, the alkali and alkaline-earth elements are lost almost entirely and silica diminishes mark-

Table 6.5
Chemical and Mineralogical Analyses of a Basalt from Kauai, Hawaii, and the Weathered Residues Derived from its Alteration[a]

| | | Increasing degree of alteration | | |
Component	Fresh rock	A	B	C
SiO_2	45.0	25.3	17.1	5.7
Al_2O_3	11.8	24.4	26.2	25.6
Fe_2O_3	4.3	29.1	32.7	40.2
FeO	9.3	0.44	0.58	0.40
CaO	9.7	<0.10	<0.10	<0.10
MgO	12.0	0.87	0.83	0.65
Na_2O	1.8	0.06	0.06	0.04
K_2O	0.62	0.03	0.03	0.12
H_2O	3.10	13.9	15.8	18.6
Others	2.55	5.20	5.71	6.04

Increasing fresh rock minerals: plagioclase feldspar, pyroxene and olivine	Kaolinite-like clay minerals, goethite, hematite, rock fragments, and minor gibbsite	Increasing alteration products: gibbsite, goethite, hematite, and minor clay minerals

[a] Data from Patterson and Roberson (1961).

edly, whereas ferric iron and water increase. Alumina remains fairly constant after an initial percentage jump. These relative losses and gains of oxide components can be illustrated by using gain-loss diagrams (Leith and Mead, 1915).

Gain-Loss Diagrams

Gain-loss diagrams are constructed by dividing the percentage of each oxide constituent in the fresh rock by its percentage in the weathered residue; this quotient is multiplied by 100 and the results plotted on a suitable scale. A gain-loss plot for the weathering of the granite gneiss and basalt is shown in Figure 6.13. It is obvious that relative to silica, alumina, and ferric iron, K_2O, MgO, CaO, and Na_2O are almost completely lost in the weathering of both rock types. This overall order of loss is what we would predict from our knowledge of mineral-soilwater reactions. The original minerals of a basalt and granite gneiss are primarily silicates. We remember that the alteration of silicate minerals generally results in the formation of more aluminous, hydrated solid phases and the release of Na, Ca, K, and Mg to the reacting water. Thus the residues resulting from the weathering of these rocks should be rich in alumina and poor in the alkali and alkaline-earth elements, as is the case. The loss of FeO relative to Fe_2O_3 represents the oxidation of ferrous iron to ferric during weathering and the formation of hydrated ferric oxides as solid residues.

Table 6.6
Chemical and Mineralogical Analyses of the Morton Granite Gneiss and the Weathered Residues Derived from its Alteration[a]

| | | Increasing degree of alteration | |
Component	Fresh rock	A	B
SiO_2	71.54	68.09	55.07
Al_2O_3	14.62	17.31	26.14
Fe_2O_3	0.69	3.86	3.72
FeO	1.64	0.36	2.53
CaO	2.08	0.06	0.16
MgO	0.77	0.46	0.33
Na_2O	3.84	0.12	0.05
K_2O	3.92	3.48	0.14
H_2O	0.32	5.61	10.39
Others	0.65	0.56	0.58
	Increasing fresh rock minerals: plagioclase feldspar, quartz, K-feldspar, biotite, and other minor minerals	Quartz, kaolinite, K-feldspar, iron oxides and hydroxides, minor plagioclase, and biotite	Increasing alteration products: kaolinite, iron oxides, and hydroxides; quartz and minor K-feldspar present as an unaltered residue

[a] Data from Goldich (1938).

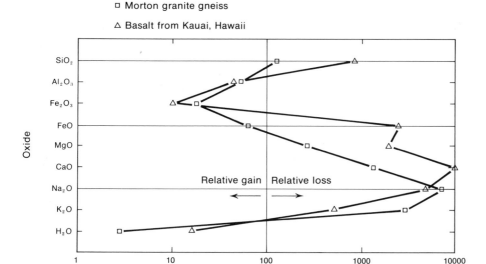

Figure 6.13 Gain-loss diagrams for a granite-gneiss and a basalt.

Notice that in the weathering of both granite gneiss and basalt, there is a large relative gain of H_2O with respect to the other constituents. The development of highly hydrous clay minerals from the primary silicates in the rock accounts for the relative increase in H_2O. The addition of H_2O to the rock minerals during weathering decreases mineral density and thus increases rock volume.

The alteration of a large number of rocks has been studied and the relative losses of oxide components determined. Prolonged weathering results in residues rich in silica, alumina, ferric oxide, and water.

Constant Oxide Calculations

Another approach to determining the losses and gains of rock constituents during weathering is to establish one of the oxides as constant and to base calculation of the loss or gain of elements on this assumption. The choice of the static oxide is not always easy. Al_2O_3 or TiO_2 usually are the oxides chosen, although Fe_2O_3 is also used. The best choice presumably is the least mobile oxide, the one that shows the greatest relative gain on a gain-loss diagram. No general rule can be put forth; the mobility of an oxide may change relative to other oxides during various stages of alteration of the primary rock. The choice must be tailored to the specific case; for example, most of the TiO_2 in basalt is present in the mineral ilmenite $(FeTiO_3)$. It has been found that some weathered residues derived from basalts contain unaltered ilmenite, in which case use of TiO_2 analyses as the reference oxide is indicated.

Table 6.7
An Example of Constant Oxide Calculations in the Weathering of a Diabase. The quartz contents of the weathered products of a lava flow (dolerite) are calculated on the basis of constant TiO_2, Al_2O_3 and Fe_2O_3 content and compared with the analyzed amounts of quartz in the weathered residues; notice the variance in quartz amount depending on the oxide considered constant [a]

	Weathered residue I	Weathered residue II	Weathered residue III	Weathered residue IV
TiO_2 constant	3.1	19.0	3.16	17.6
Al_2O_3 constant	6.5	5.3	8.2	1.4
Fe_2O_3 constant	5.5	6.5	0.43	15.3
Analyzed content	2.86	0.13	5.43	0.14

[a] Table adapted from Mohr and Van Baren (1954). The rock called dolerite by Mohr and Van Baren has a chemical composition close to that of a basalt. However, in this particular rock there was a small percentage of quartz in the fresh rock, although basalts are ordinarily quartz-free.

The quartz contents of the weathered residues of a lava flow (diabase) as calculated on the basis of constant TiO_2, Al_2O_3, and Fe_2O_3 are shown in Table 6.7 and compared with the analyzed quartz content. Notice that the calculated amount of quartz varies strikingly with the oxide that is chosen as constant.

Normative Minerals

The mineralogical changes accompanying the alteration of fresh rock may be studied in more detail by calculating so-called normative mineral compositions of the weathered residues and fresh rock. The details of the calculations need not concern us here; suffice it to say that the various oxides in the chemical analyses are apportioned among various minerals of idealized composition. By recasting the chemical analysis into compositions close to those of minerals, one gets an approximation of the actual mineralogy of the rock. Table 6.8 shows the changes in normative minerals as various rock types weather in tropical regions.

The applicability of normative mineral calculations to weathering problems is obvious. Notice that the original rock minerals, expressed as normative minerals, are nearly completely altered to aluminosilicate and aluminum oxide hydrates and also to ferric oxides in the examples of the basalt and schist. Normative quartz is still present even in the most altered residues of the granite and schist. The great susceptibility of plagioclase feldspar to weathering is apparent from this table, whereas K-feldspar and quartz appear to be less reactive.

Mineral Persistence

The differential loss of the oxides during weathering reflects the susceptibility of minerals to weathering processes. The relative persistence

Table 6.8

Examples of Normative Mineral Calculations as Applied to the Alteration of a Granite, Basalt, and Hornblende Schist Under Tropical Weathering Conditions (the alteration has been so intense that the major original rock minerals are nearly completely gone, except for quartz)[a]

Normative mineral	Rock type and weathered residue								
	Granite				Basalt		Schist		
	Fresh	Residue I	Residue II	Residue III	Fresh	Residue	Fresh	Residue I	Residue II
Quartz	31	42	37	30	—	—	4	8	23
Orthoclase	17	9	6	1	—	—	—	—	—
Plagioclase	28	2	2	1	60	5	38	—	—
Olivine	—	—	—	—	14	—	—	—	—
Augite	—	—	—	—	20	1	—	—	—
Hornblende	—	—	—	—	—	—	49	—	—
Muscovite	18	8	7	1	—	—	—	—	—
Ilmenite	1	1	1	2	3	4	9	8	7
Kaolinite[b]	3	37	44	63	—	72	—	32	44
Goethite[b]	2[c]	—	—	—	—	18[d]	—	23	12
Gibbsite[b]	—	—	2	—	—	—	—	27	2
Magnetite	—	—	—	—	3	—	—	—	6
Water	—	—	—	2	—	—	—	2	6

[a] Adapted from Mohr and Van Baren (1954).
[b] Solid alteration products.
[c] Hematite.
[d] Limonite.

of some common rock minerals under conditions of weathering has been determined by Goldich (1938), who studied in detail the chemistry and mineralogy of several soil profiles. Table 6.9 illustrates Goldich's "mineral stability series." The series is actually arranged into two series: the mafic

Table 6.9

Mineral Stability Series in Weathering (the minerals are arranged according to decreasing rate of decomposition)[a]

Mafic minerals	Felsic minerals
Olivine	
	Ca-plagioclase
Pyroxene	
	Ca-Na plagioclase
Amphibole	Na-Ca plagioclase
	Na-plagioclase
Biotite	
	K-feldspar, muscovite
	Quartz

[a] Adapted from Goldich (1938).

Table 6.10

Mineral Stability Series for Minor Minerals in Weathering Compared with a Series Based on the Frequency of Occurrence of Minor Minerals in Sandstones of Various Geologic Ages (the more persistent a mineral, that is, the more "stable" a mineral, the greater the frequency of occurrence in older rocks)

Stability under weathering conditions[a]	*Increased frequency of occurrence with age of containing rocks*[b]
1. Zircon	1. Rutile
2. Tourmaline	2. Zircon
—	3. Tourmaline
—	4. Monazite
—	5. Garnet
—	6. Biotite
—	7. Apatite
—	8. Staurolite
3. Kyanite	9. Kyanite
4. Hornblende	10. Hornblende
5. Staurolite	11. Augite
6. Garnet	12. Olivine

[a]Dryden and Dryden (1946).
[b]Pettijohn (1941).

(high in magnesium and iron) minerals and the felsic (high in feldspar and quartz) minerals. A distinction is made between the two series because in general the mafic minerals weather more rapidly than the felsic minerals.

Goldich's mineral stability series is concerned with the major rock-forming minerals. Dryden and Dryden (1946) studied the minor minerals in various fresh rocks and their weathered residues. They developed a mineral stability series for these minerals; this series is given in Table 6.10 and compared with a stability series developed by Pettijohn (1941), who studied the frequency of occurrence of minor minerals in sandstones as a function of geologic time. The more "stable" a mineral, the higher its relative percentage in older rocks. The minerals themselves need not concern us here; suffice it to say that several approaches to determining the relative susceptibility of minerals to alteration at or near the earth's surface have been employed. None of the approaches leads to completely unequivocal results, but broad generalizations which apply to a majority of rocks and weathering environments can be formulated. However, there are always exceptions to the rule.

SOIL TYPES AND CLIMATE

We have considered in some detail the reactions between rock minerals and soil waters that lead to the destruction of rock. In doing so, we characterized the general environment, or, in other words, the climate within the soil: the waters, gases, and, briefly, the biologic factors. We shall now look at the relations of these soils to climate.

Soil Profile

The weathering of rocks results in the formation of zones of materials from the surface of the soil downward. This zonation constitutes a soil profile. Generally, three zones can be recognized in a mature soil profile. The uppermost zone, which is the zone of most intense leaching, is the A horizon. Below this zone is the B horizon, consisting of reprecipitated material from above and less altered residues. The C horizon is the lowest zone and consists of partially altered parent-rock material and extends downward to the fresh rock. The zonation of the soil profile is commonly more complex than the three simple horizons described above; each horizon may be divided into subhorizons, and parts of the profile may be missing because of erosion or may be very poorly developed. Figure 6.14 illustrates the development of a soil profile on volcanic ash deposits in a tropical region. Time plays an important role in the development of the profile, and when the soil can no longer support vegetation and micro-organic life because of intense leaching, the soil profile reaches the senile stage.

Climate and Major Soil Types

Soils have been subdivided into many types and subtypes. Some scientists believe that soil types are entirely controlled by climate, and it is true

Figure 6.14 Development of a soil profile on volcanic materials in a tropical region as a function of time. Notice the complex zonation of the profile. When the soil reaches the senile stage, the soil can no longer support vegetation and microorganisms (after Mohr and Van Baren, 1954).

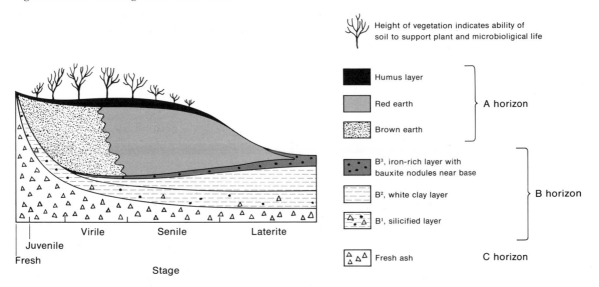

that under long-continued leaching almost any rock will yield a residue high in the nearly insoluble oxides of aluminum, iron, and titanium. On the other hand, detailed soil maps have patterns of soil types remarkably similar to geologic maps of the same area that show the nature of the underlying bedrock.

Three general soil types are recognized: pedalfer [ped(Al)(Fe)r], pedocal [pedo(Ca)(Al)], and laterite. Pedalfer soils form generally on rocks covered in vegetated areas with a plentiful supply of rain water. The rain water is charged with CO_2 in the upper few inches of the humus-rich soil due to the bacterial decomposition of plant and animal detritus. This CO_2-charged soil water attacks the primary minerals of the rock, and calcium, sodium, magnesium, and some potassium and silica are released into solution. Ferric oxides and aluminosilicate hydrates are usual residues of these processes. Ferric oxides are often concentrated in the B horizon. These processes of soil formation are named podsolization and usually occur in temperate, humid climates with a forest cover.

Pedocal soils form in regions of relatively low rainfall with brush or grass cover. The soils are often friable, sometimes black, and highly calcareous and contain montmorillonite as the primary clay mineral. Calcium and magnesium-rich carbonate concretions may be scattered throughout the profile or concentrated at a specific depth, commonly the B horizon. Because water enters these soils and then evaporates out again many times, the dissolved materials derived from the reaction of soil water with primary rock minerals concentrate in the soil solution. Eventually montmorillonite and carbonate minerals are formed from these concentrated soil solutions.

In regions of high rainfall, high temperature, and sparse vegetation, as in many parts of the tropics, bacterial decomposition of organic matter is rapid, and CO_2 evades into the atmosphere. Thus, in such regions, nearly pure rain water attacks the rock minerals. Under such conditions even quartz may be leached, and the weathered residue is rich in ferric and aluminum oxide hydrates. Soil formed by this process is called laterite.

SUMMARY

This chapter has been devoted to a rather detailed analysis of chemical weathering processes. Perhaps the best way to summarize the results obtained is by returning to the average river water draining from North America to deduce the origin of its chemical constituents.

The water composition given in Chapter 5 is reproduced here for easy reference. The concentrations of chemical species are given in milliequivalents or millimoles (SiO_2) per liter:

	HCO_3^-	SO^{2-}	Cl^-	Ca^{2+}	Mg^{2+}	Na^+	K^+	SiO_2	Total
	1.11	0.416	0.225	1.05	0.416	0.390	0.036	0.150	3.79
Adjusted	1.25								3.93

A problem arises immediately. The sum of the equivalents of cations is greater than that of the anions (1.89 to 1.75), and any given water must be electrically neutral. We shall arbitrarily adjust by increasing HCO_3^-, although we could also distribute proportionately over the anions.

In accounting for the origin of the various species, the first step is to subtract the amounts that arrived in precipitation from the atmosphere. An inspection of Table 4.7 shows that precipitation analyses at Menlo Park, California, are probably reasonably representative of average global rain. Recalculation of these analyses into milliequivalents (millimoles for SiO_2) yields (again an adjustment of HCO_3^- is required)

	HCO_3^-	SO_4^{2-}	Cl^-	Ca^{2+}	Mg^{2+}	Na^+	K^+	SiO_2	Total
Adjusted	0.039	0.028	0.096	0.040	0.030	0.087	0.006	0.005	0.331
Initial	0.066								

Because the total is only about 10 percent of the concentrations in stream water, it is clear that accurate values for constituents other than Na^+ and Cl^- are not required. Subtraction from the stream analyses leaves the following concentrations of species derived from rocks:

HCO_3^-	SO_4^{2-}	Cl^-	Ca^{2+}	Mg^{2+}	Na^+	K^+	SiO_2	Total
1.210	0.388	0.129	1.010	0.386	0.303	0.030	0.145	3.60

Next we shall subtract enough Na^+ to balance the Cl^-, on the assumption that Cl^- is present in rocks as halite, as NaCl in fluid inclusions, and as NaCl in concentrated solutions existing as films on grain surfaces. This leaves

HCO_3^-	SO_4^{2-}	Cl^-	Ca^{2+}	Mg^{2+}	Na^+	K^+	SiO_2	Total	NaCl removed
1.210	0.388	0.00	1.010	0.386	0.124	0.030	0.145	3.34	0.129

Sulfate probably is derived chiefly from gypsum or anhydrite, but some may come from oxidation of S in pyrite (FeS_2). If it were all from gypsum we could subtract out enough Ca^{2+} to balance sulfate, but the portion derived from sulfide oxidation is difficult to estimate, and even if we had a good number, which cation should be used for the balance? Fortunately, it was shown in Chapter 5 that SO_4^{2-} seems to be balanced by a combination of Ca^{2+} and Mg^{2+}, so we shall draw upon them proportionately as sources of SO_4^{2-}. The remainder is

HCO_3^-	SO_4^{2-}	Cl^-	Ca^{2+}	Mg^{2+}	Na^+	K^+	SiO_2	Total	"Gypsum" removed
1.210	0.00	0.00	0.731	0.267	0.124	0.030	0.145	2.56	0.388

We shall assume that the rest of the Ca^{2+} and Mg^{2+} has been derived from the weathering of limestones and that it comes from the minerals calcite or dolomite. In this instance each equivalent of Ca^{2+} or Mg^{2+} requires one of HCO_3^- for balance, and there remains

HCO$_3^-$	SO$_4^{2-}$	Cl$^-$	Ca^{2+}	Mg^{2+}	Na$^+$	K$^+$	SiO$_2$	Total	"Limestone" removed
0.212	0.00	0.00	0.00	0.00	0.174	0.030	0.145	0.561	0.998

We note that the ratio of Ca^{2+} to Mg^{2+} calculated as "limestone" is about 3:1, which would suggest that the ratio of limestone to dolomite weathered is about 2:1 for North America, which is a reasonable ratio. Now Na$^+$, K$^+$, and SiO$_2$ can be attributed to the weathering of feldspars. If we alter the feldspar to kaolinite, according to a reaction of the type

$$2NaAlSi_3O_8 + 2CO_2 + 3H_2O = Al_2Si_2O_5(OH)_4 + 2Na^+ + 2HCO_3^-$$
$$+ 4SiO_2,$$

two SiO$_2$ would be produced for each HCO$_3^-$, which is too much SiO$_2$ for the balance. On the other hand, if we weather the feldspar to a montmorillonite-type mineral, according to the reaction

$$2NaAlSi_3O_8 + 2CO_2 + 2H_2O = Al_2Si_4O_{10}(OH)_2 + 2Na^+ + 2HCO_3^-$$
$$+ 2SiO_2,$$

the ratio of SiO$_2$ to HCO$_3^-$ could be 1:1, which fits better.

Also, the weathering of montmorillonite to kaolinite would produce one molecule of SiO$_2$ for each molecule of HCO$_3^-$. Accordingly, we shall assume that the single best descriptive reaction is the feldspar → montmorillonite reaction and shall use up the silica in this fashion. This leaves (subtracting all K$^+$)

HCO$_3^-$	SO$_4^{2-}$	Cl$^-$	Ca^{2+}	Mg^{2+}	Na$^+$	K$^+$	SiO$_2$	Total	"Feldspar" removed
0.067	0.00	0.00	0.00	0.00	0.059	0.00	0.00	0.126	0.145

This remainder is as close to as complete a balance as could be expected from the accuracy of the original analyses, considering that we had to adjust HCO$_3^-$ in the beginning, that the minerals computed are off composition from the real ones, that many minerals not considered actually contribute to the balance, and so on. At least we get a rough outline of the relative contributions of various rock types; if we correct the equivalents of minerals weathered to weight percentages of rock types, it appears that North America's chemical contribution to the oceans comes from the following sources: 5 percent from salt beds and salt disseminated in rocks, 25 percent from gypsum and anhydrite deposits and disseminations, 40 percent from limestones and dolomites, and 30 percent from weathering of silicates, i.e., of feldspars to clay minerals or from one clay mineral to another. The ratio of suspended load to chemical load for North America is roughly 1:1 (Chapter 5); if only 30 percent of the chemical load represents silicate minerals that have been produced by weathering, 70 percent of the suspended load may just be pre-existing minerals washed into the oceans without much change.

This approximate calculation of the source materials of the chemical load of streams appears reasonable in terms of the outcrop areas of various rock types and our intuitive feelings concerning their relative erodibility; we shall try to show later that the picture developed is required from other considerations as well. Note at this stage that salt and gypsum deposits would have to be about 30 percent of all rocks exposed to weathering if their contributions to streams were proportional to their mass.

REFERENCES

Barnes, I., La Marche, V. C., Jr., and Himmelberg, G., 1964, Geochemical evidence of present-day serpentinization: *Science, 156,* 830–832.

Billings, G. K., and Williams, H. H., 1967, Distribution of chlorine in terrestrial rocks (a discussion): *Geochim. Cosmochim. Acta, 31,* 2247.

Bricker, O. P., 1968, written communication.

Bricker, O. P., and Garrels, R. M., 1965, Mineralogic factors in natural water equilibria: in *Principles and Applications of Water Chemistry,* S. Faust and J. V. Hunter, eds.: John Wiley & Sons, Inc., New York, 449–469.

Carroll, D., 1962, Rain-water as a chemical agent of geologic processes—a review: *U.S. Geol. Surv. Water Supply Paper, 1535-G.*

Clarke, F. W., 1924, The data of geochemistry: *U.S. Geol. Surv. Bull., 770.*

Dryden, L, and Dryden, D., 1946, Comparative rates of weathering of some common heavy minerals: *J. Sediment. Petrol., 16,* 91–96.

Erikkson, E., 1952, Composition of atmospheric precipitation. II. Sulfur, chloride, iodine compounds: *Tellus, 4,* 280–303.

Feth, J. H., Roberson, C. E., and Polzer, W. L., 1964, Sources of mineral constituents in water from granitic rocks Sierra Nevada California and Nevada: *U.S. Geol. Surv. Water Supply Paper, 1535-I.*

Gambell, A. W., and Fischer, D. W., 1966, Chemical composition of rainfall Eastern North Carolina and Southeastern Virginia: *U.S. Geol. Surv. Water Supply Paper, 1535-K.*

Garrels, R. M., 1967, Genesis of some ground waters from igneous rocks: in *Researches in Geochemistry,* vol. 2, P. H. Abelson, ed.: John Wiley & Sons, Inc., New York, 405–420.

Garrels, R. M., and Christ, C. L., 1965, *Solutions, Minerals, and Equilibria:* Harper & Row, New York.

Garrels, R. M., and Mackenzie, F. T., 1967, Origin of the chemical compositions of some springs and lakes: in Equilibrium concepts in natural water systems, W. Stumm, ed., *Advan. Chem. Ser., 67,* 222–242.

Goldich, S. S., 1938, A study in rock weathering: *J. Geol., 46,* 17–58.

Gorham, E., 1961, Factors influencing supply of major ions to inland waters, with special reference to the atmosphere: *Bull. Geol. Soc. Am., 72,* 795–840.

Headden, W. P., 1903, Significance of silica acids in waters of mountain streams: *Am. J. Sci., 166,* 169–186.

Johns, W. D., and Huang, W. H., 1967, Distribution of chlorine in terrestrial rocks: *Geochim. Cosmochim. Acta, 31,* 35–50.

Judson, S., and Ritter, D., 1964, Rates of regional denudation in the United States: *J. Geophys. Res., 69,* 3395–3401.

Junge, C. E., and Werby, R. T., 1958, The concentration of chloride, sodium, potassium, and sulfate in rain water over the United States: *J. Meteorol., 15,* 417–425.

Kinsman, D. J. J., 1966, Gypsum and anhydrite of recent age, Trucial Coast, Persian Gulf: *Second Symposium on Salt,* vol. 1: Northern Ohio Geological Society, Cleveland, 302–326.

Komabayasi, M., 1962, Enrichment of inorganic ions with increasing atomic weight in aerosol, rainwater and snow in comparison with sea water: *J. Met. Soc. Japan, 40,* 25–38.

Krauskopf, K., 1967, *Introduction to Geochemistry:* McGraw-Hill, Inc., New York.

Leith, C. K., and Mead, W. J., 1915, *Metamorphic Geology:* Henry Holt and Co., New York.

Miller, J. P., 1961, Solutes in small streams draining single rock types, Sangre de Cristo Range, New Mexico: *U.S. Geol. Surv. Water Supply Paper, 1535-F.*

Mohr, E. C. J., and Van Baren, F. A., 1954, *Tropical Soils:* Wiley-Interscience Publishers, New York.

Mordy, W. A., 1953, A note on the chemical composition of rainwater: *Tellus, 5,* 470–474.

Patterson, S. H., and Roberson, C. E., 1961, Weathered basalt in the eastern part of Kauai, Hawaii: *U.S. Geol. Surv. Profess. Paper, 424-C,* 195–198.

Pettijohn, F. J., 1941, Persistence of heavy minerals and geologic age: *J. Geol., 49,* 610–625.

White, D. E., Hem, J. D., and Waring, G. A., 1963, Chemical composition of subsurface waters: Data of geochemistry, 6th ed., *U.S. Geol. Surv. Profess. Paper, 440-F.*

Whitehead, H. C., and Feth, J. H., 1964, Chemical composition of rain, dry fall-out, and bulk precipitation at Menlo Park, California, 1957–1959: *J. Geophys. Res., 69,* 3319–3333.

Wolff, R. G., 1967, Weathering of Woodstock granite, near Baltimore, Maryland: *Am. J. Sci., 265,* 106–117.

7 Sedimentation in the Ocean Basins

Now that we have gained some knowledge of the materials produced during the breakdown of rocks, we can look in detail at the fate of these products of weathering as they move about the earth's surface and briefly at the major repository of these materials—the ocean. Dissolved and particulate weathered materials, from the moment of their inception, are susceptible to removal and transportation from their sites of origin and eventually to deposition and accumulation. These materials are being deposited in present-day stream beds and floodplains, swamps, lagoons, lakes, and oceans. Eventually, by compaction and cementation, accumulations of these materials will become sedimentary rocks. The same processes have occurred throughout the earth's history and are part of the complex cycle of crustal change that began when the initial crust was formed some 4½ billion years ago.

FACTORS INFLUENCING SEDIMENT DISPERSAL

Solid particles, when set in motion in stream waters or other fluids, follow laws of fluid flow. Fluids flow either laminarly or turbulently. Laminar flow occurs when fluids are moving relatively slowly, whereas turbulent flow takes place when the velocity of the fluid is great enough to overcome the viscous forces that maintain continuity of the fluid. A particle suspended in a stream flowing laminarly would follow a smooth path, whereas if the stream were flowing turbulently, the particle path would describe a series of eddies. Ground waters exhibit laminar flow; turbulent flow is characteristic of most other natural fluids—wind, water in streams, lake and ocean water currents. Turbulence spreads throughout the fluid if the Reynolds number of the flow is exceeded. The Reynolds number is a dimensionless quantity defined as

$$R_e = \frac{dvh}{n},$$

where d is the density of the fluid in grams per cubic centimeter, v is the velocity of the fluid in centimeters per second, h is the depth in centimeters, and n is the viscosity in poises. Reynolds numbers for large bodies of water range from about 300–600; to exceed these values requires only small velocities; thus turbulent flow is characteristic of most water bodies.

Particles are set in motion by wind or water currents when the drag of the moving fluid overcomes the gravitational and cohesive forces that maintain the particle at rest. However, very fine-grained particles in a stream bed or on the sea floor may remain at rest while coarser materials are being transported. This paradoxical situation arises because silt- and clay-size particles are coherent and are also so small that they do not permit development of eddies about them that initiate particle movement. Once particles are set in motion, their subsequent behavior is governed by the relative settling velocities of the particles.

The settling velocity of a particle depends on its density, sphericity, and size. The classic, and perhaps best known, formula for obtaining settling velocities is that of Stokes and is known as Stokes' law. The relationship may be formalized as

$$v = \frac{2}{9} \frac{d_1 - d_2}{n} gr^2,$$

where v is the velocity of the particle in centimeters per second, r the radius of the particle in centimeters, g is the acceleration of gravity, n is the viscosity of the fluid in poises, and d_1 and d_2 are the densities of the particle and fluid, respectively. A quartz grain with a density of 2.65 g/cm^3 and a diameter of 0.1 cm, if dropped into a quiet body of fresh water, will fall about 3.5 m in 1 sec. Care must be taken in applying Stokes' law because several assumptions underly its formulation. Perhaps the most important is that the law holds strictly only for spherical particles. In general, the less spherical the particle, the smaller the settling velocity of the particle, as compared with a sphere of the same volume. The law does not hold for large particles, whose settling velocities are not controlled by fluid viscosity but are proportional to the square root of the particle radius. In Figure 7.1, experimental data on settling velocities are compared with velocities calculated from Stokes' law and from the impact law, which holds true for large particles. Notice that Stokes' law satisfies the experimental results for particles less than 0.1 mm in diameter, whereas the impact law predicts very well the behavior of particles greater than 1 mm. The behavior of particles of intermediate size is governed by both fluid viscosity and particle size.

In summary, particles are moved by fluids—stream and ocean water currents and wind—when the forces exerted by the drag of the fluid overcome the gravitational and cohesive forces that hold the particles at rest. Fine-grained materials are difficult to set in motion, but once in suspension they may remain so for long periods of time because of vertical turbulent current velocities that exceed the settling velocities of the clay-size particles. Sand- and silt-size detritus brought into suspension settles more rapidly than clay-size materials. Some sand- and silt-size detritus is simply dragged by currents along the bottom of streams or over the ocean floor or bounces along in a series of short jumps.

The importance of these hydrodynamic relations can be illustrated by a simple example. Where a river enters the ocean, it carries a mixture of sand-, silt-, and clay-size detritus. The various sized materials become sep-

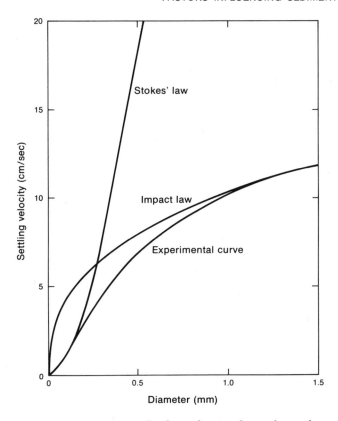

Figure 7.1 Comparison of experimentally derived curve for settling velocities in water at ordinary temperatures as a function of particle diameters with curves predicted from Stokes' law and the impact law (after Krumbein and Sloss, 1963).

arated rapidly; in general, the sands and silts settle out quickly; the clays remain in suspension and may be carried several hundred kilometers out to sea before settling to the bottom. Figure 7.2 illustrates the distances and times involved for various sized particles to settle 100 m in an ocean current moving with a horizontal velocity of 10 cm/sec. From these relations one would expect to find sands near the shores of modern ocean basins and clays farther out. Such a gross distribution of grain sizes is observed in some areas. Figure 7.3 illustrates the distribution of bottom sediment types on the Nigerian continental shelf. Notice that sands are being deposited close to shore and clays farther out, owing to the differential settling rates of the sand- and clay-size particles and the distribution of wave and current energy on the shelf.

Fine-grained materials may also be deposited near shore in areas of low current and wave energy, e.g., in lagoons behind barrier beaches, and sands may be deposited by turbidity currents (see p. 192) in the deep sea, some tens of kilometers from the coast. On some continental shelves (e.g.,

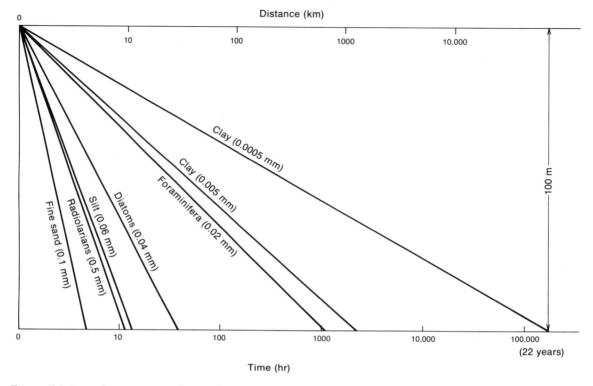

Figure 7.2 Time for various sized particles to settle 100 m in a current flowing horizontally at 10 cm/sec. Distance traveled by a particle during its settling time is also shown (American Geological Institute, 1967).

the continental shelf off the coast of the eastern United States) little deposition is occurring today because the materials transported to the shelves by rivers are trapped in back-water sites of deposition, such as lagoons and bays, or bypass the shelves owing to agitation by currents which keep the materials in suspension. A summary of relations among erosion, transportation, and deposition of sedimentary detritus as functions of current velocity and grain diameter is shown in Figure 7.4.

Notice the minimum in the critical erosion velocity curve; the erosion velocity for particles greater than about 0.3 mm in diameter increases with increasing grain size, and for particles finer than 0.3 mm the velocity increases with decreasing grain size. Thus coarse sand and gravel and silt and clay are difficult to erode, whereas fine sand is eroded more easily. Also, erosion of a sediment of mixed grain sizes could result in the removal of relatively coarse material in preference to finer particles.

As wind, river, and ocean currents move detritus about, the original heterogeneous materials that were initially available at the sites of weathering are selectively sorted according to particle sizes, shapes, and densities and segregated into sedimentary masses of similar texture, structure,

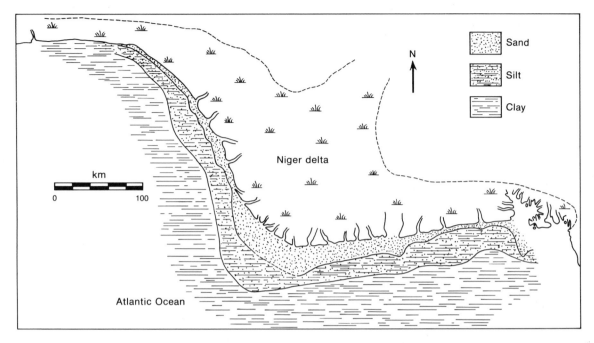

Figure 7.3 Bottom sediment type and distribution on the Nigerian continental shelf and upper slope. Notice decrease of sediment grain size seaward (after Allen, 1964).

and composition. The textures of some sedimentary rocks were pictured in Chapter 2, p. 50. The structure of a sedimentary rock refers to the larger features of the rock, e.g., bedding, ripple marking, cross-bedding (Figure 7.5). Textures and structures are commonly used to determine the genesis of a particular rock body. For example, the texture of a sandstone containing primarily well-rounded, spherical grains of quartz of about the same size may indicate derivation from a pre-existing sandstone and stream transportation over a long distance or extensive reworking by marine waves and currents of the quartz grains that make up the sandstone. The same sandstone may contain structural evidence, such as ripple marking or cross-bedding, that indicates deposition in a current-agitated environment and supports the textural evidence of reworking.

Figure 7.6 schematically illustrates the changes in the properties of stream sediment as it moves downstream. Notice that the average particle diameter decreases downstream, whereas particle roundness (smoothing of edges and corners of particle) and sphericity (degree of approach of particle shape to a sphere) increase. These changes are due primarily to sorting of the sedimentary debris by the stream but are also due to abrasion of the particles during transportation. Because of the relative chemical and mechanical stabilities of minerals, and the sorting of these minerals according to size, shape, and density, sedimentary deposits are not only segregated

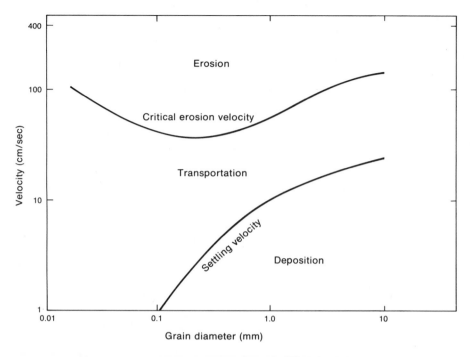

Figure 7.4 Relations among erosion, transportation, and deposition for material of uniform grain size as a function of velocity and grain diameter (after Hjulström, 1955).

on a mechanical basis but also on a chemical basis. For example, quartz is a mechanically and chemically resistant mineral; it is not easily broken down into particles smaller than sand and silt size and it does not weather rapidly chemically. Therefore, the mineral quartz tends to segregate into

Figure 7.5 Some representative depositional structures found in sedimentary rocks. A, Mud cracking (Courtesy of W. D. Keller, Department of Geology, University of Missouri, Columbia, Missouri). B, Ripple marking. C, Cross-bedding.

A   B   C

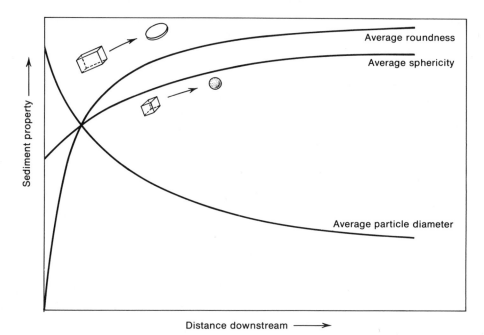

Figure 7.6 Generalized changes of sediment properties due to selective sorting and abrasion in a stream. Roundness refers to smoothing of edges and corners; sphericity refers to degree of approach to a sphere.

sedimentary deposits of sand and silt size, and consequently these deposits tend to be high in SiO_2. Potassium feldspar, on the other hand, is resistant to chemical weathering in most surface environments but is less resistant mechanically than quartz; during transportation and dispersal it is abraded into silt- and clay-size particles. Thus K-feldspar tends to concentrate in silts and clays, and these deposits tend to be higher in K and Al than the sand-size deposits. In contrast, plagioclase feldspars are susceptible to chemical alteration; by-products of this alteration are the clay minerals. These minerals concentrate in clay-size sedimentary deposits, which are generally higher in Al, Na, and K than sandstones.

DEPOSITIONAL ENVIRONMENTS OF THE OCEAN BASINS

Introduction

The oceans are the predominant repository for products of weathering. The ocean basins are brimful; their waters cover the margins of the continents and extend into the continents in great inland seas—about 7 percent of the continents are covered by marine waters of a few meters to 200 m deep.

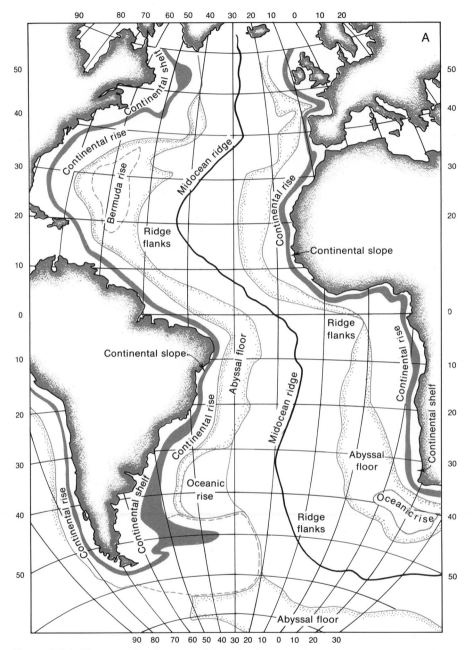

Figure 7.7 A, Physiographic diagram of the Atlantic Ocean basin (after Heezen et al., 1959).

These shallow water continental shelves and inland basins are major areas of accumulation of sedimentary materials. Figure 7.7 illustrates the major divisions of the ocean basins—continental margins (continental shelves, slopes, and rises), ocean-basin floors (plains, rises, and trenches), and

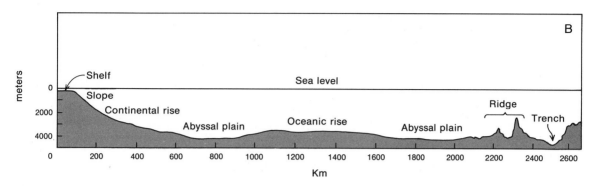

Figure 7.7 B, Generalized topographic profile of the ocean basins. Vertical scale greatly exaggerated.

oceanic ridges—and also shows the physiographic subdivisions of the Atlantic Basin.

Continental Shelves

General

The continental shelves vary in width and generally have smooth, gently seaward-dipping surfaces. Carbonate and terrigenous sediments are accumulating in various physiographic regions of the shelf. Fine-grained terrigenous materials are deposited in the nearshore, low-energy environments of lagoons, parts of bays and estuaries, tidal flats, and the swamps and marshes of deltas and on the offshore, low-energy regions of shelves. Sands accumulate in the higher-energy environments of beaches, distributary channels of deltas, river-mouth bars, and on the nearshore, high-energy regions of some shelves.

Carbonate sediments accumulate in shallow water regions of the shelves where the influx of terrigenous materials is small. Modern carbonate sediments are produced either by direct precipitation of carbonate minerals from sea water (warm, shallow, marine waters are oversaturated with respect to calcite) or by the breakup and accumulation of the skeletons and tests of carbonate-secreting organisms; the latter process appears to be much more important today. Because carbonate-secreting organisms thrive in warm waters and these regions are also most favorable for direct chemical precipitation of carbonate minerals from sea water, the major areas of shallow water carbonate accumulation today are in the tropics and subtropics. However, warm water is not a necessity for carbonate accumulation; modern calcareous deposits are also found in cold or temperate regions.

Modern provinces of shallow water carbonate sedimentation exhibit a range of environments in which carbonate-secreting organisms live. A spectrum of grain sizes can be produced by mechanical disintegration of the skeletons of organisms. Some green algae (e.g., *Penicillus*) produce fine, needle-like aragonitic fragments that accumulate as carbonate mud; the green alga, *Halimeda*, produces primarily sand-size debris upon breakdown and the red alga, *Lithothamnion*, which encrusts the surfaces of reefs, may be broken apart into gravel-size clasts under strong wave attack. Corals produce gravel- and sand-size debris or much finer materials depending on the type of coral and the physical environment. Mollusks, echinoderms, forams, brachiopods, and other organisms that secrete skeletal edifices also supply carbonate detritus in a spectrum of grain sizes to form deposits of carbonate mud, silt, sand, and gravel. It is likely that in some cases organisms that are capable of boring into the hard skeletal substrates of other organisms weaken the skeletal framework of these organisms and promote mechanical breakup by waves and currents. The sponge *Cliona* and the mollusk *Lithophaga* are important boring organisms; large coral heads are commonly found riddled with holes produced by *Lithophaga*. The important point is that organisms themselves, through breakup of their skeletons, supply detritus that can be worked upon by waves and currents in the same way as terrigenous materials.

Environments

In the following sections we shall look briefly at a few environments of deposition [1] found on the shelves to provide some idea of the types of shelf and nearshore sediments and the processes of sedimentation. The Nigerian continental shelf (Figure 7.3) has already been described. It exhibits a sediment distribution pattern consistent with the distribution of the intensity of wave and current agitation; sands are being deposited in current-agitated nearshore areas and clays in deeper, less-agitated water.

An investigation by Creager (1963) of a portion of the continental shelf area of the Chukchi Sea off the northwest coast of Alaska provides a detailed example of the relationship between sediment distribution and physical properties. Figure 7.8 illustrates the distribution of sediments, directions of bottom current movement, bathymetry, and sorting [2] of sediments in this high-energy embayment of the sea. Because of strong current agitation, the embayment is characterized by medium- to coarse-grained sediments. The sediments generally decrease in grain size from the shores of the embayment toward its center. Gravelly sediments are found in the Point Hope-Cape Thompson area and at the Bering Strait. The current through the Bering Strait is the main source of sediments in the embayment; a minor source is the cliffs near Cape Thompson. The current entering the Bering Strait moves so rapidly that sedimentation does not occur in the Strait, but a coarse, poorly sorted residual deposit is found on the floor of the Strait. As the Bering Strait current moves northeastward into the embayment, its velocity decreases and sand is deposited in the shallow water area north of Cape Prince of Wales. The sand is further cleaned by the re-

---

[1] Geographically restricted areas, having variable sets of physical, chemical, and biological conditions; usually described in geomorphic terms.

[2] Sorting is an index of the range in particle size of a sediment. Poor sorting indicates a wide range and excellent sorting a small range.

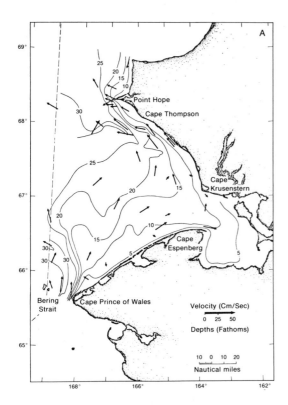

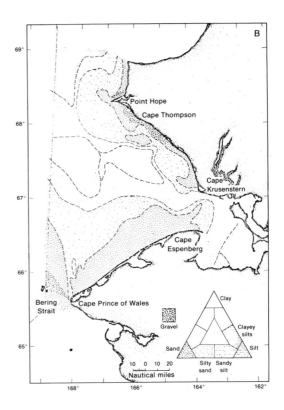

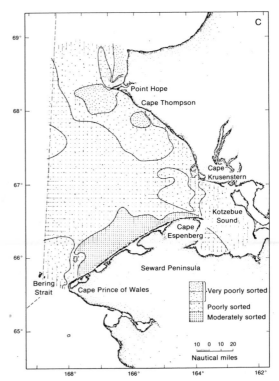

Figure 7.8 Distribution of physical properties of the continental shelf area of the Chukchi Sea off the northwest coast of Alaska. A, Bathymetry and near-bottom current pattern. B, Areal distribution of clay, silt, sand, and gravel and their mixtures. C, Sorting of sediment (after Creager, 1963).

moval of finer materials by wave action resulting in a moderately sorted sand. The current is deflected westward along the north shore of the embayment; water piled up against this shore causes an increase in current velocity such that only coarse sediments, locally derived from the marine erosion of cliffs, accumulate in this area. The sand and gravel deposited are poorly sorted because of the nearness of the source and the types and poor sorting of the source rocks. Finer materials accumulate in the deeper, central portion of the embayment owing to decreased current agitation.

The Chukchi Sea embayment provides an example of shelf processes in an area where no large rivers enter the sea. Where rivers enter the sea, thick prisms of sediment may accumulate on the continental shelf in the form of deltas. The morphologies of several of the world's deltas are shown in Figure 7.9. Generally, deltas exhibit a seaward bulging of the shoreline where the major river channel and associated distributory channels that cross the low-lying land of the delta enter the sea. The major river channel may change its course with time, owing to decrease in channel grade because of filling of the channel with deposits and breakthrough of the natural levee by the stream, which then occupies a new course; for example, the Yellow River of China flowed northeast between 1856 and 1938 into the Gulf of Pechili (Figure 7.9) but left this channel in 1938 to flow southeast into the Yellow Sea, returning in 1946 to its original northeast course.

A delta consists of a myriad of sedimentary deposits formed in a variety of environments. Sands are deposited in thin sheets interbedded with clay- and silt-size detritus near the seaward margin of a delta and are sedimented in the channels of the distributaries crossing the deltaic plain, along beaches, or as bars at the mouths of rivers. Fine-grained materials accumulate in marshes, swamps, and lakes of the low-lying land between distributary channels, and clays are deposited at the farthest seaward margin of a delta. The environments of deposition of a deltaic complex are not fixed geographically but migrate with time. The large deltas of the world have built slowly seaward because of an abundant supply of sediment, slow subsidence of the regions in which the deltas are forming, and low rate of removal of river-supplied detritus by ocean waves and currents. Thus the clays deposited at the seaward edge of many deltas are slowly being covered by shallow water, sands, silts, and silty clays. Lateral shifting of distributaries and consequent changes in the loci of sediment accumulation may result in a shift in the emergent and submergent parts of a delta. The many environments represented in a delta and the temporal and spatial variation of these environments produce a wedge of deposits exhibiting a complex vertical and horizontal distribution of sedimentary rocks.

Figure 7.10 is a block diagram of part of the Mississippi Delta in the northern Gulf of Mexico and shows the various generalized types of deposits that have accumulated in the delta in the last 500 years. These deposits are all of shallow water origin and illustrate the fact that to accumulate this wedge of sediments, the Gulf of Mexico shelf must have gradually sunk

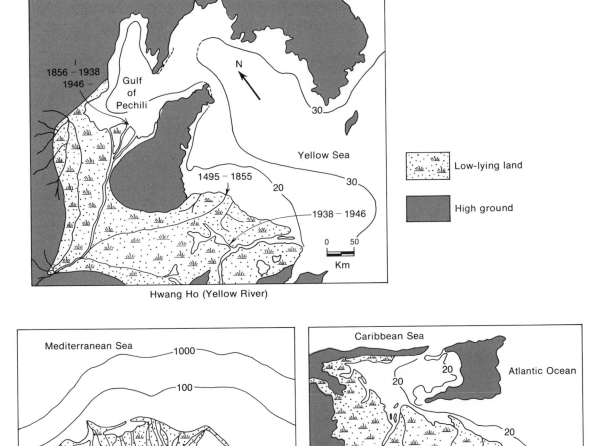

Figure 7.9 Morphologies of some major deltas of the world. Depth contours are in meters. The Hwang Ho River has changed its major channel several times during recorded history; the dates of these changes are shown (after Dunbar and Rodgers, 1957, and Scruton, 1960).

or the sea must have risen. The cross section in Figure 7.11 shows that for nearly 200 million years, shallow water marine sediments have been accumulating in a portion of the Gulf Coast basin. Certainly the sea has not

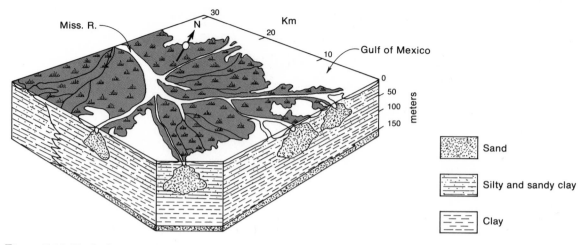

Figure 7.10 Block diagram of part of the Mississippi delta illustrating the heterogeneity of deltaic deposits (after Fisk et al., 1954).

risen some 3500 m to permit accumulation of these sediments! On the contrary, such large areas of ocean basins in which shallow water sedimentary rocks have accumulated for long periods of time are evidence of slow sinking of parts of the crust. The North American continent is the source for the materials that enter the Gulf Coast Basin and has performed isostatically much the same as a raft from which sand is being unloaded into a nearby boat. As the land is denuded and slowly rises, the adjacent basin sinks. However, whether the basin sinks owing to the weight of the overlying sedimentary rock mass alone or whether it is slowly depressed by an

Figure 7.11 Diagrammatic cross section of the Gulf of Mexico. Sedimentary rocks found in this great sedimentary basin are mainly of shallow water origin (after King, 1959).

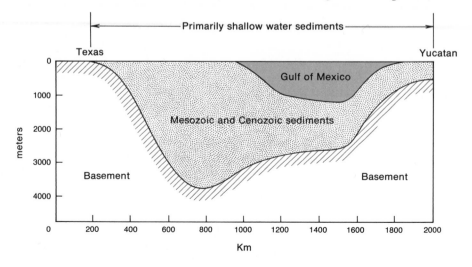

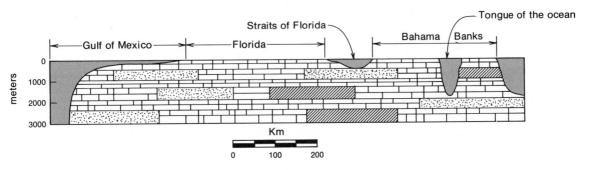

Figure 7.12 Generalized cross section of the Florida-Bahamas area, a region of essentially continuous carbonate deposition since the Cretaceous. ⊞, Carbonate deposits. ⊡, Quartz silt and sand deposits. ▨, Evaporite deposits.

external force, e.g., convection currents, is a major question concerning the earth's history.

Areas of extensive carbonate accumulation also sink slowly with time. Figure 7.12 is a cross section of the upper part of the crust in the region of Florida and the Bahama Islands. The section illustrates that parts of this region of the crust have been areas of shallow water carbonate deposition for more than 70 million years. The present land of Florida began to emerge in the middle Miocene, and prior to that time Florida had been a shallow water marine basin accumulating carbonate deposits.

The present-day Straits of Florida separate the Florida shelf from the Bahama Banks, a region of modern carbonate sedimentation. Until Miocene time, an area including the eastern edge of the present Gulf of Mexico, present-day Florida, and the Bahamas had been a region of carbonate sedimentation, primarily because the detritus carried off the North American continent into the Gulf of Mexico was not transported to the east by ocean currents but was trapped in the deltas of the major rivers entering the Gulf or was transported westward by marine currents. Organisms flourished in the clear, relatively warm waters of the eastern Gulf and the Bahamas and provided skeletal detritus that makes up the carbonate sediments. At times, marine waters within this region were sufficiently restricted from the general open ocean circulation to evaporate to the extent that evaporite deposits were formed; at other times, quartz sand and silt were transported into the region by marine currents and accumulated in thin sheet and lens-like deposits. The bulk of limestones and dolomites in the geologic column accumulated under comparable shallow water marine conditions in inland seas and on continental shelves.

Continental Slopes and Rises

Along the seaward margin of a continental shelf, the shelf gives way to a topographic surface inclined some 3–6° seaward—the continental slope.

The slope may be a relatively smooth, nearly featureless surface or may be notched by submarine canyons or dissected into a complex submarine valley and ridge topography. The canyons resemble gullies and valleys formed on land and generally run down the slope and head at the seaward margin of the continental shelf. The origin of these canyons has been a subject of hot debate for the past 20 years. It is likely that they cannot be explained by one process only. Some may have been excavated by stream erosion at a time in the earth's past when sea level was lower and may have since been modified by marine processes. Others probably have been scoured by undersea marine currents of dense muddy water that have moved down the continental slope. These turbidity currents can be produced in the laboratory and are known to occur in lakes where fresh, muddy waters enter the lake and flow along the bottom. A turbidity current has never been observed in the oceans, although the breaking of telegraph cables during an earthquake in the North Atlantic Ocean has been attributed to a very large turbidity current that was initiated by the shock of the quake. Large "waterfalls" of sediment-laden water have been observed cascading down submarine canyon walls off the coast of southern California. These cascades may transform into turbidity currents lower down at the mouths of the canyons. Turbidity currents are apparently capable of scouring submarine materials to form canyons and of depositing sand-size debris at depths of 1000–5000 m or more. Sedimentary deposits resulting from presumed turbidity current flows commonly are poorly sorted and contain both clay- and sand-size detritus. Sedimentary rocks termed *graywackes* (Chapter 8), with the same grain size and compositional characteristics as modern turbidity current deposits, are found throughout the geologic column; much of this kind of rock is thought to have been deposited by turbidity currents.

In some areas of the ocean, turbidity current deposits, instead of forming delta-like masses, intercalate with fine-grained sediments to form a thick prism of terrestrially derived sediment at the base of continental slopes. This wedge of sediments, known as the continental rise, extends at a lesser inclination than the continental slope out to the ocean-basin floor at depths of 3500–5500 m.

The proportion of sand-size detritus in continental rise sediments is small; most of the sediment is lutite of terrigenous origin that bypasses the continental shelf because strong, irregular currents on the shelf prevent deposition of the slow settling clay-size detritus. Eventually, the materials settle from deeper, slower moving currents. The thickest parts of the sediments of the continental rise are found beneath the axes of the fastest moving deep-ocean currents, and the sediments become thinner in the direction of decreasing current velocity. In effect, the deep currents shape the continental rise. An interpretation of the development of the continental rise off eastern North America is shown in Figure 7.13, adapted from Heezen et al. (1966).

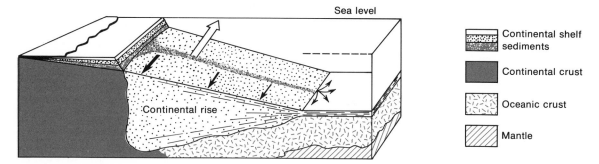

Figure 7.13 Development of the continental rise. The black arrows show the direction of flow and relative velocity of the deep-sea currents and the white arrow the direction of flow of the Gulf Stream. A small turbidity current is shown flowing down the continental rise and out onto the ocean-basin floor (after Heezen et al., 1966).

Ocean-Basin Floor and Oceanic Ridges

In some areas of the world adjacent to and on the oceanward side of long, narrow submarine ridges and island arcs, a deep depression instead of a rise lies at the base of the continental slope. These trenches reach depths of 7,500–11,000 m. The greatest depth recorded is in the Marianas Trench near the island of Guam in the Pacific Ocean. A depth of 11,030 m was recorded by echo-sounding in 1957 by the Russian research vessel *Vityaz*.

The worldwide distribution of oceanic ridges is shown in Figure 7.14. The longest continuous ridge forms a great submarine mountain chain that extends for some 60,000 km down the middle of the Atlantic basin, into the Indian Ocean basin, and westward between Australia and Antarctica to enter the South Pacific basin. Other ridges are found in the Arctic and North Pacific. The Mid-Atlantic Ridge is the best known of these submerged mountain chains. It extends nearly continuously down the middle of the Atlantic basin and rises some 3500 m above the adjacent sea floor.

Figure 7.14 Worldwide distribution of oceanic ridge systems (after Wilson, 1963).

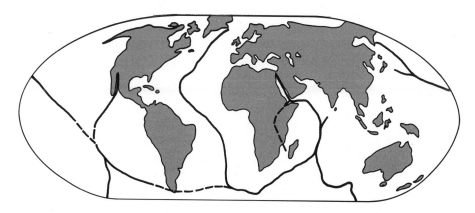

The central part of the ridge is characterized by a deep trench-like depression, commonly referred to as a rift valley. This rift may represent a series of major fractures in the earth's crust through which new materials for the crust are derived from the upper mantle (see Chapter 12).

The ocean-basin floor lies at depths of about 4000–6000 m and consists of (1) very gently sloping plains with interspersed hills rising some tens of meters to a few hundred meters above the abyssal plain, (2) broad rises that cover areas of hundreds of square kilometers, with elevations several hundreds of meters above that of the abyssal plain, and (3) isolated peaks, termed sea mounts, that rise some 1500 m above the sea floor. The flatness of the abyssal plains suggests that these surfaces have formed by continued deposition from turbidity flows moving down the continental slopes and spreading out in thin sheets across the sea floor. However, it is likely that even at these depths, ocean-bottom currents have helped to develop this featureless topography.

Deep-Sea Sediments

Deep-sea (pelagic) sediments have been sampled more extensively during the past two decades than in all preceding oceanographic investigation. Our increased scientific interest in the ocean floor and the increased number and availability of oceanographic vessels and improved sampling equipment have led to this more vigorous program of deep-ocean investigation. The sediments of the deep ocean have been accumulating slowly and continuously; because of this and because they contain materials from all the erosional agents, as well as cosmic additions, they have information stored in them that can be interpreted in terms of the history of the continents as well as that of the ocean itself.

Turbidity currents cannot reach many areas of the central oceans because of the barriers posed by the ocean-margin trenches and ridges. In these areas only fine-grained materials accumulate. They consist of terrestrially derived clay minerals that have been redistributed by ocean currents; windborne detritus that falls on the ocean surface and slowly settles to the bottom; minerals precipitated from sea water; submarine volcanic materials; remains of the skeletons of marine plankton, such as foraminiferans, diatoms and radiolarians; and extraterrestrial dust. Figure 7.15 summarizes the distribution of deep-sea sediments in the Pacific Ocean basin.

Organic accumulations (calcareous and siliceous "oozes") of the shells of plankton occur in the North and South Pacific Oceans and near the equator. These are regions in which plankton are abundant in the overlying nutrient-rich surface waters. Calcareous oozes are composed primarily of the tests of foraminiferans and coccoliths. Coccoliths are made up of tiny rhombohedrons of calcite that are derived from the skeletons of the family of algae called *Coccolithophoridae* and constitute about 20 percent of the calcium carbonate in the deep sea. Foraminifera are protozoa that have shells made of calcite, aragonite, or quartz sand cemented by calcite. The

1. Diatoms, radiolaria
   Fine-grained silicates
   Glacially rafted debris

2. Carbonates
   Fine-grained silicates

3. Diatoms, radiolaria
   Carbonates
   Volcanics
   Fine-grained silicates
   Phillipsite

4. Carbonates
   Fine-grained silicates
   Phillipsite

5. Diatoms
   Carbonates

6. Glacially rafted debris

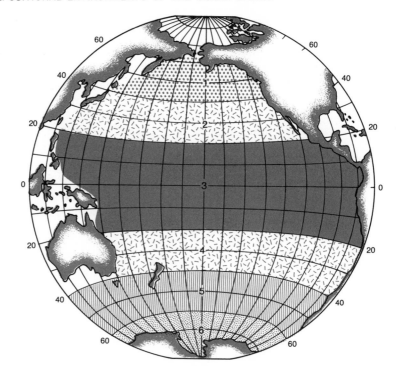

Figure 7.15 Schematic map of the types and distribution of deep-sea sediments in the Pacific Ocean basin. Boundaries of sediment types are more transitional than illustrated (modified from Peterson, 1967).

most common planktonic forams have skeletons of calcite. Calcareous oozes are primarily the remains of calcitic forams. Pteropods are mollusks with aragonitic tests and are found in sediments deposited in water less than 3500–4000 m deep. The term *Globigerina* ooze is commonly applied to deep-sea calcareous oozes, although these oozes contain not only the tests of the foram genus *Globigerina* but also other genera of foraminifera as well as remains of the coccoliths. Siliceous oozes are composed of tests of microscopic algae termed diatoms and two major groups of protozoans termed radiolarians and dinoflagellates. All of these organisms secrete tests of opaline (amorphous) SiO_2.

When the planktonic organisms die, their tests fall slowly toward the bottom of the ocean. The rate of accumulation is directly proportional to the numbers of organisms per unit area that die and inversely related to the rate of solution of the tests as they fall through the water column or lie exposed on the sea floor. The calcium carbonate concentration in deep-water-bottom sediments is a result of the relative accumulation rates of calcium carbonate and clay. The highest calcium carbonate concentration values are associated with oceanic ridges and regions of high organic productivity, as in surface waters of the equatorial Pacific.

It has been observed that over large areas of the Atlantic and Pacific ocean basins, below a depth of about 4500 m, the calcium carbonate concentration of sediments drops abruptly to insignificant values. This depth is called the "snow line" or "compensation depth." The reasons for this pattern of calcium carbonate deposition are not known, and there is even some question as to whether the correlation between depth and calcium carbonate concentration exists over large areas of the ocean. However, it is known that in some places the snow line is not a direct result of the saturation state of ocean water because the sea water may be undersaturated with respect to calcium carbonate at depths below 300–500 m. Consequently, the snow line, if it were a result only of the depth in the ocean at which calcium carbonate begins to dissolve, should be at 300 and not 4500 m. It has been observed that organic materials coat calcium carbonate particles suspended in ocean water. Chave (1965) suggested that these coatings may inhibit reactions between carbonate particles and sea water. It is possible that the compensation depth is the depth at which most of these coatings are removed by organisms utilizing the organic materials for food, permitting dissolution of the carbonate detritus in the undersaturated sea water. Any remaining coatings may be lost at the sediment surface, as bottom-dwelling, sediment-ingesting organisms feed on the organic material and expose the calcium carbonate to solution. The overall result of removal of these protective organic coatings is to prevent accumulation of carbonate detritus in the deeper parts of the ocean basins.

The mechanisms that control the snow line, and perhaps even the concept of an ocean-wide compensation level, are matters of intense debate today. Field experiments (e.g., Peterson, 1966) have been performed to obtain the amount of material lost per unit time by calcium carbonate spheres suspended at various ocean depths by a wire line hung over the side of a ship. Initial results in the northeast Pacific Ocean show that the material is lost at all depths below a few hundred meters, but that the rate of solution increases sharply at 3800 m, approximately the depth of the snow line level. These results suggest that the compensation level is not directly controlled by the saturation state of sea water but is a rate-controlled process, perhaps regulated by the removal of organic materials from grain surfaces by biological processes.

The equatorial region of biogenic deposits in the Pacific Ocean reveals the position of the equator during geologic time. Because of the increased addition of skeletal remains to the sediments at the equator, the sediment thickness in the equatorial region is greater than the thickness of contemporaneous strata north and south of the equator. If, during the past, the equator had been displaced from its present position, the thickness maximum should also have been displaced. The record of the sediments obtained from deep-sea cores below and near the present equator show that the equator has not changed position for the last 500,000 years. However, now that longer cores have been obtained that penetrate a greater thickness of deep-sea sediments, they may reveal a shift in the thickness maximum and consequently a shift in the position of the equator, and, conversely, of the geographic poles.

Fine-grained silicate minerals are the chief materials that accumulate in broad belts between the equatorial and polar regions of organic-rich sediments in both the North and South Pacific. These clay minerals are primarily of terrestrial origin, although in some areas of the South Pacific, the greater bulk of clay minerals have formed by alteration of volcanic debris.

The terrestrial origin of the clay minerals is well demonstrated in Figure 7.16, which shows the concentration of kaolinite in the less than 2-μ (0.002 mm) size fraction of the sediments in the ocean. The highest kaolinite concentrations are found in sediments of the equatorial region. Intense weathering of rocks in equatorial regions produces kaolinite that is transported to the oceans by rivers or as airborne dust and is further dispersed by ocean currents to produce the distribution of Figure 7.16.

In some areas, the clays are associated with abundant volcanic materials, with manganese oxide particles and nodules, and with an interesting silicate mineral termed *phillipsite* (a zeolite rich in K, Na, and Si). Some of this phillipsite probably formed from alteration of volcanic materials, primarily from minute angular fragments of volcanic glass.

The high concentration of manganese in deep-sea deposits is a continuing problem. Manganese concentrations form in areas of slow sedimentation (the rate of accumulation of deep-sea clays is about 0.1–5 mm/1000 years) and they commonly contain Fe and minor amounts of Cu, Ni, Co, Zn, and Pb. Manganese concentrations at the water-sediment interface have been attributed to precipitation of manganese from the sea water; to diffusion of reduced manganese up through the pore waters of the sediments, with precipitation occurring at the oxygenated sediment-water interface; and locally, as well as much more rapidly than by the two preceding mechanisms, by addition of manganese during periods of active volcanism when Mn, Fe, Ba, and other elements are leached from hot submarine lava by reaction with sea water. The manganese accretions someday may be mined, although this is not economically favorable now.

The quartz content of deep-sea sediments near 30° N and S latitudes in the Pacific is very high. The quartz particles average about 0.01 mm in diameter and are thought to be of eolian origin. The quartz maxima in the deep sea are probably due to fallouts of materials from the stratospheric jetstreams that circle the globe in these two latitudinal zones (Rex and Goldberg, 1958). Clay minerals and carbonate minerals are also delivered to the oceans by the wind.

Because of the slow sedimentation rates of deep-sea sediments, cosmic material is more easily recognizable than in other deposits. Microscopic spherules of nickel-iron are the most easily identified component of cosmic material. It appears that these spherules represent the solidified droplets of molten iron-nickel splashed off meteorites entering the earth's atmosphere (Fredriksson, 1958).

Atlantic pelagic deposits are similar to those of the Pacific, although there are some significant differences (Revelle et al., 1955). The average carbonate content of Atlantic deposits is lower than that of the Pacific, and the area dominated by clay minerals is less, although the rate of clay min-

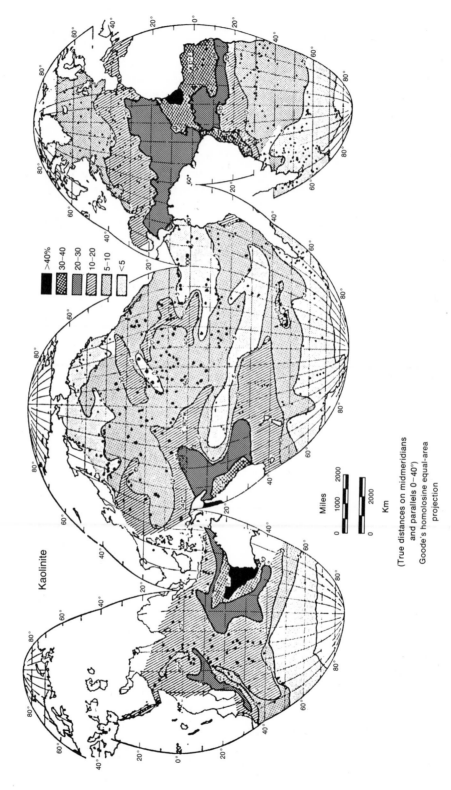

Kaolinite

>40%
30–40
20–30
10–20
5–10
<5

Miles
0   1000   2000

Km
0   2000

(True distances on midmeridians
and parallels 0–40°)
Goode's homolosine equal-area
projection

Figure 7.16 Distribution of kaolinite concentrations in the less than 2-μ size fraction of sediments in the world ocean (Griffin et al., 1968).

eral deposition is higher. The Atlantic near-surface deposits, in general, are lower in Fe and Mn, in phillipsite, in the remains of siliceous organisms, and in volcanic materials. Some of these differences are perhaps related to the greater dilution of organic detritus and chemical precipitates by terrigenous material in the Atlantic than in the Pacific. However, the differences between the Atlantic and Pacific pelagic sediments are not completely explained by simple dilution. Recent deep drilling in the Atlantic has uncovered thick sections of zeolitic and volcanic materials, suggesting that the sediments of the two ocean basins are not as dissimilar as was previously thought.

The deep-sea sediments someday will provide us with the knowledge necessary to interpret the history of the ocean basins. So far we have only extensively sampled a few tens of meters of the pelagic sediment thickness —the upper veneer of a sedimentary rock column that averages about 1 km in thickness. Based on an average pelagic sedimentation rate of 1 mm/1000 years, it would take a maximum of 10^9 years to accumulate all of the existing deep-sea sediment mass, an interval of time equal to only about 25 percent of geologic time. However, it is likely that the above average rate is a minimum; if it were on the order of 1 cm/1000 years, it would take only 10^8 years to accumulate the present thickness of pelagic sediments. Deep drilling in the ocean basins during 1969 and 1970 indicates that there are no significant areas in the ocean basins underlain by sediments with ages greater than 2×10^8 years.

Within recent years, sedimentary layers that occur within the middle of the deep-sea sediment column in the Atlantic Basin have been found at the surface of the ocean floor east of the Bahama Islands. The layers are Cretaceous in age. This relatively young age and the thin column of pelagic sediment immediately below these layers throughout most of the Atlantic Basin suggest that it may have taken less than 200 million years to accumulate the present thickness of deep-sea sediments (Ewing et al., 1966). In fact, no rocks older than Jurassic have been found in the ocean basins up to the time of writing of this book. Deep-sea sedimentary deposits, other than those found in present ocean basins, are exceedingly scarce in the geologic record. Do these observations mean that there were no deep-sea basins in the past, or that the deep-sea sediments were never uplifted to form part of the sedimentary record of the present continents? In other words, are the present ocean basins young features of the earth's crust, and if so, how and when did they originate? Our answer lies in the deep sedimentary record of the ocean basins. In Chapter 12 we shall discuss this question in more detail and present a model that has been employed to explain these enigmas.

SUMMARY

Solid particles are moved by wind and water currents and come to rest when the relative settling velocities of the particles are greater than the upwardly directed vertical velocity component of the current. The materials transported to the ocean are deposited in various depositional environments found on continental shelves, slopes, and rises and on the deep-sea floor. Fine-grained detritus accumulates in regions of lower wave and current agitation than does coarser debris. Thus clays are usually found farther from land and in deeper water than sands. However, fine-grained materials are also deposited in shallow, quiet water environments, e.g., lagoons, and sands are found in the deep sea. Carbonate sediments are forming today in regions having a low rate of supply of terrigenous detritus and sources of calcareous skeletal particles.

The thick wedges and prisms of shallow water sediments found at the margins of ocean basins are indicative of slow sinking of parts of the earth's crust owing to the weight of the overlying sedimentary burden or to stresses generated by mechanisms that bow the crust downward. The thin layer of pelagic sediments beneath the deep ocean suggests that the ocean basins, as we know them today, may be relatively young features of the earth's surface.

REFERENCES

Allen, J. R. L., 1964, The Nigerian continental margin: bottom sediments, submarine morphology and geological evolution: *Marine Geol., 1,* no. 4, 289–332.

American Geological Institute, 1967, *Investigating the Earth:* Houghton Mifflin Company, Boston.

Chave, K. E., 1965, Carbonates: Association with organic matter in surface seawater: *Science, 148,* 1723–1724.

Creager, J. S., 1963, Sedimentation in a high energy, embayed continental shelf environment: *J. Sediment. Petrol., 33,* 815–830.

Dunbar, C. O., and Rodgers, J., 1957, *Principles of Stratigraphy:* John Wiley & Sons, Inc., New York.

Ewing, J., Worzel, J. L., Ewing, M., and Windisch, C., 1966, Ages of horizon A and the oldest Atlantic sediments: *Science, 154* (3753), 1125–1132.

Fisk, H. N., McFarlan, E., Jr., Kolb, C. R., and Wilbert, L. J., Jr., 1954, Sedimentary framework of the modern Mississippi delta: *J. Sediment. Petrol, 24,* 76–99.

Fredriksson, K., 1958, A note on investigations of cosmic spherules and other small meteoritic particles: *Astronomical Notes,* Univ. of Goteborg, *21.*

Griffin, J. J., Windom, H., and Goldberg, E. D., 1968, The distribution of clay minerals in the world ocean: *Deep-Sea Res., 15,* 433–459.

Heezen, B. C., Tharp, M., and Ewing, M., 1959, The floors of the oceans: *Geol. Soc. Am. Spec. Paper, 65.*

Heezen, B. C., Hollister, C. D., and Ruddiman, W. F., 1966, Shaping of the continental rise by deep geostrophic contour currents: *Science, 152,* 502–508.

Hjulström, F., 1955, Transportation of detritus by moving water: In Recent marine sediments, P. D. Trask, ed., *Soc. Econ. Paleontologists Mineralogists Spec. Pub., 4,* 5–31.

King, P. B., 1959, *The Evolution of North America:* Princeton Univ. Press, Princeton, N.J.

Krumbein, W. C., and Sloss, L. L., 1963, *Stratigraphy and Sedimentation,* 2nd ed.: W. H. Freeman and Company, San Francisco.

Peterson, M. N. A., 1966, Calcite: Rates of dissolution in a vertical profile in the Central Pacific: *Science, 154,* 1542–1544.

Peterson, M. N. A., 1967, written communication.

Revelle, R., Bramlette, M., Arrhenius, G., and Goldberg, E. D., 1955, Pelagic sediments of the Pacific: in Crust of the earth, A. Poldervaart, ed., *Geol. Soc. Am. Spec. Paper, 62,* 221–236.

Rex, R. W., and Goldberg, E. D., 1958, Quartz contents of pelagic sediments of the Pacific Ocean: *Tellus, 10,* 153–159.

Scruton, P. C., 1960, Delta building and the deltaic sequence: in *Recent Sediments, Northwest Gulf of Mexico,* F. P. Shepard, F. B. Phleger, and T. H. van Andel, eds.: American Association of Petroleum Geologists, Tulsa, Okla., 82–102.

Wilson, J. T., 1963, Continental drift: *Sci. Am., 207,* 86–95.

8 | # Sedimentary Rocks

Materials transported to the ocean cannot continuously accumulate in the ocean basins. Storage in the oceans of the dissolved constituents of streams would change present sea water composition markedly in a few million years. The solids deposited by streams would fill the ocean basins in a few hundred million years if the present rate were continued. In this chapter the immediate fate of the products of weathering will be considered. To do so, it is necessary to look at the final abode of weathered materials, the sedimentary rocks, to provide information about the mechanisms and sinks for removal of these materials.

Sedimentary rock types were discussed in a general way in Chapter 2; however, little emphasis was placed on the details of their mineralogical and chemical compositions. In the following sections the chemistry and mineralogy of sedimentary rocks are presented in terms of the oxide composition and normative mineral composition, respectively, of average rock types. Although the use of averages does not permit description of the wide range of sedimentary rock compositions and may provide the feeling that rocks are easily "pigeonholed," comparison of average rock types helps in deducing the mechanisms by which the products of weathering are segregated and enables discussion of these materials in terms of the sediments in which they finally come to rest. The procedure by which normative mineral compositions are calculated is given in Appendix C.

The total mass and composition of the sedimentary lithosphere have been estimated by several authors; a detailed analysis of these estimates is given in Chapter 9. To provide a framework for the following discussion of sedimentary rocks, the mass and lithology of the major sedimentary units of the earth's crust as estimated by Horn (1966) are given in Table 8.1. Horn's totals are not in agreement with our final estimates of totals, for a variety of reasons explained later, but his estimates of relative masses and rock types in the *subdivisions* of the sedimentary units are the most accurate and detailed breakdown available.

The sedimentary lithosphere has been subdivided into four major units. The continent-shield unit includes the relatively undeformed sedimentary rocks found in the interior, stable parts of the continents; the mobile belt-continental shelf sedimentary mass is composed of the sediments underlying the present continental shelves and the thick wedges of strongly deformed sedimentary rocks found near the edges of continents; the hemipelagic unit includes the thick prisms of sedimentary rocks of the oceanic slope and rise; and the pelagic unit comprises the sediments beneath the deep-ocean floor.

Table 8.2 provides some perspective for the confidence that should be placed in the estimates of the percentages of the various major lithologies

Table 8.1
Estimates of Mass and Lithologic Properties of Major Sedimentary Units[a]

	Sedimentary unit			
	Continent-shield	Mobile belt-continental shelf	Hemipelagic	Pelagic
Volume (units of 10^6 km³)	127.0	395.0	315.0	241.0
Porosity (%)	25.3	28.1	20.0	48.1
Solid phase mass (units of 10^{20}g)	2561.5	7668.1	6804.0	3377.1
Liquid phase mass (units of 10^{20}g)	329.0	1136.6	645.1	1187.0
Total mass (units of 10^{20}g)	2890.5	8804.7	7449.1	4564.1
Sedimentary proportions (%)				
Lutite	57.0	59.0	95.0	90.0
Sandstone	32.0	36.0	—	—
Carbonate	8.0	2.0	5.0	10.0
Evaporite	3.0	3.0	—	—

[a]Data from Horn (1966).

in the sedimentary rock mass, whether based on measurements of stratigraphic sections or on geochemical calculations. The differences in percentages of rock types obtained by measurement as opposed to those stemming from geochemical calculations raises a major geologic problem that is discussed in detail in Chapter 9.

SANDSTONES

Sandstones are made up of the chemically and mechanically resistant minerals produced during the weathering and erosion of granular crystalline rocks (e.g., granite) or the products of erosion of pre-existing sandstones. Sandstones constitute about 30 percent of the geologic column based on stratigraphic measurements. Modern pelagic and hemipelagic sediments (continental slope and rise sediments) contain little sand-size detritus.

Table 8.2
Estimates of the Percentages of the Three Major Lithologies in the Sedimentary Lithosphere

	Measured			
	Leith and Mead (1915)	Schuchert (1931)	Kuenen (1941)	Krynine (1948)
Shale	46	44	56	42
Sandstone	32	37	14	40
Carbonate	22	19	29	18

Many classifications of sandstones based on genetic and descriptive properties have been proposed. Most investigators subdivide sandstones into four major groups: quartzites, arkoses, lithic sandstones, and graywackes. The chemical and mineralogical compositions of sandstones reflect the initial source rock type and the intensity of the processes of weathering, dispersal, deposition, diagenesis, and metamorphism. Quartzites are the extreme in the sedimentary differentiation of an original granular source rock to chemically and physically inert materials, principally coarsely crystalline quartz and chert. The high quartz content of the average quartzite is apparent from the chemical analysis which shows 95 weight percent SiO_2. Of the remaining 5 percent, only CaO and CO_2, present as calcite cement, are above the 1 percent level.

Arkoses are relatively high in K_2O and Na_2O, which are present in feldspars. K_2O generally exceeds Na_2O, showing that K-feldspars are more abundant than Na-feldspars. CaO and CO_2 are high in some analyses of arkoses and are attributable to calcite cement. In contrast to the values found in graywackes, ferrous iron and MgO are low in arkoses. There is not enough CaO to make normative Ca-feldspar.

Lithic sandstones are somewhat more complex chemically and mineralogically than quartzites but contain less feldspar than arkoses and less matrix material than graywackes. Rock fragments are a major component of lithic sandstones. Silica is tied up in quartz, in feldspars, in silicate minerals comprising rock fragments, and in clay minerals. The molar ratio of CO_2 to CaO is sufficient to make normative $CaCO_3$, and only a small excess of CO_2 is left to form $MgCO_3$. Therefore the carbonate mineral present is almost entirely calcite, rather than dolomite. The relatively high amount of normative kaolinite reflects Al_2O_3 and SiO_2 tied up in the minerals of various shaly rock fragments and in the clay minerals of the matrix. Normative K-feldspar and Na-feldspar show the presence of feldspar detritus. The magnesium in excess over that necessary to satisfy the CO_2 is present chiefly as clay minerals in the rock fragments or matrix; some FeO is probably present in ferromagnesian minerals in the rock fragments. The iron is found in clay minerals of the rock fragments and matrix and as hydrous iron oxides derived from mobilization of iron originally within or sorbed on clay minerals and organic matter.

			Calculated			
Mead (1907)	*Clarke* (1924)	*Goldschmidt* (1933)	*Holmes* (1937)	*Wickman* (1954)	*Horn* (1966)	*Authors* (1969)
82	80	⎧ 91 ⎫	70	83	73	74
12	15	⎨ ⎬	16	8	20	11
6	5	⎩ 9 ⎭	14	9	7	15

Graywackes are the most chemically and mineralogically complex sandstones. Normative Ca-feldspar is present, reflecting the anorthite component of the detrital plagioclase feldspars. Plagioclase is easily destroyed during chemical weathering of crystalline rocks; thus graywackes, as a group, are a result of mild chemical weathering of nearby source rocks. The predominance of Na-feldspar over K-feldspar and the high Na_2O/K_2O ratio also indicate a low intensity of chemical weathering. The iron present is primarily in the reduced state; both the iron and MgO are contained in the chlorite-like clay minerals of the graywacke matrix.

Pettijohn (1963) estimated the chemical composition of the average sandstone based on weighted average compositions of the four major sandstone groups. Figure 8.1 gives the average sandstone composition and its norm. Sandstones are a repository for quartz, and the segregation of quartz is indicative of the tendency of granular igneous rocks to break down chemically and mechanically into relatively well differentiated end products—shales, carbonate rocks, and sandstones—with unique chemical and mineralogical compositions.

Figure 8.1 Chemical composition and normative mineral composition of average sandstone (from Pettijohn, 1963).

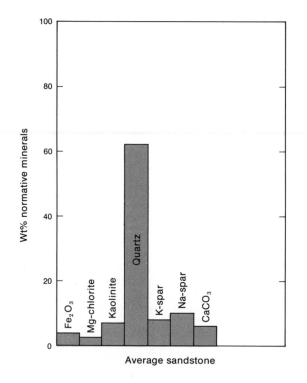

SiO_2	77.6
TiO_2	0.4
Al_2O_3	7.1
Fe_2O_3	1.7
FeO	1.5
MnO	—
MgO	1.2
CaO	3.1
Na_2O	1.2
K_2O	1.3
H_2O	2.1
P_2O_5	0.1
CO_2	2.5
SO_3	0.1

Lutites are fine-grained sedimentary rocks and include, among others, the rocks known as schist, slate, shale, argillite, siltstone, and claystone. Lutites are the most abundant of all sediments, consisting of about 50 percent of the measured stratigraphic column. Pelagic and hemipelagic sediments contain 90–95 percent lutite.

Figure 8.2 illustrates the normative mineral and chemical composition of some average lutites. Silica is the dominant oxide found in lutites and is contained in detrital feldspars, in clay minerals, in detrital silica (mainly quartz), and as silica derived from siliceous organisms such as diatoms and radiolarians. The alumina present is found mainly in feldspars and clay minerals, although highly aluminous lutites contain simple aluminum oxides or hydroxides such as gibbsite. The alkali metals are present mainly in feldspars or clay minerals. Some K_2O is found in muscovite, and some of the Na in the analytical Na_2O may be present as NaCl. CaO is present mainly in carbonate minerals. In most lutites, the molar ratio of CO_2 to

Figure 8.2 Chemical composition and normative mineral composition of some average lutites. A, From Clarke (1924). B, From Eckel (1904). C and D, From Shaw (1956).

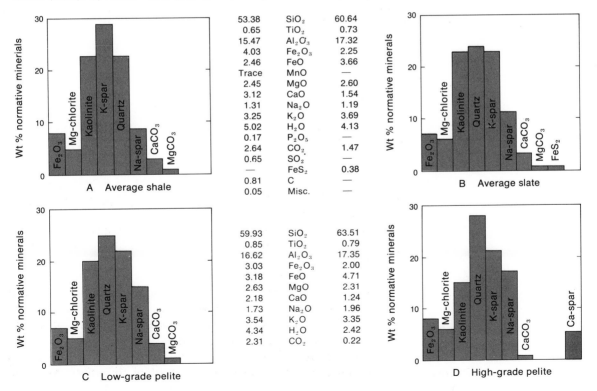

CaO is sufficient to form only normative $CaCO_3$. In those few lutites with excess CO_2, some of the MgO is tied up in carbonate minerals, primarily dolomite. In CO_2-deficient lutites, the CaO is also contained in feldspars, whereas nearly all of the MgO is in the clay minerals chlorite, montmorillonite, and illite. In some lutites CaO is found in sulfate minerals, chiefly gypsum. The FeO and Fe_2O_3 are contained in iron-bearing oxides, silicates, or sulfides. Pyrite and Fe-chlorites are common Fe-bearing minerals in lutites.

Comparison of the chemical compositions and mineralogical norms between low-grade "pelites" (clays, shales, and slates) and high-grade (strongly metamorphosed) pelites (phyllites and schists) shows the changes that accompany the transformation of fine-grained sediments into metamorphic rocks. In general, as can be seen from Figure 8.2, these changes are minor and show that the metamorphism is essentially isochemical—there is little transfer of materials into or out of the rock. Some H_2O is lost, and the iron is partially reduced during metamorphism, according to reactions typified by

$$6Fe_2O_3 + C = 4Fe_3O_4 + CO_2.$$

In low-grade pelites, there is enough CO_2 to satisfy all the CaO and some MgO as normative carbonate minerals, whereas in high-grade pelites, there is a deficiency of CO_2 and the CaO is found in normative Ca-feldspar. The percentage of normative kaolinite decreases significantly in metamorphosed pelites. Therefore, it is probable that the general reaction

$$CaCO_3 + Al_2Si_2O_5(OH)_4 = CaAl_2Si_2O_8 + CO_2 + 2H_2O$$

obtains during the metamorphism of fine-grained sediments. In accord with this reaction, the anorthite content of plagioclase increases as the intensity of metamorphism increases.

CARBONATES

Carbonate sediments are composed primarily of the carbonate minerals calcite (including a spectrum of magnesian calcites containing up to 20–30 mole percent $MgCO_3$), aragonite, and dolomite.

Carbonate rocks constitute about 22 percent of the geologic column, based on the measurement of stratigraphic sections. Pelagic and hemipelagic sediments contain only about 5 percent carbonate.

The chemical and normative mineral compositions of the average limestone and of a nearly pure dolomite are given in Figure 8.3. As can be seen by a comparison of the histograms, the dolomite content of the average limestone is low. The major diluents in carbonate rocks are SiO_2 and Al_2O_3. The SiO_2 is present mainly as fine-grained detrital quartz and various forms of silica derived from silica-secreting organisms. The Al_2O_3 is present in clay minerals, which also contain most of the minor K_2O and

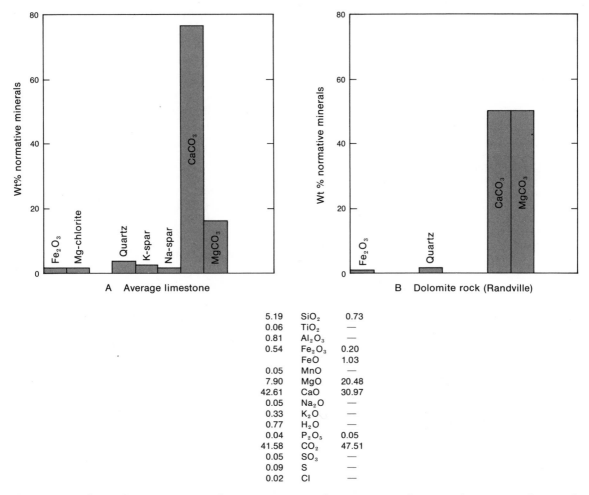

5.19	SiO$_2$	0.73
0.06	TiO$_2$	—
0.81	Al$_2$O$_3$	—
0.54	Fe$_2$O$_3$	0.20
	FeO	1.03
0.05	MnO	—
7.90	MgO	20.48
42.61	CaO	30.97
0.05	Na$_2$O	—
0.33	K$_2$O	—
0.77	H$_2$O	—
0.04	P$_2$O$_5$	0.05
41.58	CO$_2$	47.51
0.05	SO$_3$	—
0.09	S	—
0.02	Cl	—

Figure 8.3 Chemical composition and normative mineral composition of average limestone and a nearly pure dolomite rock. A, From Clarke (1924). B, From Bayley (1904).

Na$_2$O. Authigenic albite is widespread but not abundant. The Fe$_2$O$_3$ is mainly in the clay minerals or is found as hydrous iron oxides. Some carbonates contain gypsum or anhydrite; these minerals are responsible for the relatively high sulfur concentrations in the rock.

EVAPORITES

Evaporites constitute about 3 percent of the total sedimentary rock mass. They are repositories for Ca, S, Na, and Cl, as well as ubiquitous O, and consist mainly of the minerals anhydrite, gypsum, and halite, commonly associated with a complex of chloride, sulfate, and borate minerals. Evaporites are most abundantly deposited in arms of the sea that have limited communication with the open ocean. The evaporation of normal sea water that continuously enters these basins leads to the precipitation of evaporite minerals. The sequence of minerals found in evaporite deposits follows closely the experimental results discussed in Appendix B; i.e., carbonates and sulfates precipitate first and chlorides second. However, the sequence is commonly complex owing to seasonal changes in evaporation rate, post-depositional alteration, and other factors.

MODERN SEDIMENTS

In the previous sections, the chemistry and mineralogy of sedimentary rocks were presented without consideration of possible variations of these parameters during geologic time. The temporal changes will be discussed in detail in Chapters 9 and 10; here we shall discuss the compositions of sediments deposited during the last few thousand years. With these data the compositions of recently deposited materials can be obtained and the pathways and processes of disposal of the products of weathering evaluated.

Graphical Representation of Chemical Composition

The chemical compositions of modern sediments can be conveniently represented by a plot in which the logarithms of the weight ratios of SiO_2/Al_2O_3 and $(Na_2O+CaO)/K_2O$ form the coordinates of the graph (Figure 8.4). The former ratio provides a numerical representation of the distinction between sandstones and cherts, which are relatively high in silica, and lutites, which are more aluminous. The latter ratio provides an estimate of the argillaceous or calcareous nature of a sediment. A lutite has a low $(Na_2O+CaO)/K_2O$ ratio, because of its mineralogical composition of quartz, clay minerals, and feldspars (primarily K-feldspar), whereas a carbonate sediment has a high ratio, reflecting its relatively high CaO and low K_2O content.

The average compositions of a large number of sediments were originally plotted on Figure 8.4; however, to maintain clarity, only a few representative compositions are shown. The compositions of igneous rocks are represented by the banana-shaped shaded area; siliceous rocks (granite) plot high on the left and iron-magnesium rich rocks (basalt) low on the right. The stippled area represents the compositions of modern argillaceous sediments (lutites). Although there are significant differences in the compositions of shallow and deep-water sediments, as a first approximation these modern sediments can be conveniently subdivided into three major groups:

1. Argillaceous sediments, relatively high in Al_2O_3 and low in CaO; the Na_2O and CaO contents cover a small range and depend on the extent to which the sediment has been segregated into the minerals most character-

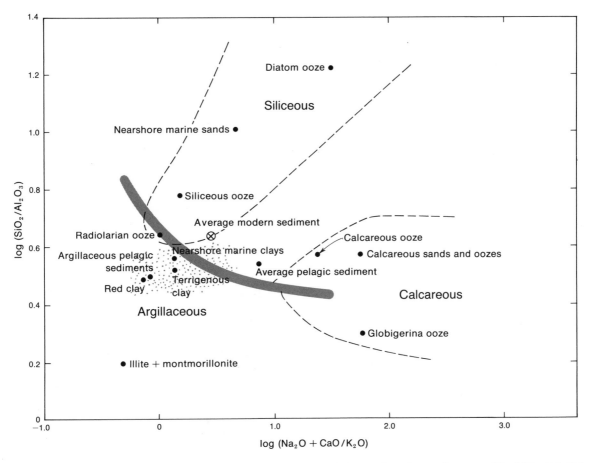

Figure 8.4 Chemical compositions of some average modern sediments plotted as a function of $\log(SiO_2/Al_2O_3)$ and $\log(Na_2O+CaO)/K_2O$. The shaded, banana-shaped area denotes the range in composition of igneous rocks with more silicic-, sodium-, and potassium-rich rocks to the left grading into less silicic-, iron-, and magnesium-rich rocks to the right. The stippled area shows the approximate range in composition of argillaceous sediments.

istic of lutites—clay minerals, quartz, and K-feldspar. Notice that no modern sediment approaches the composition of a pure clay mineral mixture of illite plus montmorillonite; more siliceous minerals are always present along with minor amounts of carbonate minerals. These sediments include the red clays of the deep sea and the lutites of nearshore and deeper-water environments that become shales and slates upon lithification.

2. Siliceous sediments containing a relatively high SiO_2 content owing to the presence of quartz sand and silt or skeletal remains of silica-secreting organisms. These sediments, including the radiolarian and diatom oozes of the deep sea, become sandstones and cherts upon lithification.

3. Calcareous sediments, which are relatively high in CaO and low in

Na_2O and K_2O. *Globigerina* ooze is a deep-sea representative of this group. Upon lithification these sediments become limestones.

This subdivision could be criticized from the point of view that modern sediments have a complete spectrum of compositions and should not be separated into groups. However, except near boundaries where one sedimentary type mixes with another on the sea floor, the three major sediment groups are remarkably well segregated on the basis of composition. Mixed sediments are found, however, but generally they reflect the independence of processes that govern the accumulation of sediments in a particular environment. For example, in the Gulf of California, Calvert (1966) has shown that the production of diatoms is greatest in the surface waters of the central part of the Gulf. This is also the region of the Gulf in which the sediments contain a high proportion of opal. Diatoms are the source of the opal; it accumulates in the central Gulf owing to the death of the diatoms in the surface waters and the relatively rapid raining down of their skeletons through the Gulf water onto the sea floor. Independent sources of sediment for the central Gulf are the rivers draining the mainland of Mexico. These rivers supply terrigenous materials that are distributed by marine currents and eventually settle on the Gulf floor, mixing with the biogenic opal. Thus a mixed sediment is produced from two independent processes.

Table 8.3
Chemical and Mineralogical Composition of Modern Marine Invertebrates[a]

Type of organism	*Most common mineralogy*	$CaCO_3$ (%)
Foraminifera	Calcite	77–90
Calcareous sponges	Calcite	71–85
Madreporian corals	Aragonite	98–99
Alcyonarian corals	Calcite	73–99
Echinoids	Calcite	78–92
Crinoids	Calcite	83–92
Asteroids	Calcite	84–91
Ophiuroids	Calcite	83–91
Bryozoa	Calcite, aragonite	63–97
Calcareous brachiopods	Calcite	89–99
Phosphatic brachiopods	Chitinophosphatic	?–8
Annelid worms	Calcite, aragonite	83–94
Pelecypods	Calcite, aragonite	98.6–99.8
Gastropods	Calcite, aragonite	96.6–99.9
Cephalopods[b]	Aragonite	93.8–99.5
Crustaceans	Calcite, calcium phosphate	29–83
Calcareous algae	Calcite, aragonite	65–88

[a] Data from Clarke and Wheeler (1917), Chave (1954a), Thompson and Chow (1955), and Lowenstam (1961).

The mineralogical and chemical compositions of each major sediment group will be considered in detail in the following sections.

Calcareous Sediments

Modern carbonate sediments contrast sharply in their chemistry and mineralogy with ancient carbonate rocks (Figures 8.5 and 8.6). Modern shallow water calcareous sediments consist predominantly of aragonite and magnesian calcites. Calcite, although ubiquitous in modern sediments, is present in relatively small amounts, except in deposits where calcite-secreting coccoliths or foraminiferans are important sediment contributors. Most of the carbonate in modern marine sediments is derived from the skeletons of marine organisms. Table 8.3 summarizes the compositions of modern calcareous organisms. Notice that except for some crustaceans and brachiopods, the skeletons of marine invertebrates are composed predominantly of aragonite and calcite containing a range of magnesium. The magnesium and strontium contents of skeletal carbonates depend on the mineralogy and biochemistry of the skeletal species considered and on the temperature of the environment in which the organism grew. Skeletal calcite tends to have higher magnesium and lower strontium contents in solid solution in

		Range in chemical composition		
$MgCO_3$ (%)	$SrCO_3$ (mean %)	P_2O_5 (%)	SiO_2 (%)	$(Al,Fe)_2O_3$ (%)
1–16	0.363	Trace	Trace–15	Trace–4
5–14	0.116	?–4	Trace–8	1–6
0.1–0.8	1.33	0–Trace	0–1	0–0.6
0.3–16	0.385	Trace–3	0–2	Trace–1
4–16	0.334	Trace–0.7	Trace–10	0.1–5
7–16	0.244	Trace–0.4	Trace–2	0.1–1
9–16	0.225	—	—	—
9–17	0.257	—	—	—
0.2–11	0.383	Trace–1.1	0.2–17	0.1–2
0.5–9	0.190	Trace–0.2	0.1–0.5	Trace–0.5
2–7	—	29–36	0.5–0.9	0.3–1.2
6–17	0.707	—	—	—
0–3	0.258	Trace–0.2	0–0.4	Trace–0.5
0–2	0.234	Trace–0.4	0–2	Trace–2
Trace–0.3	0.492	Trace	0–0.2	Trace–0.1
1–16	0.573	3–20	0–1	Trace–9
7–29	0.322	Trace–0.2	Trace–4	Trace–2

[b] Egg case of *Argonauta argo* is calcite and contains 7 percent $MgCO_3$.

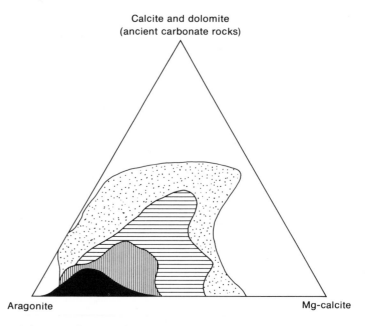

Figure 8.5 Schematic diagram showing mineralogical composition of modern shallow-water calcareous sediments. Ancient carbonate rocks contain primarily calcite and dolomite. The number of sediment samples per unit area increases from the stippled pattern to the shaded area (data from Land, 1967).

the crystalline edifice than skeletal aragonite. Furthermore, the magnesium content of skeletons appears to decrease with increasing structural complexity (phylogenetic level) of the organism. Some species that precipitate both calcite and aragonite exhibit an increase in the aragonite/calcite ratio with increasing environmental temperature. The magnesium content of skeletal calcites generally increases with increasing temperature, whereas the currently available data for the strontium content show that strontium may increase or decrease with temperature elevation. These variations in the mineralogy and magnesium and strontium contents of carbonates as a function of temperature, along with their variations in O^{18}/O^{16} ratios, can be used to interpret the environmental temperature of growth of an organism that is now fossilized. This approach has been used by Lowenstam (1961) on fossil brachiopods. He analyzed the strontium and magnesium contents and the O^{18}/O^{16} ratios of recent and ancient articulate brachiopods. The relationship between O^{18}/O^{16} ratios and the $MgCO_3$ and $SrCO_3$ contents of some fossilized samples as old as the early Permian were similar to recent species, suggesting that the environmental conditions of growth were much the same as today.

It appears that most of the calcium carbonate being transported to the ocean by rivers today is being deposited as oozes in the deep sea, whereas in shallow water areas the accumulation of carbonate is quantitatively un-

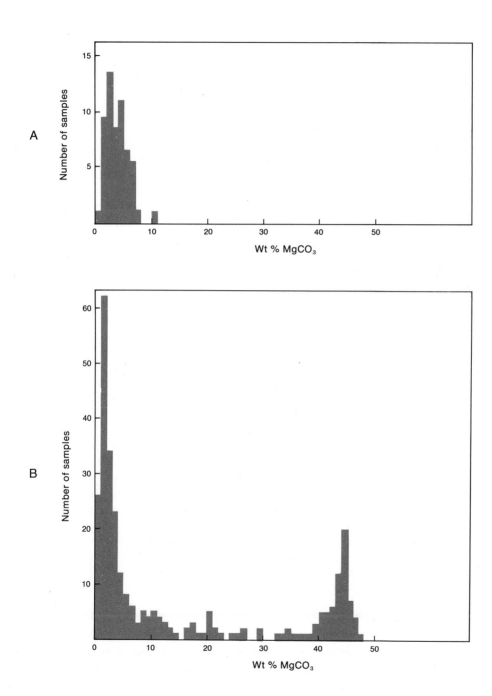

Figure 8.6 Magnesium content of (A) modern shallow-water calcareous sediments and (B) ancient carbonate rocks (after Chave, 1954b).

known and of minor importance. The geologic record is replete with carbonate rocks interpreted as being deposited in shallow water; thus the marine locus of major carbonate accumulation has changed with time. Perhaps the advent of planktonic foraminifera in the Cretaceous Period brought about the change from predominantly shallow water to deep water accumulation. Modern shallow water calcareous deposits average about 5 mole percent $MgCO_3$, whereas ancient carbonate rocks show two maxima in the distribution of $MgCO_3$ (Figure 8.6), one of 2 percent and the other of 45 percent. The similarity of the low magnesium end of the ancient carbonate rock histogram to that of the modern sediment histogram suggests that the $MgCO_3$ of the low magnesium end, like that of modern sediments, is of biogenic origin. The magnesium making up the mode at the high $MgCO_3$ end of the ancient rock histogram is probably derived from another source, perhaps from shales as they are progressively leached during geologic time (Chapter 9).

Siliceous Sediments

Siliceous sediments are a home for the quartz and dissolved silica produced during the weathering and erosion of rocks. Modern siliceous sediments are of two major types: quartz-rich sands and coarse silts, and biogenic deposits (siliceous oozes). Figure 8.7 illustrates the normative mineral and chemical compositions of these sediments and emphasizes the silica-rich nature of the deposits. In siliceous ooze, the silica is present as opaline skeletons of diatoms and radiolarians, whereas in the sands and silts it is present mainly as quartz. In both types of deposits, clay minerals and minor amounts of feldspar account for some of the silica. Notice that both the sands and siliceous oozes are diluted by argillaceous and calcareous components; the average siliceous ooze is somewhat higher in clay minerals and lower in calcium carbonate than the nearshore sands. This difference is reflected in the slightly higher CaO and normative $CaCO_3$ and the lower normative kaolinite of the sands than the siliceous oozes. Both types of deposits have more CaO than is necessary to satisfy the CO_2 in the analysis, suggesting that minor amounts of calcium-bearing silicates are present.

We do not wish to imply that all siliceous sediments are like the ones illustrated in Figure 8.7. Both types vary in composition, because of dilution by carbonate or clay minerals. Sands deposited by turbidity currents in the deep sea may contain a matrix of mud partially enclosing the larger sand grains. The mud is composed largely of finely comminuted clay minerals, quartz, and feldspars. This muddy matrix causes these sands to have lower SiO_2, higher Al_2O_3, and higher normative kaolinite and feldspars than clean sands.

Upon lithification siliceous oozes become cherts; the original siliceous skeletal fragments can be so thoroughly reorganized by solution and redeposition within the rock that the original diatom and radiolarian skeletons

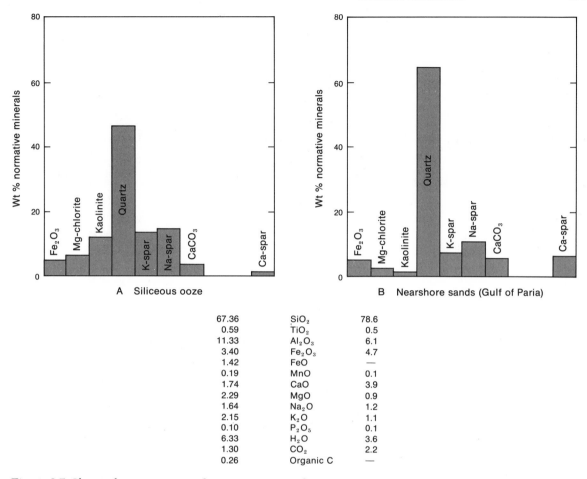

67.36	SiO$_2$	78.6
0.59	TiO$_2$	0.5
11.33	Al$_2$O$_3$	6.1
3.40	Fe$_2$O$_3$	4.7
1.42	FeO	—
0.19	MnO	0.1
1.74	CaO	3.9
2.29	MgO	0.9
1.64	Na$_2$O	1.2
2.15	K$_2$O	1.1
0.10	P$_2$O$_5$	0.1
6.33	H$_2$O	3.6
1.30	CO$_2$	2.2
0.26	Organic C	—

Figure 8.7 Chemical composition and normative mineral composition of representative modern siliceous sediments. A, From El Wakeel and Riley (1961). B, From Hirst (1962).

are no longer recognizable. Thus a siliceous sediment can be transformed to an aggregate of microcrystalline quartz containing no visual evidence of the original biogenic nature of the deposit.

Argillaceous Sediments

Argillaceous sediments include the fine-grained alumina-rich materials that accumulate in all water depths. Figure 8.8 illustrates the compositions of three major types of argillaceous sediment: nearshore marine clays, blue and green muds (predominantly continental slope and rise deposits), and red clays that occur in the deep oceans. These sediments are higher in Al$_2$O$_3$ and lower in SiO$_2$ and contain more normative kaolinite than siliceous sediments. These differences are reflections of the clay mineral-rich

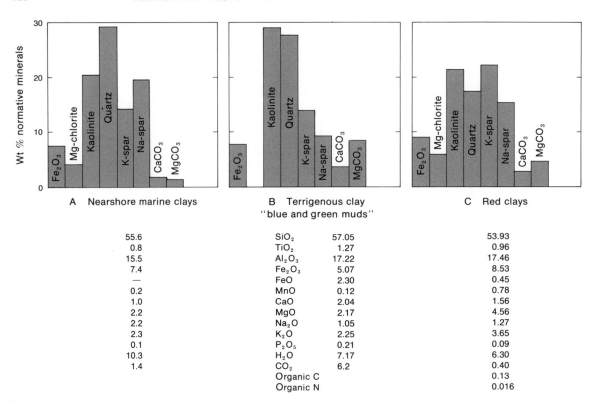

Figure 8.8 Chemical composition and normative mineral composition of representative modern argillaceous sediments. A, From Hirst (1962). B, From Clarke (1924). C, From El Wakeel and Riley (1961).

	A	B	C
SiO$_2$	55.6	57.05	53.93
TiO$_2$	0.8	1.27	0.96
Al$_2$O$_3$	15.5	17.22	17.46
Fe$_2$O$_3$	7.4	5.07	8.53
FeO	—	2.30	0.45
MnO	0.2	0.12	0.78
CaO	1.0	2.04	1.56
MgO	2.2	2.17	4.56
Na$_2$O	2.2	1.05	1.27
K$_2$O	2.3	2.25	3.65
P$_2$O$_5$	0.1	0.21	0.09
H$_2$O	10.3	7.17	6.30
CO$_2$	1.4	6.2	0.40
Organic C			0.13
Organic N			0.016

nature of argillaceous deposits and lack of quartz or siliceous skeletal debris. Even though these three types of sediments are deposited at different depths, they are remarkably similar in composition. The weight percent of SiO$_2$ and Al$_2$O$_3$ is nearly the same in all three. The total iron contents differ by only about 1.5 percent and the iron is mainly in the ferric form. The MgO and K$_2$O contents of the red clays are significantly higher than in the shallow water argillaceous sediments; this difference may be partially due to the abundance of the minerals phillipsite and montmorillonite in the red clays. Both minerals are thought to be in part of authigenic (newly formed) origin in deep-sea sediments. Red clays are also slightly higher than shallow water sediment in manganese, reflecting the presence of manganese pellets and nodules. The slightly higher CaO content of blue and green muds, and the fact that the molar content of CO$_2$ in the analysis equals the sum of molar CaO and MgO, shows the presence of tests of calcareous plankton and of fine-grained carbonate detritus of shallow water origin that has been transported to the slope and rise by marine currents.

The major chemical constituents of sedimentary rocks are found in a limited number of minerals. Modern carbonates contain primarily metastable phases, whereas ancient carbonate rocks consist dominantly of the minerals calcite and dolomite. Modern clays and sands have compositions that are similar to their ancient counterparts. Small changes in the chemistry and mineralogy of lutites with time are evident, however, and will be discussed in the following chapter.

The chemical composition of sediment deposited within the last few thousand years is difficult and perhaps at this stage in our knowledge somewhat precarious to estimate. As pointed out in Chapter 5, chemical analyses of the composition of river-borne suspended sediment transported to the ocean are few, and no analyses of the composition of suspended sediment of Southeast Asian rivers, which supply about 80 percent of the world suspended load, are available. In addition, averages for particular sediment types are only roughly known. However, to provide some feeling for the composition of recently deposited sediments, an average (Table 8.4) was obtained by using the compositions of Mississippi silt and of Gulf of Paria marine sands and muds to represent the detrital portion of shelf sediments, calcareous sands and oozes to represent shelf carbonates, blue and green muds to represent slope and rise sediments, and oceanic pelagic sediment to represent deep-sea deposits. The various sediment compositions were weighted by area and sedimentation rate. The resultant composition given in Table 8.4 should be viewed with caution because of the aforementioned difficulties. It is interesting to note that we know less about modern sediment compositions than we do about their ancient counterparts.

Table 8.4

Chemical Composition in Weight Percent of Average Modern Sediment

Oxide	Percent	Percent (water-free)
SiO_2	59.9	64.0
Al_2O_3	13.8	14.7
Fe_2O_3[a]	6.4	6.8
CaO	4.2	4.5
MgO	2.1	2.2
Na_2O	1.3	1.4
K_2O	2.0	2.1
CO_2	4.0	4.3
H_2O	6.3	

[a]Total Fe as Fe_2O_3.

REFERENCES

Bayley, W. S., 1904, The Menominee iron-bearing district of Michigan: *U.S. Geol. Surv. Monograph 46.*

Calvert, S. E., 1966, Accumulation of diatomaceous silica in the sediments of the Gulf of California: *Bull. Geol. Soc. Am., 77,* 569–596.

Chave, K. E., 1954a, Aspects of the biogeochemistry of magnesium. 1. Calcareous marine organisms: *J. Geol., 62,* 266–283.

Chave, K. E., 1954b, Aspects of the biogeochemistry of magnesium. 2. Calcareous sediments and rocks: *J. Geol., 62,* 587–599.

Clarke, F. W., 1924, Data of geochemistry: *U.S. Geol. Surv. Bull., 770.*

Clarke, F. W., and Wheeler, W. C., 1917, The inorganic constituents of marine invertebrates: *U.S. Geol. Surv. Profess. Paper, 124.*

Eckel, E. C., 1904, On the chemical composition of American shales and roofing slates: *J. Geol., 12,* 25–29.

El Wakeel, S. K., and Riley, J. P., 1961, Chemical and mineralogical studies of deep-sea sediments: *Geochim. Cosmochim. Acta, 25,* 110–146.

Goldschmidt, V. M., 1933, Grundlagen der quantitativen Geochemie: *Fortschr. Mineral., 17,* 112–156.

Hirst, D. M., 1962, The geochemistry of modern marine sediments from the Gulf of Paria: I. The relationship between the mineralogy and the distribution of the minor elements: *Geochim. Cosmochim. Acta, 26,* 309–334.

Holmes, A., 1937, *The Age of the Earth:* Thomas Nelson & Sons, Camden, N. J.

Horn, M. K., 1966, Written communication.

Krynine, P. D., 1948, The megascopic study and field classification of sedimentary rocks: *J. Geol., 56,* 130–165.

Kuenen, Ph. H., 1941, Geochemical calculations concerning the total mass of sediments in the earth: *Am. J. Sci., 239,* 161–190.

Land, L. S., 1967, Diagenesis of skeletal carbonates: *J. Sediment. Petrol., 37,* 914–930.

Leith, C. K., and Mead, W. J., 1915, *Metamorphic Geology:* Holt, Rinehart and Winston, Inc., New York.

Lowenstam, H. A., 1961, Mineralogy, O^{18}/O^{16} ratios, and strontium and magnesium contents of recent and fossil brachiopods and their bearing on the history of the oceans: *J. Geol., 69,* 241–260.

Mead, W. J., 1907, Redistribution of elements in the formation of sedimentary rocks: *J. Geol., 15,* 238–256.

Pettijohn, F. J., 1963, Chemical composition of sandstones—Excluding carbonate and volcanic sands: in Data of geochemistry, 6th ed., M. Fleischer, ed., *U.S. Geol. Surv. Profess. Paper, 440-S.*

Schuchert, C., 1931, in The age of the earth: *Bull. Natl. Res. Council, 80,* 10–64.

Shaw, D. M., 1956, Geochemistry of pelitic rocks, Part III: Major elements and general geochemistry: *Bull. Geol. Soc. Am., 67,* 919–934.

Thompson, T. G., and Chow, T. J., 1955, The strontium-calcium atom ratio in carbonate-secreting marine organisms: in Papers in marine biology and oceanography, *Deep-Sea Res. Suppl., 3,* 20–39.

Wickman, F. E., 1954, The "total" amount of sediment and the composition of the "average igneous rock": *Geochim. Cosmochim. Acta, 5,* 97–110.

9

The Origin and Mass of Sedimentary Rocks

A major goal of geologists is to produce geologic maps that show the ages of the rocks at the earth's surface. The complex patterns exhibited by most maps, whatever the size of the area they depict, indicate the complicated geologic history of almost every part of the earth's crust. In many areas the rocks mapped include sedimentary, metamorphic, and igneous types, and the events deduced from their relationships commonly include repetitious sequences of sedimentary deposition, folding, intrusion of igneous rocks, uplift, erosion, and submergence. Every area of the continents has been at one time covered by the sea, and there are some places that show a clear record of having been submerged at least 20 separate times.

Presumably there was a time in the earliest history of the earth when mapping would have shown only the igneous rocks of the primeval crust. What we see now is the result of more than 4 billion years of the interaction of sun-driven erosional and depositional processes, which tend to destroy all rocks above the sea and deposit them as sediments, with restoring forces driven by the earth's internal energy.

Despite the fact that rocks of various types and ages are now intimately intermixed, the bulk composition of the crust should be the same today as it was in the beginning, if the crust can be considered as a closed chemical system, that is, a system with no significant additions during geologic time of material from the mantle or from outer space. Because additions have been made to it, we know that the crust is *not* a fixed chemical entity as a function of time. For example, molten rock is being added today from volcanoes such as those in Hawaii, where it is known that some of the lavas are derived from depths of as much as 100 km and thus from the mantle. When the lavas are erupted, gases escape from them. We are not very confident of the composition of the gases present in the mantle, as compared with those picked up as the lavas move toward the surface, but there is undoubtedly a contribution of volatiles from the mantle.

Addition of cosmic material at present is at such a low rate that it would have little effect on the bulk composition of the crust, even over a period of several billion years. Although the earth probably accreted rapidly from cosmic material, significant additions during the 3.5 billion years for which we have a rock record do not show up in the crust we can now observe.

The chemical compositions of the igneous rocks that have intruded the crust over the past 3 billion years show little change with time, so that although the crustal chemical system is not closed, and the mass of crust may have increased significantly since initial solidification, there probably has been little change in the ratios of the major chemical elements of the crust during this time. The "average igneous rock" of today, determined by weighting the various types of igneous rock on the basis of their relative volumes, is apparently similar compositionally to one that geologists would have calculated had they been mapping a billion years ago.

The water and other volatiles of the crustal system have been derived from solidification of igneous rocks; we are unable to determine when these were added to the crustal system, but it is likely that if the composition of the average igneous rock has remained fairly constant, the composition of the released gases may also have remained about the same.

The sedimentary rocks preserved today have been derived either from igneous rocks or pre-existing sedimentary rocks. Thus the ultimate source of sedimentary rocks is igneous rocks. Through reaction with atmospheric constituents, igneous rocks are granulated, dissolved, or chemically altered. The solid debris then accumulates as sandstones or shales; dissolved constituents are precipitated inorganically or through the intervention of organisms. The average composition of sedimentary rocks should be the same as that of average igneous rocks, including their volatile constituents, except for those constituents that have remained in solution in the oceans, lakes, rivers, or the pore waters of rocks. In this chapter we shall investigate the chemical relations between igneous and sedimentary rocks and show that although there are compositional variations of sedimentary rocks that are functions of their ages, the bulk composition of the total sedimentary mass now in existence is in accord with an origin from igneous rocks with an average composition close to that of the average igneous rock of today. In addition, we estimate the total mass of rocks of sedimentary origin, which will give us a value for the mass of igneous rock that has been converted into sediments.

CHEMICAL RELATIONS OF IGNEOUS AND SEDIMENTARY ROCKS

It is possible, by considering the compositions of existing sedimentary and igneous rocks, to deduce approximately the chemical changes that are required for the transformation. Alteration of igneous to sedimentary rocks involves the formation of carbonate minerals, which tend to segregate into limestones, removal of sodium into the oceans, and retention of potassium in shales. These chemical relations are the basis for the graphical portrayal of the compositions of sedimentary and igneous rock types already used in Chapter 8 with regard to modern sediments. The weight ratio $(Na_2O+CaO)/K_2O$ is a measure of the degree to which a particular sedimentary rock has become "differentiated." A shale that has been strongly segregated into clay minerals, feldspars, and quartz should be low in sodium and calcium, whereas a limestone formed far from suspended materials should be an almost pure calcium rock, with extremely low sodium and potassium, and hence a high ratio of $(Na_2O+CaO)/K_2O$. The ratio SiO_2/Al_2O_3 is a helpful variable to use in distinguishing between sandstones, with their high quartz content and low Al_2O_3 and more highly aluminous lutites.

Figure 9.1 shows a logarithmic plot of a variety of rock compositions, using weight percent SiO_2/Al_2O_3 as ordinate and $(Na_2O+CaO)/K_2O$ as abscissa. These parameters were used to show the composition of modern sediments in Chapter 8; comparison of Figure 9.1 with Figure 8.4 shows that there are no gross differences between modern and ancient sedimentary rocks. Only a few analyses for particular kinds of rocks are labeled, but the areas enclosed by dashes show the compositional regions defined by plotting many sets of analyses. Typical igneous rocks fall along a smooth curve, with the siliceous rocks such as rhyolite and granite high on the left and grading off to the right toward diorites and basalts. Sandstones plot in the high silica part of the diagram and

Figure 9.1 Relation between the compositions of igneous rocks and those of sedimentary rocks.

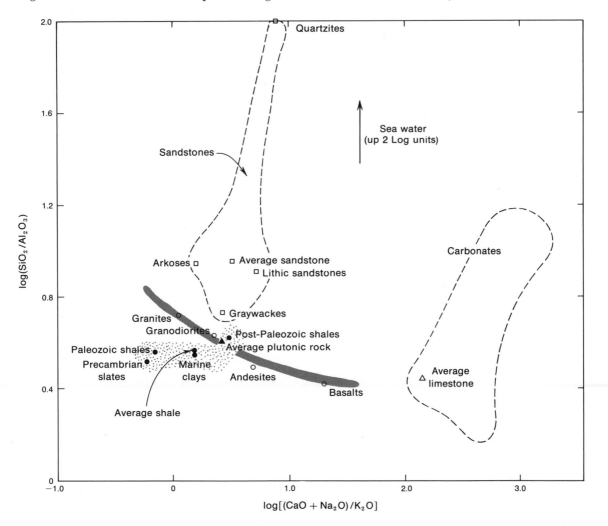

show a range along the abscissa, depending chiefly on whether they are argillaceous or carbonate-rich. The shaly rocks, including metamorphosed sediments, fall within a well-defined area. They have a small range of the ratio of silica to alumina. The broad spread along the abscissa can be attributed largely to variations in CaO. Indeed, the CaO in shales is almost invariably balanced by exactly enough CO_2 to make $CaCO_3$, indicating that calcium in most shales is present as carbonate minerals and is not contained in clay minerals or unaltered silicates of igneous origin.

The chemical basis for distinguishing three major sedimentary rock types emerges from the diagram. The areas enclosing the analyses of sandstones, shales, and carbonate rocks show no overlap. In terms of differentiation from an igneous parent of average composition, shales show postassium enrichment and sodium and calcium loss; sandstones are silica-enriched; and limestones represent isolation of calcium and some magnesium, plus the addition of much CO_2. The positions of the average compositions of each rock type, in terms of their distances on the graph from the "parent" igneous rock, are indicative of the fact that lutites comprise most of the mass of sedimentary rocks. A small proportion of sandstone, plus an even smaller one of limestone, would suffice to bring the average shale composition to that of the average igneous rock, in terms of the five oxides shown.

In Figure 9.2 the compositions of lutites relative to igneous rocks are shown by plotting in terms of the ratios Na_2O/Al_2O_3 and K_2O/Al_2O_3. The irregular solid line is the lower limit of most igneous rock compositions. All of the shaly rocks, regardless of their age or degree of metamorphism, are distinctly lower in Na_2O/Al_2O_3 than any of the igneous rocks. Furthermore, except for one set of schist analyses, the lutite compositions lie on a trend line that intersects the boundary of the lower igneous limit at approximately right angles. The various sets of analyses that have been used are a hodgepodge of composite analyses; some represent an age, some a degree of metamorphism, some a place and an age, but there is no obvious correlation among any of these variables and position on the trend line.

A projection of the shaly trend line up to the igneous rock limit intersects the limit among the various estimates of the "average igneous rock," reinforcing the conclusion that whereas sedimentary rocks are derived from all kinds of igneous rocks and from each other as well, calculations based on a single ingneous progenitor probably have some validity. Brotzen's average rock is closest to the projection of the shale trend line, so we shall use it in future calculations.

Some important relations among igneous rocks are shown; for example, although there are significant differences between average Precambrian, Paleozoic, Mesozoic, and Cenozoic basalts, there is no obvious time trend. Mesozoic and Cenozoic basalt compositions lie between those of the Paleozoic and those of the Precambrian. On the other hand, Precambrian granites appear to be a little richer in K_2O than more recent ones.

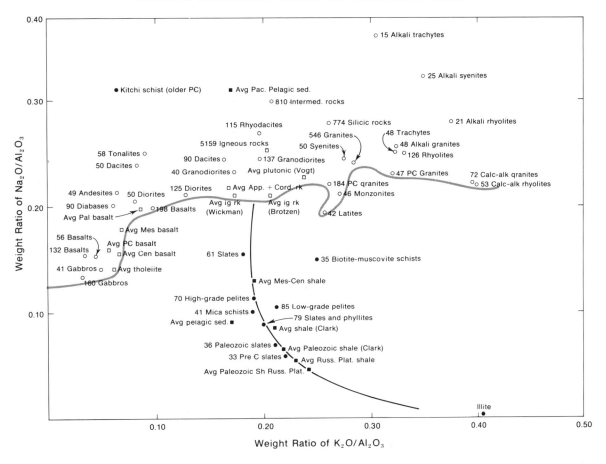

Figure 9.2 Compositions of lutites compared with those of igneous rocks in terms of their contents of Na₂O and K₂O.

Two analyses of "sedimentary rocks" are shown that require us to consider a fourth sedimentary rock type. The Kitchi schist, from the older Precambrian of Upper Michigan (≈2500 million years old), and the average pelagic sediment of the Pacific lie high and to the left on the diagram, with compositions more sodic than almost any of the igneous rocks. These two analyses, representing materials of extremely different ages, are of rocks that lie in a borderland between igneous and sedimentary rocks. Although they are not plotted specifically, there are many more analyses available of similar materials that plot in the same general region or a little lower. These are the volcanogenic sediments. These sediments were originally basaltic or slightly more silicic lavas that were fragmented and altered but retained their original basaltic imprint. There are many places today, especially in the areas of oceanic trenches and volcanic islands, where volcanic debris is blown into the air or lavas flow to the sea. In these places a great deal of this volcanogenic material becomes intermingled with other rocks.

In the Cretaceous rocks of Puerto Rico, for example, a jumble of volcanic debris and coral reef material is found, almost a necessary accompaniment of a volcanic island or island chain projecting abruptly above the ocean floor. This type of situation apparently has occurred many times in the past, especially near regions of extensive volcanism at continental margins. In a sense, then, there are two kinds of sediments: those resulting from the standard processes of weathering and erosion of the continents and those that are fragmented and altered volcanic products. The Kitchi schist, for example, is a complex mass of basaltic flows and fragmented debris in which the original minerals of the basalt—chiefly plagioclase (Na, Ca, Al, Si) and pyroxene (Ca, Mg, Fe, Si)—have been changed to a mixture of albite (Na, Al, Si), chlorite (Mg, Fe, Al, Si), and epidote (Ca, Al, Si). This mineral assemblage represents an increase in sodium, an addition of H_2O, and a loss of Ca, Mg, and Fe.

Rocks such as the Kitchi, as we shall see, do not enter into the usual *geochemical* calculations of the mass of sedimentary rocks, which are arbitrarily limited to "normal" sediments: shales, sandstones, limestones, and evaporites. Yet rocks such as the Kitchi are debris derived from igneous rocks and have chemical requirements for their formation. Furthermore, they are included in the measured, as opposed to the geochemically calculated, mass. Our procedure will be to make the usual calculations and then modify them to include the formation of the fragmental volcanics.

To assess the chemical changes involved in the formation of sedimentary rocks from igneous rocks, we need to know the chemical variations of the various sedimentary rock types as a function of time, as well as the distribution of the sedimentary mass with time, so that a properly weighted average composition can be obtained.

Chemical Changes in Shaly Rocks as a Function of Age

The data required for an assessment of chemical differences in rocks as a function of their ages are just beginning to be sufficiently numerous to permit the determination of major trends that are not entirely lost in the noise of the variations from Period to Period. We find that for purposes of comparison it is necessary to lump together whole Eras. Figure 9.3 illustrates this point. Detailed studies of the Russian Platform by Ronov and his co-workers (cf. Vinogradov and Ronov, 1956 a,b) have been most useful, especially after it was discovered that the major chemical trends there are similar to those of the sedimentary rocks of North America, and hence perhaps of global rather than local significance. Rocks ranging in age from Precambrian to Recent, with formations representing every intervening Period, are present on the platform, so it has been an important site for studying time variations of the chemistry and mineralogy of various rock types. We plotted the data for the weight ratios of Na_2O/Al_2O_3 in shaly rocks versus those of the weight ratios of Mg/Ca in carbonate rocks. It can be seen from Figure 9.3 that the points fluctuate widely throughout the

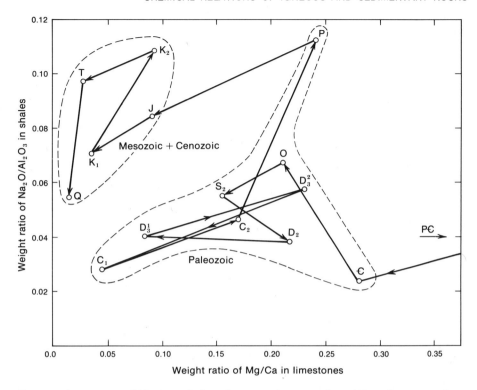

Figure 9.3 Variation of Na_2O in shales relative to the ratio of Mg/Ca in limestones for the rocks of the Russian Platform. The capital letters beside the points refer to the standard geologic periods; subscripts or superscripts denote subdivisions of the periods (data from Vinogradov and Ronov, 1956a, b).

Paleozoic Era, that the Permian Period has unique ratios, and that there is great variability within the Mesozoic and Cenozoic Eras. However, all the points for the Paleozoic can be enclosed in an area completely distinct from the area enclosing Mesozoic and Cenozoic data. Therefore, although there is no obvious time trend for the points within the Paleozoic, or within the Mesozoic and Cenozoic, it can be stated that the ratio of Na_2O/Al_2O_3 for the most recent 200 million years is about twice that for the preceding 400 million years, whereas the ratio of Mg/Ca in limestones is only about one third or one fourth. This particular plot is fairly representative of quite a few others, all suggesting that analyses of sedimentary rocks representing time spans of era length should be grouped together if systematic secular variations are to be seen.

The same kind of information is supplied by Figure 9.4, which shows weight percent of Na_2O, MgO, and K_2O plotted versus weight percent of Al_2O_3 for a variety of shaly rocks: sediments, sedimentary rocks, and metamorphosed sediments. In this instance, the rocks are grouped on the basis of their degree of compaction and recrystallization, rather than on

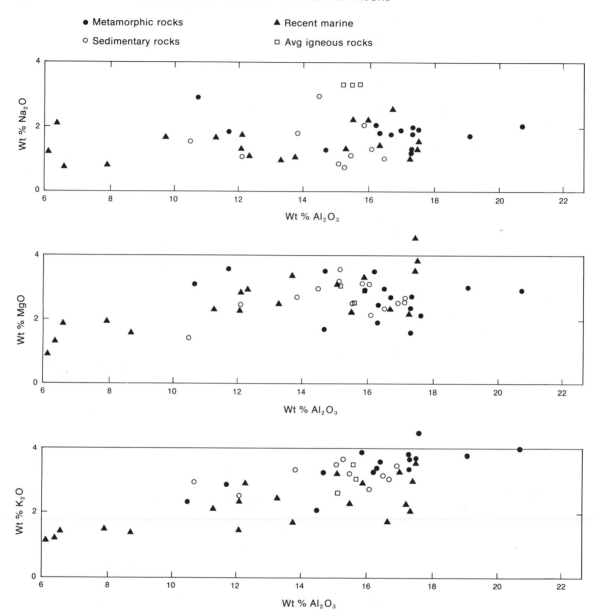

Figure 9.4 Weight percents of Na₂O, MgO, and K₂O plotted against weight percent of Al₂O₃ for shaly rocks. The square symbols show the compositions of three individual estimates of the average igneous rock.

the basis of age. In general, sedimentary and metamorphosed rocks range widely in composition but show almost complete overlap in terms of the variables chosen; sediments cover nearly the same ranges of composition as do the other two types but include lower values of all other oxides shown. Also, a few of the composite analyses of metamorphosed sediments show uniquely high values of Al_2O_3. It is also apparent that the ratio of Na_2O and MgO to Al_2O_3 diminishes with increasing Al_2O_3, whereas the change in the ratio of K_2O to Al_2O_3 is slight.

In Figure 9.5 we show four sets of composite analyses of shaly rocks, an estimate of the average composition of modern sediments, that of Mesozoic and Cenozoic shales, of Paleozoic shales, and of Precambrian slates, schists, and phyllites. The ratios of the various metal oxides to Al_2O_3 were chosen as the ordinate because of the relative immobility of alumina. The chemical trends shown can be interpreted in two chief ways: Either they represent initial differences in composition of the shales, or they have been produced by selective postdepositional addition or removal of some of the constituents. Because little Al_2O_3 would be expected to move in or out of the rocks, the plot used permits assessment of the second alternative, as well as being satisfactory for representation of composition, even if there has been little postdepositional compositional change.

Lack of data forced us to lump together all analyses of Precambrian shaly rocks; however, many of the analyses used were from "middle" Pre-

Figure 9.5 Chemical trends in shaly rocks as a function of age.

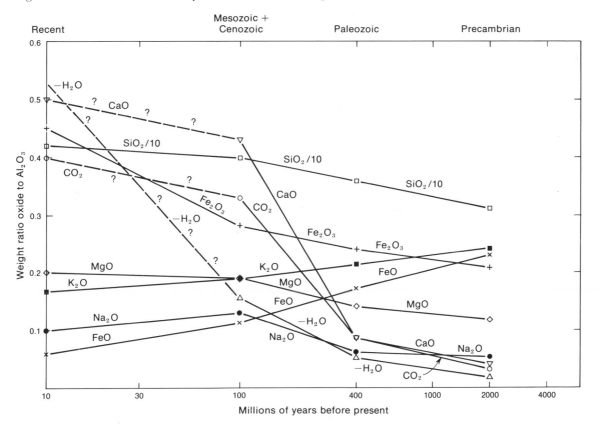

cambrian, which is our only justification for assigning an age of 2000 million years to these older rocks.

The most striking aspect of the diagram is that there is but one reversal in trend, namely the decrease in Na_2O/Al_2O_3 between Mesozoic-Cenozoic and Recent, and it is a change so minor that it may not be real. Second, all oxides except K_2O and FeO diminish with respect to Al_2O_3 with increasing rock age. Water, CO_2, and CaO behave similarly, going from high to very low values. MgO, Na_2O, and SiO_2 form a second behavioral group and change much less than water, CO_2 and CaO. There is a reciprocal relation between FeO and Fe_2O_3; young rocks are high in oxidized iron and old rocks are low, but total iron oxide remains almost constant.

In Figure 9.6 the data are plotted so as to show differences of oxide contents relative to the average Paleozoic shale. The graph emphasizes the uniformity of behavior of the various oxide groups previously delineated and also that the differences in compositions of shales increase exponentially from the Precambrian to the present.

The H_2O plotted is so-called "minus" water, pore water that is removed by drying at $110°C$; its loss is to be expected as a result of burial, compaction, and heating. The covariance of CO_2 and CaO with H_2O is suggestive

Figure 9.6 Enrichment and depletion of oxide components of Mesozoic-Cenozoic and of Precambrian lutites relative to Paleozoic lutites.

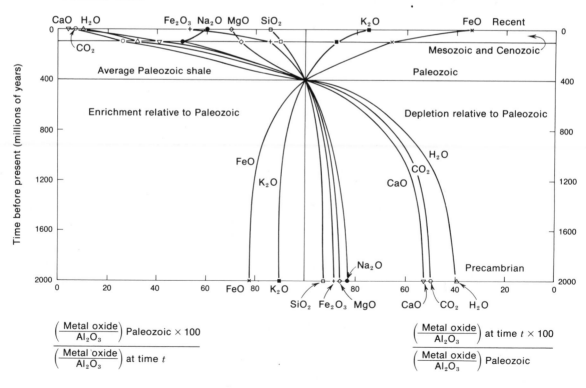

of a loss of original calcium carbonate through time in response to the same kinds of environmental changes that remove pore water. The alternative to postdepositional loss of calcium carbonate would be secular changes in the sites of original calcium carbonate deposition resulting in more initial mixing of shale and carbonate minerals in Mesozoic and Cenozoic sediments than in those of older rocks—in other words, a decrease in the effectiveness of sedimentary differentiation processes.

The reciprocal variation of FeO and Fe_2O_3 also has alternative explanations. The increase in the proportion of Fe_2O_3 in younger lutites can be interpreted as a primary feature of the shales that reflects more and more effective oxidation of iron during weathering, perhaps related to an increase in atmospheric oxygen or to more effective bacterial oxidation processes in the soil. On the other hand, burial of sediments high in Fe_2O_3 with accompanying organic matter could result, through time, in oxidation of the organic matter and reduction of the iron.

The chemical trends are reflected in mineralogical differences. Figure 9.7 shows the relative percentages of clay mineral types from Cambrian to Recent. The more restricted chemistry of Paleozoic shales permits only two important clay minerals, illite and chlorite, with minor kaolinite and expanded clays. In younger rocks, kaolinite and mixed-layer clays add up to about 50 percent of total clay mineralogy. The increase of illite in older

Figure 9.7 Variations in relative percentages of clay minerals in shales as a function of time (data from Weaver, 1967).

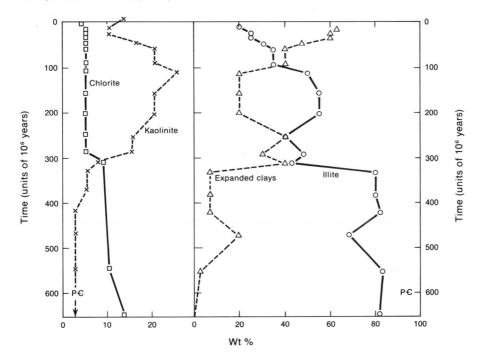

rocks reflects their relative enrichment in K_2O; illite is a micaceous mineral with an approximate composition $[K_{0.6} Mg_{0.3} Al_{2.2} Si_{3.5}O_{10}(OH)_2]$. The expanded clays are Na-Mg-Al silicates, with more silica than occurs in illites.

Increased percentages of expanded clays in younger rocks account for increases in SiO_2, MgO, and Na_2O, the three oxides that were found to behave so similarly on the gain-loss diagram (Figure 9.6). The marked mineralogical change at about the end of the Paleozoic is readily apparent, as are the large fluctuations over short time intervals in the Mesozoic and Cenozoic. The low percentages of kaolinite and expanded clays in Paleozoic shales are characteristic of Precambrian shales as well. If the relative percentages of clay minerals are primary depositional features, then the conditions of the most recent 200–300 million years are unique in geologic time in that they have resulted in the formation of a greater variety of important clay mineral species. The alternative explanation, based on postdepositional change, would require reactions within the rocks of the type

(low silica) kaolinite + (high silica) expanded clay + potassium = (intermediate silica) illite + MgO, Na_2O, and SiO_2 (lost from the shale).

A study of the ages of shales by the potassium-argon method shows that some important changes in chemistry take place after deposition. Recent illitic marine sediments, such as those in the Atlantic basin, give ages of several hundreds of millions of years by this method. Such ages are consistent with the ages determined by the same methods for the source-rock shales being eroded from the continents. Mesozoic sediments also yield ages of 200–400 million years, which are, of course, much closer to their true ages. Paleozoic shales yield age determinations characteristically a little younger than their true ages. Thus, if Paleozoic shales, when deposited, reflected the ages of their source rocks, something has happened since deposition to cause the radioactive clock to reset. If Recent sediments, like those of the Paleozoic, are eventually to give ages younger than their true ages, instead of ages hundreds of millions of years older, it will be necessary for the clay minerals to lose their present argon content and then to start retaining argon some time in the future. Because the K/Ar ages of Paleozoic shales are about 80 percent of their true ages, it can be estimated that at least 20–50 million years were required before the illites settled down and stopped leaking argon. The process may have taken much longer, depending on the actual mechanism; some of the original illite may have retained its argon, whereas new illite may have formed slowly over 2 or 3 hundred million years at the expense of kaolinite and expanded clays, resulting in a bulk age younger than the true age.

Whatever the reasons for variations of the compositions of shales as a function of their ages, the differences are sufficient to cause difficulties in obtaining a chemical analysis for the average shale. One widely used value has been obtained by taking a mean of the analyses of Paleozoic and Mesozoic-Cenozoic shales, on the assumption that roughly equal masses are involved and indirectly that Precambrian shales or their metamorphosed compositional equivalents can be neglected. As we shall see later, the mass of Precambrian sedimentary rocks may be nearly equivalent to the Paleozoic and Mesozoic-Cenozoic total. If so, reference to Figure 9.5 shows that the average composition of Paleozoic shales may be closer to a true composite than a 1:1 mixture of Paleozoic and Mesozoic-Cenozoic. The errors involved in using the average Paleozoic are small, whether we mix Mesozoic-Cenozoic, Paleozoic, and Precambrian on a 1:1:1 mass basis, or even in a 1:1:2 ratio. The major chemical effect is a change in the percentages of CaO, CO_2, and H_2O. The greater the mass of older rocks mixed with the post-Precambrian in obtaining an average, the lower the content of these three oxides. Table 9.1 lists the composition of the average Paleozoic shale.

Chemical Changes in Carbonate Rocks as a Function of Age

The composition of the average carbonate rock is given in Table 9.1. Again, there is a difficult problem in weighting the analytical data because of compositional change with age. Recent carbonate rocks are nearly pure calcium carbonate, with Ca/Mg weight ratios of about 50:1, whereas Precambrian carbonates approach the composition of the mineral dolomite, with a Ca/Mg ratio 1.7:1. Figure 9.8 shows the variation in the Mg/Ca ratio for North American and for Russian Platform carbonate rocks as a function of time. The similarity of the general trends, as well as the Period-to-Period "noise," is well illustrated.

Carbonate rocks less than 100 million years old have the low average magnesium content expected of carbonate deposits formed by accumulation of the skeletal debris of organisms. With increasing age the Mg con-

Table 9.1
Chemical Analyses of "Average Rocks" (wt %)

	A Average igneous rock (after Brotzen, 1966)	B Average limestone (after Clarke, 1924)	C Average shale (after Clarke, 1924)	D Average sandstone (after Pettijohn, 1963)	E Average sedimentary rock (authors)
SiO_2	63.5	5.19	61.90	78.0	59.7
Al_2O_3	15.9	0.81	16.90	7.2	14.6
Fe_2O_3	2.9	0.54	4.20	1.7	3.5
FeO	3.3		3.00	1.5	2.6
MgO	2.9	8.00	2.40	1.2	2.6
CaO	4.9	43.00	1.49	3.2	4.8
Na_2O	3.3	0.05	1.07	1.2	0.9
K_2O	3.3	0.33	3.70	1.3	3.2
CO_2	—	41.90	1.54	2.6	4.7
H_2O (110°C)	—	0.21	3.90	2.2	3.4

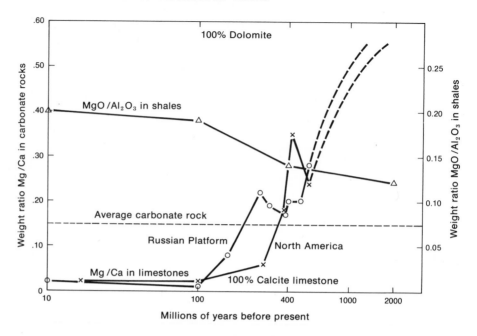

Figure 9.8 Magnesium to calcium weight ratios in carbonate rocks from the Russian Platform and from North America as a function of age. Also plotted is the weight ratio MgO/Al₂O₃ for average shaly rocks as a function of their ages (data from Vinogradov and Ronov, 1956a,b, and Chilingar, 1956).

tent rises irregularly, and the trend apparently extends back into the older Precambrian, as suggested by the dashed lines on Figure 9.8. Voluminous research on the "dolomite problem" has shown that the reasons for high Mg content of carbonates are diverse and complex. Some dolomitic rocks are primary precipitates: others were deposited as $CaCO_3$ and then converted entirely or partially to dolomite before deposition of a succeeding layer; still others were dolomitized by migrating underground waters tens or hundreds of millions of years after deposition.

The mass of Precambrian carbonate rocks is small compared to the mass of younger ones; the average limestone is a composite of largely Phanerozoic limestones and thus may be a little low in its MgO content. However, an adjustment for the poorly known mass and composition of Precambrian carbonates does not seem justified at this time.

The time variation of the weight ratio of MgO/Al_2O_3 in shaly rocks also is plotted on Figure 9.8 and shows the reciprocal relation to the Mg/Ca ratio that would be required by distribution of Mg between shales and carbonate rocks if there has been a fairly constant total amount of Mg deposited.

Chemical Changes in Sandstones as a Function of Age

There seem to be few marked changes in the major element composition of sandstones with time. Composite analyses of Paleozoic, Mesozoic, and Cenozoic sandstones of the Russian Platform show no particular trends; their average composition looks very much like that of Pettijohn's estimate of the world average. The Russian samples are lower in SiO_2 (70 percent versus Pettijohn's 78 percent) but similar in their contents of other oxides.

The above-mentioned difference in SiO_2 percentages comes from the inclusion of analyses of quartzites in the Pettijohn estimates; they contain about 95 percent quartz. The change from sandstone to quartzite is accomplished by the diminution of constituents other than SiO_2 to trace levels.

Despite the lack of documentation of time differences in sandstones by chemical analyses, the possibility of major differences in the proportions of the four sandstone types with time must be considered. Very old geologic terrains, especially those in the plus 2500-million-year range, are now being studied intensively, and according to several investigators (cf. Figure 10.2) the percentages of graywackes and arkoses are increased relative to quartzose sandstones. If so, the average sandstone composition might be shifted from the analysis given in Table 9.1 to a composition somewhat less siliceous and higher in the other oxides.

Chemical Changes in Volcanogenic Sediments as a Function of Time

Little is known about the composition of the "average volcanogenic sediment" and even less about compositional variations of these rocks with time. In some areas the volcanic material is little changed from igneous rock composition, even in very old rocks; elsewhere Recent volcanic sediments are completely converted to clay minerals. Some general features of the alteration processes that seem to be time-independent can be summarized. The original rock may be a glass or may be an aggregate of fine-grained minerals. The same elemental changes apparently take place when an initially glassy rock is altered, even before formation of new minerals. Plagioclase feldspar, usually with a molar ratio of Na/Ca of about 1, makes up roughly half the nonglassy rocks. Pyroxenes (Ca, Mg, Fe silicates) and olivines (Mg, Fe silicates) in various proportions make up most of the remainder. In either case, glass or crystals, alteration is dominantly a leaching and carbonatization process. Mg, Ca, and Fe are lost; K is gained, or perhaps only increases relatively in the residue; Na is conserved or gained. The pyroxenes, by reaction with CO_2, form carbonate minerals such as calcite, dolomite, or siderite ($FeCO_3$). The oxidation state of the iron is variable; in some instances the Fe is converted almost entirely to Fe_2O_3. Conservation of Na is achieved by conversion of the Na,Ca-feldspar to a nearly pure albite or similar Na, Al silicate, such as the zeolite analcime. The presence of a micaceous mineral, usually illite, points to the conservation or increase of K. Mg that is not converted to a carbonate mineral is found in chlorite (Mg, Fe, Al, Si), as is some of the residual ferrous iron. The schematic chemical reaction is

Basaltic sediment $+ CO_2 + H_2O =$ Na-feldspar $+$ clay minerals $+$ Fe, Mg,
Ca carbonates $+$ Fe oxides $+ SiO_2$.

Chemical analyses of volcanic sediments indicate that in some instances the carbonate minerals have been leached away; in others they remain in situ. The SiO_2 released also behaves erratically; it may accumulate locally with the rock to form masses and stringers of chert, or it too may be transported in solution away from the place of its release from the primary rock.

There are insufficient analyses available to us to permit determination of the degree to which Na is conserved; the Kitchi schist shows clear-cut addition of Na from an external source. If so, Cl is probably the balancing anion for the added Na, and the schematic chemical reaction is

$$NaCl + Na,Ca\text{-}feldspar = Na\text{-}feldspar + CaCl_2.$$

The volcanogenic sediments require CO_2 and probably some O_2 for their conversion from their igneous parents. The carbonate minerals formed may stay in place, or they may migrate out of the sediments, presumably to become limestones somewhere else. The chemical influence of the volcanogenic sediments must have been greater in the early days of the earth than it is now; the relative percentages of such sediments increase as we go back in time.

Figure 9.9 shows schematically the mineralogic trends of the major rock types as a function of their ages.

CONVERSION OF AVERAGE IGNEOUS ROCK INTO AVERAGE SEDIMENTARY ROCK

The preceding pages should have sufficed to illustrate the difficulties of choosing a chemical composition for each major rock type that represents a properly composited sample, weighted in terms of the relative masses of different compositions. However, a good test of the analyses finally chosen can be made by attempting to convert the average igneous rock into limestone, shale, and sandstone. If a mixture of these three rock types can be made that will yield the proportions of oxides found in the average igneous rock, the sedimentary rock analyses chosen to represent each type can be validated. Additionally, calculated proportions of the three types can be compared with the observed ones. The test is quite sensitive to the chemical analyses chosen for the individual rock types because the sum of *each* oxide component in the three sedimentary rock types should add up to that in the igneous rock. If a mixture is chosen so that total K_2O, for example, adds up to that in the igneous rock, the proportions of all other oxides in the individual rock-type analyses must be such that they also sum correctly.

The procedure that we used to get a balance can be followed by reference to Table 9.2. We started with 1 kg of average igneous rock (column I). Then 70 percent of the CaO, or 34 g of the original 49 in the igneous rock, was assigned to limestones. This is equivalent to deciding upon the proportion of limestone in the sedimentary rock-type mixture. Then, from

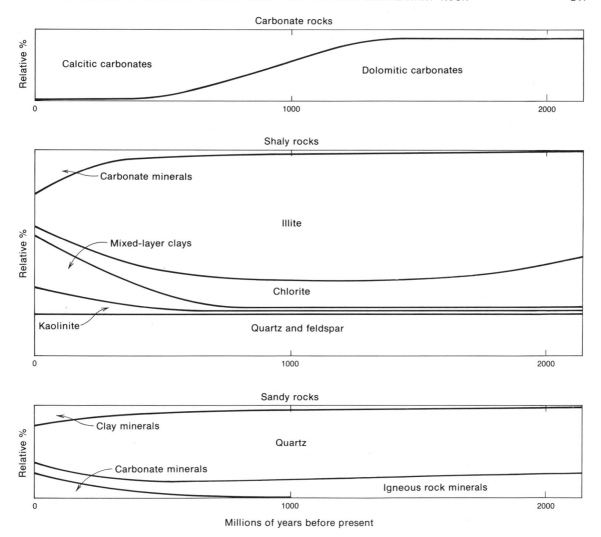

Figure 9.9 Schematic diagram summarizing major changes in the mineralogy of sedimentary rocks as a function of their ages.

the chemical analysis of the average limestone, we calculated the number of grams of each other oxide component required by 34 of CaO. These requirements (column II) were subtracted from the igneous rock to yield a remainder, from which to make shale and sandstone. Carbon dioxide is required to make limestone; because it is not present, except in traces, in the igneous rock analysis, the 34 g of CO_2 were carried forward into the igneous rock remainder as a deficit.

In the next step (column III), 97 percent of the K_2O in the igneous rock remainder was assumed to be present in shales. Again this particular amount is equivalent to selecting the proportion of shale in the sedimen-

Table 9.2
Conversion of Igneous to Sedimentary Rocks (analyses in grams)

	I Average igneous rock (after Brotzen, 1966) (g/kg)		II Average limestone (assuming 70% of CaO of igneous rock)		III Remainder		IV Average shale (assuming 97% of K₂O in igneous remainder)
SiO_2	635		4.0		631		535
Al_2O_3	159		1.0		158		146
Fe_2O_3	29		—		29		36
FeO	33		—		33		26
MgO	29		6.0		23		21
CaO	49	70% →	34		15		13
Na_2O	33		—		33		9
K_2O	33		—		33	97% →	32
CO_2	—		34		−34		13
H_2O	—		—		—		34
Total	1000		79.0				865

tary rock mixture. As before, the chemical analysis of the average shale was employed to calculate the number of grams of other oxides required by the 32 g of K_2O taken from the igneous rock remainder, and these amounts were subtracted from the igneous remainder to provide a second remainder. This second remainder, if the assumed proportions of limestone and shale were correct, should correspond to the composition of the average sandstone. In column VI, the composition of the average sandstone was calculated, adjusting the amounts of the various oxides to the SiO_2 content of the remainder in column V.

Finally, column VII is the test of the validity of the average analyses chosen and of the proportions of rock types mixed. Of the oxides in the original average igneous rock, all except Na_2O balance out to within a few percent of their original amounts, a result as close as could be expected when the difficulties of choosing averages are considered. The deficiency in Fe_2O_3 is balanced by the excess of FeO, indicating that the average sediment is somewhat more oxidized than the average igneous source rock. The oxygen required is about ½ g/kg of igneous rock converted to sediments.

The excess in Na_2O is certainly significant and, if the balances of the other oxides are used as a measure, is somewhere between 21 and 25 g/kg of rock. This excess Na_2O, which disappears in the conversion of igneous to sedimentary rocks, has long been the basis for a variety of geochemical studies. It is found today converted to NaCl dissolved in the oceans and in the pore waters of sediments or in deposits of solid NaCl in evaporites. Just as CO_2 was required to convert CaO and some of the MgO of the average

V Remainder		VI Average sandstone	VII Remainder, found in ocean, pore- waters, evaporites	
96	100% →	96	0	
12		9	3	
−7		2	−9	HCl required to convert oxides
7		2	5	to chlorides, 27 g
2		1	1	
2		4	2	CO$_2$ required for carbonate
24		1	23	minerals, 50 g
1		2	−1	
−47		3	−50	
−34		2	−36	
		122.0		

igneous rock to carbonate minerals, so HCl must be invoked to convert the excess Na$_2$O to NaCl. The reaction is

$$Na_2O + 2HCl = 2NaCl + H_2O$$

$$23 \text{ g} \quad 27 \text{ g} \quad 43 \text{ g} \quad 7 \text{ g}$$

A "perfect" balance would show a residue of oxides in ratios similar to those in sea water, but the amounts of oxides other than Na$_2$O in sea water are so small that they are lost in the errors of the calculation.

The mass balance calculation permits us to determine the weight ratios of the three major sedimentary rock types. Note that the sum of the masses of the sediments is a little greater than that of the original igneous rock mass, 1066 versus 1000 g, because added H$_2$O and CO$_2$ more than balanced the Na$_2$O loss to the oceans. According to the calculations, limestones, sandstones, and shales occur in the ratio of 8:11:81. These ratios are in fair accord with Horn's estimate (1966), based on an elaborate computer program of element balances, of 7:20:73. The chemical composition of the 8:11:81 mixture has been given in Table 9.1, column E.

Our results are comparable to those of other investigators who have made similar calculations, but they emphasize two important points. First, the calculated proportions of rock types do not agree with those obtained by integration of field measurements of rock types. The obvious difference is the small percentage of limestone obtained by calculation. Second, the balance calculation is a demanding test of the relative proportions of the three rock types, so we can conclude that the average igneous rock would indeed give rise to limestones, sandstones, and shales in the relative quantities calculated.

A solution to the limestone problem is that there is a neglected contribution to the mass of limestones, sandstones, and shales that produced chiefly limestone. The preceding description of the formation of volcanogenic sediments showed that the chemical changes involved in their origin from igneous material might provide the necessary carbonate minerals to add to the limestone mass to achieve agreement between calculated and observed masses.

Mineralogic Changes

So far only chemical changes, keyed on the use of oxide components, have been used to investigate the conversion of igneous to sedimentary rocks. Investigation of mineralogic changes helps in understanding the processes of mineral formation and segregation that give rise to the sedimentary rock types.

Because the average chemical analyses are not represented by any individual real rocks, it is necessary to calculate their normative mineralogies. The details of calculation of normative minerals from chemical analyses is given in Appendix C.

Calculation of the normative minerals for the average igneous rock was done in the following way. All of the sodium, plus enough alumina and silica to balance, was assigned to Na-feldspar. All of the potassium was disposed of as K-feldspar. Calcium was also assigned to feldspar until all the alumina was used up. Excess calcium was assigned to $CaSiO_3$. The rest of the cations were calculated as metasilicates, with the exception of Fe_2O_3, which was left as hematite. The residual silica was considered to be present as quartz. Although the normative mineral composition obscures the actual complex compositions of constituent minerals, the general picture is satisfactory (Table 9.3). The average igneous rock is composed of about 16 weight percent quartz, has plagioclase feldspar in excess over orthoclase, and if the sodium and calcium feldspar constituents are combined to make a single plagioclase molecule, its composition (approximately $Na_{60}Ca_{40}$) is that to be expected in an igneous rock body of intermediate composition.

Determination of the normative composition of the average sedimentary rock is a difficult task. We chose to recast the chemical analysis into the minerals calcite, dolomite, quartz, illite, chlorite, montmorillonite, orthoclase, and albite. To make the calculation it was necessary to assign particular compositions to the minerals illite, chlorite, and montmorillonite, each of which has a wide range of compositions in nature. Furthermore, a unique distribution of mineral percentages could be obtained only by arbitrarily setting the K-feldspar percentage. All the CO_2 in the chemical analysis was assigned to calcite and dolomite, using up all the CaO and a part of the MgO in this way. Then enough K_2O was reacted with Al_2O_3 and SiO_2 to make 6 percent of K-feldspar. The arbitrary value of 6 percent was

Table 9.3

Normative Mineralogy of the Average Igneous and the Average Sedimentary Rock (wt %)

	Average igneous	Average sedimentary
Ferrosilite ($FeSiO_3$)	6	
Enstatite ($MgSiO_3$)	7	
Wollastonite ($CaSiO_3$)	3	
Anorthite feldspar ($CaAl_2Si_2O_8$)	18	
Albite feldspar ($NaAlSi_3O_8$)	28	6
K-feldspar ($KAlSi_3O_8$)	19	6
Hematite (Fe_2O_3)	3	4
Quartz (SiO_2)	16	35
Calcite ($CaCO_3$)		7
Dolomite [$CaMg(CO_3)_2$]		4
Illite [$K_{0.6}Mg_{0.3}Al_{2.2}Si_{3.5}O_{10}(OH)_2$]		27
Chlorite [$Mg_2Fe_3.Al_2Si_3O_{10}(OH)_8$]		7
Montmorillonite [$Na_{0.33}Al_{2.33}Si_{3.67}O_{10}(OH)_2$]		3

chosen as reasonable from what is known about actual shale mineralogies. The remaining K_2O, plus the required MgO, Al_2O_3, and SiO_2, was assigned to illite. All the FeO, plus the required MgO, Al_2O_3, and SiO_2, was assigned to chlorite. As it turned out, this process also used up the remaining MgO. The Na_2O was then distributed between albite and montmorillonite in such a way as to use up the Na_2O and the remaining Al_2O_3. Excess silica was calculated as quartz and Fe_2O_3 as hematite.

The results of these normative calculations show how drastic the mineralogic changes are in the igneous to sedimentary rock conversion process, whereas the chemical analyses differ markedly only in the addition of CO_2 and a little water and in the loss of Na_2O. Of the original igneous minerals, all are gone or drastically reduced in amount in sedimentary rocks except for quartz and hematite. The plagioclase feldspars are almost entirely destroyed, the Ca goes into calcite, and the Na is lost to the oceans. K-feldspar is definitely more resistant to alteration than plagioclase. The quartz content of sedimentary rocks is about twice that of the igneous source, suggesting that the original quartz is transferred unchanged into sediments, and is augmented by quartz formed from the silica released when feldspars and other silicates are converted into clay minerals. Hematite, like quartz, is almost inert and tends to be transferred unchanged. There is a slight increase in hematite in the sedimentary rocks because of oxidation of FeO to Fe_2O_3 during conversion.

The large acid requirement for the formation of sedimentary rocks is illustrated by asking whether the process could be duplicated in the laboratory. According to the preceding discussion, one could perform the following experiment. Take 1 liter of 0.75 normal HCl and dump into it 1 kg of powdered average igneous rock (about 370 cm³). Stir vigorously and simultaneously bubble CO_2 through the solution. Permit the reaction to continue until HCl is neutralized and 1.14 moles of CO_2 has been consumed. Admit about 400 cm³ of oxygen during the reaction to oxidize a little of the ferrous iron to ferric oxide.

To our knowledge, no one has performed this experiment. Some preliminary work indicates that neutralization of all the HCl would take place in a few months, but that the reaction products would require many, many years to organize into the proper mineral species.

The Total Mass of Sedimentary Rocks

So far we have dealt with the conversion of a unit mass of igneous rock to sedimentary rock. Now we can turn to making geochemical estimates of the total mass of sedimentary rocks. The total mass has been estimated both geophysically and geochemically. The physical estimate requires knowledge of the masses of the major sedimentary units: continent-shield, mobile belt-continental shelf, hemipelagic, and pelagic. The mass of each of these units is a function of areal extent, average thickness, average mineral density, and average porosity. The average thickness of each major sedimentary unit is difficult to determine and is based primarily on seismic evidence. Pelagic sediment thickness has been most difficult to estimate because of (1) lack of seismic data from the deep ocean and (2) difficulty in determining the nature of the seismic reflectors in oceanic sediments. The average pelagic sediment thickness is probably about 1 km and certainly does not exceed 2 km.

Poldervaart (1955) estimated the total sedimentary mass as $17,000 \times 10^{20}$ g, and Horn (1966) obtained $20,400 \times 10^{20}$ g. Gregor (1968), leaning heavily on Ronov's work, estimated the Phanerozoic sedimentary mass at about $18,000 \times 10^{20}$ g, whereas Ronov himself (1968) puts the total mass at about $24,000 \times 10^{20}$ g.

This considerable range in the estimates illustrates the incompleteness of our current knowledge. In general, mass estimates have been getting larger and larger as more information about the earth is available. One of the most difficult aspects of measurement procedures is to make proper allowances for rocks that have maintained their sedimentary compositions but have become intimately intermixed with igneous material. Not only are detailed mapping and many chemical analyses required, but because the rocks are largely Precambrian in age, they are commonly covered by younger deposits.

Geochemical methods are based upon the igneous rock-sedimentary rock conversion and rely upon some measured parameter, such as the total chloride representing HCl that has been neutralized by the destruction of igneous rock minerals. Ideally, the geochemically calculated mass should be larger than the measured mass, for it includes sedimentary rocks that are missed by measurement.

Before we can move on to calculations of total rock masses, it is necessary to reconcile the calculated and measured ratios of limestones, shales, and sandstones. According to several investigators these ratios in Post-Precambrian rocks are about 20:65:15. Precambrian sedimentary rocks are notably low in carbonate rocks, which average perhaps 5–10 percent of the sum of the three rock types. At this juncture the relative mass of Precambrian sedimentary rocks has not been estimated, but the amount of any composition required to get an overall ratio of 8:81:11, by mixing with 20:65:15, is prohibitive.

The volcanogenic sediments should have been included in the determination of the average igneous rock, and the chemical requirements for their alteration from their parent rocks also should have been assessed, but we do not have adequate analytical data. For the moment, we shall assume that the postdepositional alteration of the volcanogenic rocks, with subsequent leaching of $CaCO_3$ and $MgCO_3$, which become part of the limestone mass, has yielded a mass of limestone comparable to that from the average igneous rock. The proposed interrelations of the "average igneous rock system" and the "volcanogenic sediment system" are shown in Figure 9.10.

The numbers suggested for the masses of volcanogenic rocks are crude estimates, but they are in general accord with Ronov's estimate (1968) of the relative mass of such rocks. From the values given in Figure 9.10, the volcanogenic rocks make up about 26 percent of the total sedimentary mass. Also, the 100 g of limestone calculated to be derived from 500 g of volcanogenic rock is possible, at least, as judged by calculations of the chemical relations between fresh and altered basaltic rocks.

All of the foregoing discussion of rock compositions and ratios illustrates the intricacies and value judgments required in mass balance calculations on a global scale. The best we can do is to make a model that is internally fairly consistent and to use it with great caution as a temporary "best fit" of the data.

Mass Calculations

We can now estimate the relative masses and the total mass of sedimentary rocks, using the numbers from Figure 9.10. First we shall employ an adaptation of the so-called "chloride method," which depends on an estimate of the total Cl in the crust that represents HCl neutralized by reaction with igneous rocks. Total Cl in the crust, exclusive of that in igneous rocks, is given by Horn (1966) as 560×10^{20} g. Reference to Figure 9.10 shows that 27 g of HCl, corresponding to 26 g of Cl, produce 1500 g of sedimentary rock. The calculation is

$$\frac{560 \times 10^{20} \times 1500}{26} = 32{,}000 \times 10^{20} \text{ g.}$$

A similar calculation can be done, utilizing a quantitative estimate of the Na released to the oceans, to evaporite deposits, and to pore waters, by

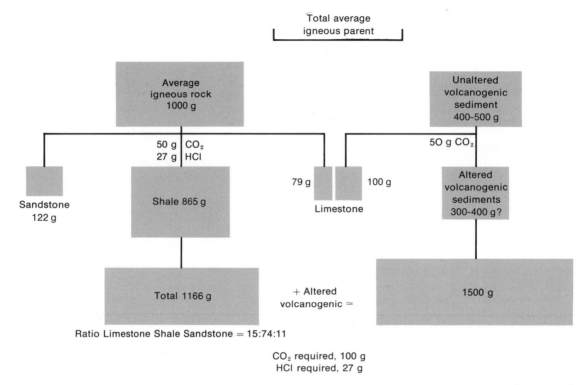

Figure 9.10 Flow sheet of the conversion of igneous rocks into sedimentary rocks by reaction with CO_2 and HCl.

the alteration of igneous rocks to sedimentary rocks, but the Cl method is preferable, because Cl is a direct measure of the acid neutralized, whereas Na can be bound or freed without acid consumption by various exchange reactions with other cations.

The value of $32,000 \times 10^{20}$ g is substantially larger than the preceding values, almost 30 percent larger than Ronov's estimate (1968). It does gain some support, however, if we accept Gregor's value (1968) of about $18,000 \times 10^{20}$ g for the Phanerozoic mass, in that our estimate leaves a substantial body of Precambrian sediments, whereas Ronov's estimate of about $24,000 \times 10^{20}$ g leaves only 6000×10^{20} g for rocks older than Cambrian. Such a relatively small mass implies more destruction of old rocks than most geologists are willing to admit.

The figure $32,000 \times 10^{20}$ g also leads to a consistency in carbonate rock relations. If we accept the Phanerozoic ratios of limestone, shale and sandstone of about 20:65:15 proposed by several investigators, we can inquire, from the overall ratios of 15:74:11 from Figure 9.10, concerning the ratios in Precambrian rocks if we know the relative total rock masses of Precambrian and post-Precambrian rocks. The calculation is given in Table 9.4.

The calculated ratios of various rock types for the Precambrian are com-

Table 9.4

	Lime-stone	Shale	Sand-stone	Volcano-genic	Total
			(units of 10^{20}g)		
Total	3500	17300	2600	8600	32,000
Post-Precambrian	2600	8600	2000	4800	18,000
Precambrian (difference)	900	8700	600	3800	14,000
Precambrian percent (ratios)	6	62	4	27	
Ronov (1968)	4	37	14	40	
Precambrian percent of sandstone, limestone, shale	8	86	6 (shale + sandstone = 92)		
Ronov (1968)	7	67	26 (shale + sandstone = 93)		

pared with Ronov's estimates (1968) of the same rock types, which he obtained by measurement.

The calculations show that, whatever the actual mass numbers, the mass of Precambrian sediments relative to post-Precambrian is substantial —of the order of 1:1 as opposed to ratios such as 1:3 or 1:4. The total mass of $32,000 \times 10^{20}$ g is, of course, based on Gregor's value of $18,000 \times 10^{20}$ g for the post-Precambrian, and his estimate is subject, in our opinion, to a probable error of ± 15 percent.

Our calculations so far have neglected a sedimentary rock type that makes up some 2–3 percent of the total mass—the evaporite deposits, composed of salt, gypsum, or anhydrite, and intermixed material of various kinds. Holser and Kaplan (1966) estimate that there are about 400×10^{20} g of $CaSO_4$ in evaporites; the amount of Na (Horn, 1966) is 95×10^{20} g, corresponding to 210×10^{20} g of NaCl. The total mass of $CaSO_4$ and NaCl is about 600×10^{20} g; we can add another 300 or 400×10^{20} g for intermixed material and come up with a total mass of about 1000×10^{20} g. This mass, about 3 percent of the total of other rocks, is restricted almost entirely to Phanerozoic rocks.

SUMMARY

The current situation in regard to the mass of sedimentary rocks, and their relative proportions, according to our estimates, can be summarized as follows.

1. The total mass of sandstones plus limestones plus shales plus volcanogenic deposits is of the order of $32,000 \times 10^{20}$ g.

2. The Precambrian mass is almost as large as the post-Precambrian mass—$14,000 \times 10^{20}$ versus $18,000 \times 10^{20}$ g.

3. The percentage of carbonate rocks in the post-Precambrian mass is more than twice that in the Precambrian mass; the total carbonate rock is about 3 times as much.

4. The percentage of shaly rocks is higher in the Precambrian than in the post-Precambrian, but the total masses are about the same.

5. There is a discrepancy between the percentage of sandstones that we calculate and the percentage observed; our estimates are considerably lower and are probably wrong. There may well be a significant contribution to sandstones by silica released from the alteration of volcanogenic deposits that would increase the sandstone mass as it apparently increases the carbonate rock mass. However, inasmuch as sandstones, in terms of chemical composition, are essentially shales plus silica, it is of interest that our total mass of sandstone plus shale is in accord with measured ratios. The implication is that the SiO_2 content of the average shale that we have chosen is too high; not enough SiO_2 was left over for sandstones.

6. The major barrier to a geochemical calculation of sedimentary rock-type *ratios*, which are more trustworthy than estimated *masses*, is a lack of knowledge of the chemical compositions of the altered volcanogenic rocks, as well as a lack of knowledge of the average chemical compositions of the weathered parent rocks of the existing volcanogenic materials.

7. Previous geochemical calculations of the total mass of sedimentary rocks have neglected what appears to be a highly significant contribution to the mass of limestones from the alteration and leaching of Ca and Mg from volcanogenic rocks. The assumption that the mass of carbonate rocks derived from alteration of volcanogenic rocks is about the same as that derived from the average igneous rock is a possible solution to the long-standing problem of the discrepancy between measured carbonate rock percentages and calculated percentages.

8. The percentages of volcanogenic sediments and perhaps also of gray-wackes may be much higher in rocks of the early Precambrian than in later time.

9. The formation of 1500 g of sedimentary rocks requires about 100 g of CO_2 and 27 g of HCl. These numbers place the total neutralized CO_2, now in the form of carbonate minerals, at about 2100×10^{20} g and the neutralized HCl, now in the oceans, in pore waters of rocks, and in evaporite deposits, at about 560×10^{20} g.

10. The sedimentary rock mass can be regarded as having two types of sources. One is the classic source of continental erosion in which material weathered from the continents is carried to the oceans by streams and deposited or precipitated to make limestones, sandstones, and shales. The other is from the alteration of volcanogenic material, which may take place in many environments, including the deeps of the seas themselves. The products are the residue of the volcanic material, which remains in situ, plus a dissolved fraction that can move away and eventually join the chemical precipitates derived from the land. The characteristic changes from fresh to altered volcanic material involve loss of Ca, Mg, and Fe, conservation or gain of Na, and addition of K and H_2O.

11. The numbers we offer, and those of previous investigators, must be

used with the knowledge that they may include large errors. Their values lie in their yielding a working model expressive of our current knowledge.

REFERENCES

Brotzen, O., 1966, The average igneous rock and the geochemical balance: Geochim. Cosmochim. Acta, *30*, 863–868.

Chilingar, G. V., 1956, Relationship between Ca/Mg ratio and geologic age: *Bull. Am. Assoc. Petrol. Geologists, 40*, 2256–2266.

Clark, F. W., 1924, Data of geochemistry: U.S. Geol. Survey Bull. *770*, 841 pp.

Gregor, C. B., 1968, The rate of denudation in Post-Algonkian time: *Koninkl. Ned. Akad. Wetenschap. Proc., 71, 22*.

Holser, W. T., and Kaplan, I. R., 1966, Isotope geochemistry of sedimentary sulfates: *Chem. Geol., 1*, 93–135.

Horn, M. K., 1966, Written communication.

Pettijohn, F. J., 1963, Chemical composition of sandstones—Excluding carbonate and volcanic sands: in Data of Geochemistry, 6th ed., M. Fleischer, ed., U.S. Geol. Survey Prof. Paper *440-S*, 21 pp.

Poldervaart, A., 1955, Chemistry of the earth's crust: in Crust of the earth, A. Poldervaart, ed., *Geol. Soc. Am. Spec. Paper, 62*, 119–144.

Ronov, A. B., 1968, Probable changes in the composition of sea water during the course of geological time: Invited Lecture, VII International Sedimentological Congress, Edinburgh, Scotland.

Vinogradov, A. P., and Ronov, A. B., 1956a, Evolution of the chemical composition of clays of the Russian Platform: *Geochemistry, 2*, 123–129.

Vinogradov, A. P., and Ronov, A. B., 1956b, Composition of the sedimentary rocks of the Russian Platform in relation to the history of its tectonic movements: *Geochemistry, 6*, 533–559.

Weaver, C. E., 1967, Potassium, illite, and the ocean: *Geochim. Cosmochim. Acta, 31*, 2181–2196.

10 | # Sedimentary Rock Mass-Age Relations and Sedimentary Cycling

There are two major points that were made in Chapter 9 that will be analyzed further here. First, individual sedimentary rock types show secular variations in their chemical compositions. Second, the proportions of the different rock types also vary with time. The major problem raised with respect to the first point was whether the secular variations in chemical composition were dominantly a result of initial differences in mineralogy or whether the changes were chiefly postdepositional. The second relation also has two similar basic interpretations: Are the variations in rock-type ratios primary, or have they too been changed since the time of formation of a given sedimentary mass? The questions are alike and the problems are definitely interrelated.

The data on secular variation of chemical composition with age already have been presented. Now we need the information available on the distribution of sedimentary mass as a function of the ages of the rocks, as well as the same information for individual rock types. Then simple models of erosion, deposition, and sediment distribution can be devised and tested against what we know of the actual age distribution of sediment types and the chemical variations of each type.

SEDIMENTARY MASS AS A FUNCTION OF AGE

The only direct information we have about the distribution of sedimentary mass as a function of the ages of the rocks covers the interval from the Devonian to the Jurassic. Figure 10.1 shows our estimate of the mass distribution of sedimentary rocks as a function of time based on the assumption that the maximum thickness of strata, as given by Kay (1955), is proportional to the mass of rock within a given Period. Also illustrated are the estimates of Ronov (1959) based on measurement of the volumes of sedimentary rocks for the Devonian through Jurassic Periods and those of Gregor (1968), which are dependent on Ronov's volume data and Holmes' maximum thickness data (1965). The graph deserves a few extra words of explanation. We estimated the mass of sedimentary rocks of each Period that remain today and then divided these estimated masses by the duration of the Periods (in millions of years). This calculation provided the average mass of rocks that remain today per million years of the given Period. Gregor (1968) termed this ratio the "survival factor." These average values for the individual Periods were then plotted as a histogram of rock mass per unit time versus age. The area of each block of the histogram is proportional to the mass of rock of that Period that remains today. Differences between our values and those of Ronov and Gregor indicate the approximate error inherent in the values available for estimating the height of each block.

However, the estimates agree on the basic pattern of mass-age distribu-

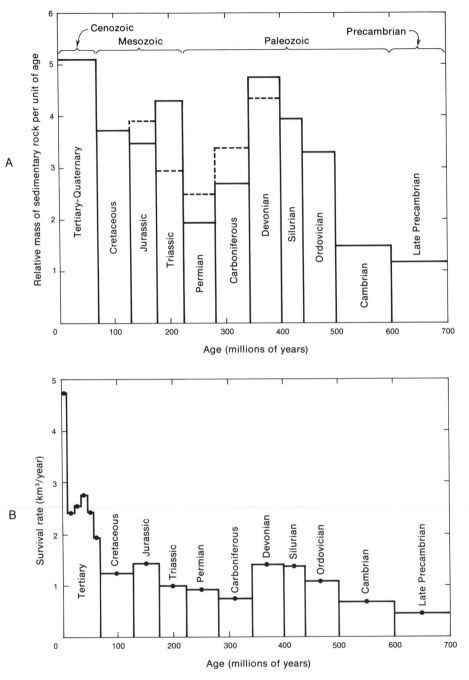

Figure 10.1 Distribution of sedimentary rock mass as a function of rock ages. The diagram above is our estimate; the dashed lines show Ronov's estimate (1959) for the periods from Devonian through Jurassic. Below is Gregor's estimate (1968). All three sets of estimates show low mass per unit time about 300 million years ago and high mass per unit time today and during mid-Paleozoic.

tion, and the following conclusions are undoubtedly valid, despite the differences in estimates of mass for individual Periods. The Cambrian Period, the immediately preceding late Precambrian Period, the Carboniferous Period, and the Permian Period have much less rock preserved per year of their respective time spans than do other Periods.

There is a clear-cut minimum of mass per unit time between 200 and 300 million years ago, a maximum at about 400 million years ago, and a low which may or may not be a minimum during the interval between 500 and 700 million years ago. The center of gravity of the mass distribution shown is at about 300 million years; inasmuch as the total time portrayed is 700 million years, the gross mass distribution is almost uniform with age, but with somewhat more mass remaining per unit time for the most recent 350 million years than for the preceding 350 million. The Mesozoic-Cenozoic mass is somewhat larger than the Paleozoic mass, but this difference is attributable almost entirely to the large mass of the Tertiary and Quaternary.

According to our preceding estimate, the mass of Precambrian sedimentary rocks is a little less than the total of Paleozoic plus Mesozoic-Cenozoic ($14,000 \times 10^{20}$ versus $18,000 \times 10^{20}$ g), but we know next to nothing about its distribution with age. We know the mass is spread over a total of at least 3 billion years and that the amount remaining per unit time diminishes with increasing age. We are also fairly certain that there are maxima and minima in the spread of $14,000 \times 10^{20}$ g of rock over the 3-billion-year time span. Sedimentary rocks with ages of about 1 billion years are found in many areas. There are relatively few occurrences of rocks in the 600–800-million-year range, so the low value for the late Precambrian shown on Figure 10.1 may be a minimum.

Figure 10.2 emphasizes, by its schematic presentation of mass distribution, the small mass of Precambrian rock that remains today spread over the time span of 3 billion years as compared to the mass available for investigation for the 600 million years from Cambrian to Recent. It is apparent that mass per unit time diminishes in an exponential fashion from the present to the distant past. The sedimentary rocks are crowded toward the front of geologic time.

Information on the distribution of rock types within the total sedimentary rock mass has been assembled by Ronov (1964). Figure 10.3 is taken from his work. He expressed his results in terms of the relative volume percentages of the various types as a function of age. The conversion of his volume percentages for a given rock type into the mass of that type remaining for a given time interval requires knowledge of rock densities and of the total rock mass for the given time interval. Our lack of information on total mass-age relations for the Precambrian precludes construction of diagrams showing mass per unit time for the individual rock types as a function of age. In assessing the trends shown on Ronov's diagram, it must be remembered that the total rock mass diminishes with increasing age and that a given volume percentage of rock 2 billion years old represents

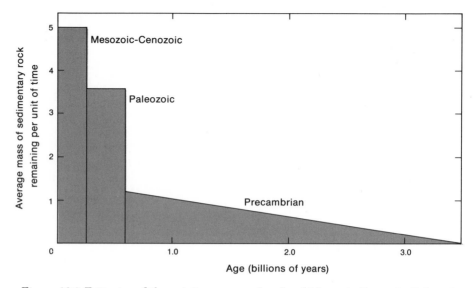

Figure 10.2 Estimates of the existing masses of rocks of Mesozoic-Cenozoic, Paleozoic, and Precambrian ages, plotted as blocks whose areas are proportional to rock masses. The shape of the Precambrian block has been drawn schematically to emphasize the diminution of preserved mass with increasing age.

much less rock than the same volume percentage for rocks only 200 million years old.

Among the major trends shown by Ronov are the relative constancy of the percentages of lutites. He maintains that the relative percentage of sandstones is almost constant, too, but increases the proportion of gray-wackes at the expense of quartz sands in the older rocks. Ronov shows a marked increase in the percentage of submarine lavas with increasing age, which is of interest relative to the discussion in Chapter 9 on the origin of carbonate rocks by alteration of volcanogenic sediments. In the early days of the earth, alteration of lavas may have been the dominant process in the genesis of sediments. Carbonate rocks increase strikingly in their relative percentages from almost zero 3 billion years ago to a significant fraction of the total column in post-Precambrian time; as illustrated, dolomite rocks dominated the scene in the late Precambrian, and calcite rocks are restricted almost entirely to the post-Precambrian. Evaporites behave like carbonate rocks, only more so; they are almost entirely found in the post-Precambrian.

An interesting aspect of the Ronov diagram is the significant relative percentage of jaspilites in the younger Precambrian, rocks that he reduces to zero at about the end of the Precambrian. These are massively layered cherty rocks, commonly associated with iron deposits. The chert-iron association is a unique rock type of the Precambrian.

As illustrated in Chapter 9, our crude attempts to calculate relative masses of rock types are in general accord with, and influenced by, Ronov's

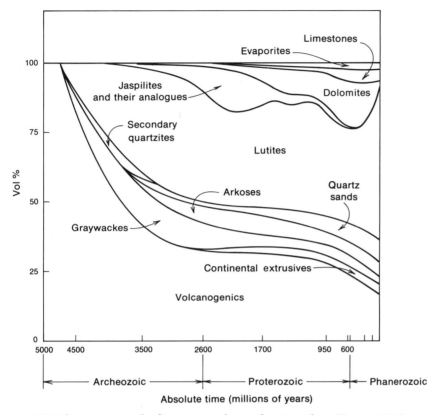

Figure 10.3 Volume percent of sedimentary rocks as a function of age (Ronov, 1964).

work. We have constructed Figure 10.4 not only to try to show the general trends indicated by Ronov but also to include the mass factors as well.

SEDIMENTARY CYCLING

Now that we have rough estimates of the total mass of preserved sedimentary rocks, as well as some notion of the distribution of that mass with age and rock type, we can construct models of the distribution, making assumptions concerning rates of deposition, destruction, and sediment accumulation throughout geologic time, and see which of the models gives the best fit to the little we know.

The distribution of sedimentary mass with age, despite large variations, shows little trend with time for the past 600 million years. As already pointed out, the center of mass is about midway between the Cambrian and the Present, and when we realize that we can expect an increasing percentage of sediments *deposited* to be destroyed in one way or another as the sediments get older and that we are looking at the sediments *preserved*, it

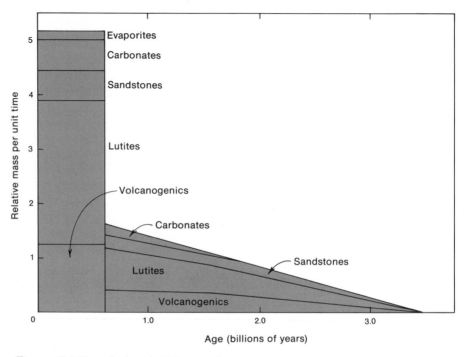

Figure 10.4 Ronov's data (1964) on rock types as a function of age, replotted to show the relation between percentages of types and their relative masses.

becomes apparent that depositional rates for the past 600 million years may well have averaged out about constant, or even have decreased a little. The constant rate required to *deposit* the $18,000 \times 10^{20}$ g of post-Precambrian sediments, making no allowance for subsequent destruction, is $18,000 \times 10^{20}/600 \times 10^6 = 30 \times 10^{14}$ g/year.

On the other hand, the present-day sediment flux is about 250×10^{14} g/year. If the present rate of deposition were continued for 600 million years and there were no erosion, the accumulation would be about 8 times what we see preserved during the past 600 million years. Today's rate of erosion is probably much higher than the average for the geologic past; we know that the continents are higher, and the area of land exposed is greater, than during much of the rest of geologic time. Gregor (1968) has shown that for the Tertiary Period, some 70 million years in length, the existing sedimentary mass corresponds to an annual sediment flux of about 50×10^{14} g/year, whereas the rate over the Pleistocene and Recent is about 100×10^{14} g/year. All of these numbers are, of course, minimal rates, based on surviving sediments, but if we arbitrarily consider a rate of 50×10^{14} g/year over 3.5 billion years, the total sedimentary mass deposited would be

$$50 \times 10^{14} \times 3.5 \times 10^9 = 175,000 \times 10^{20} \text{ g.}$$

This rough estimate implies that the total mass of sediments deposited through time is 5 or 6 times greater than that remaining $(32,000 \times 10^{20}$ g) and fits also the sedimentary rock mass distribution, which is consistent with the destruction of most of the Precambrian sediments that initially were deposited.

We shall now consider several simplified models of the erosional-depositional-accumulational history of sediments. There are two chief types of models: one in which a mass of sediments comparable to that preserved today was formed early in the earth's history and has been continuously destroyed and redeposited and a second in which the preserved mass increased steadily with time.

CONSTANT MASS MODELS

The underlying basis for constant mass models is an early degassing of the earth, with the emission of all the water to the hydrosphere and all the CO_2, HCl, and other acid gases that reacted with primary igneous rocks to form sedimentary rocks. This model implies that since that time there has been no increase in the total mass of sediments because there have been no new acid gases released to create them, although the sediments have been turned over by erosion, and destroyed by metamorphism, with concomitant recycling of CO_2 and HCl. This view, of course, is an extreme one, but it should provide us with one set of limiting conditions.

The assumptions of constant mass models are formally

1. A mass of sediments equal to the existing mass was formed very early in the earth's history from primary igneous rock.

2. This mass has been destroyed (by erosion or by metamorphism) and deposited at a constant rate through time.

3. The probability of a given mass of sediments being destroyed is proportional to the ratio of the given mass to the total mass.

Figure 10.5 shows the fraction remaining today of the sediments of the past as a function of various rates of deposition relative to the total mass. The calculation of the curves could have been done in several ways; we chose to develop them as a series, because the physical significance of the operation is more obvious in its relation to the model. (See Gregor, 1968, for the use of exponential functions in solving a similar problem.)

The 4 billion years of geologic time were divided into 100 units, designated $\Delta t_1, \Delta t_2, ..., \Delta t_{100}$. The mass of sediments, assumed constant, is designated M and the mass deposited or destroyed as ΔM, which can be set at any desired fraction of M.

During the most recent time interval, Δt_1, ΔM_D sediments are deposited and ΔM_E destroyed. But those that are destroyed are taken from the whole mass M, so the fraction of ΔM_D that is destroyed is $\Delta M_E/(M+\Delta M_D)$ and the fraction remaining is $1-\Delta M_E/(M+\Delta M_D)$. Because ΔM_D is numerically equal to ΔM_E, the fraction remaining reduces to $M/(M+\Delta M)$. These relations are best visualized by a simple sketch, as shown in Figure 10.6.

If we go back two time intervals and consider the sediments deposited during Δt_2, the fraction remaining at the end of Δt_2 is again $M/(M+\Delta M)$,

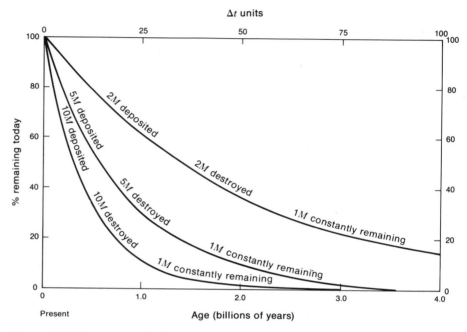

Figure 10.5 Percentage of sediments originally deposited that remain today, according to the constant mass model.

and the fraction remaining at the end of Δt_1 is simply $[M/(M+\Delta M)]^2$. Consequently, the fraction remaining of the sediments deposited during any chosen time interval, Δt_n, is $[M/(M+\Delta M)]^n$.

This is perhaps an unfamiliar form of the exponential curve for which we can say that the rate of change of the fraction remaining is inversely proportional to age—a so-called radioactive breakdown type of curve (see Chapter 3).

The shapes of the curves are quite sensitive to the ratio of total sediment deposited to the total mass existing. Clearly a ratio of 2:1 is too small, giving too much very old sediment remaining today, whereas a ratio of 10:1 would leave too little. Further discussion of constant mass models will be postponed until the linear accumulation models are considered.

LINEAR ACCUMULATION MODELS

The major alternative to the constant mass model is one in which water, HCl, and CO_2 are continuously being degassed from the interior of the earth, and as a result the mass of sediments existing at any one time continually increases with time. A linear rate of degassing would appear to represent the extreme alternative condition to a constant rate, because most hypotheses support a model with a high early degassing rate, fol-

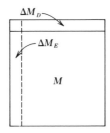

Figure 10.6

lowed by a continuously decreasing rate, which probably is somewhat ir-regular. The assumptions of the linear accumulation models are as follows:

1. The mass of sediments has grown linearly through time from zero to the currently existing mass.

2. Deposition, destruction, and accumulation have all gone on at con-stant rates, so that deposition rate minus destruction rate equals accumula-tion rate.

3. As before, the probability of a given mass of sediments being de-stroyed is proportional to the ratio of the given mass to the total mass.

In this model, the oceans also have grown linearly with time. Note, however, that the model assumes that the ratios of $H_2O/CO_2/HCl$ in the released gases are constant and that the oceans and the sediment mass grow together.

The calculations for the linear accumulation models are similar to but somewhat more complicated than for constant mass models. If the present mass of sediments is M, then we can again divide geologic time in $100\Delta t$ units and accumulate $0.01\,M$ units of mass during each Δt interval to yield the present mass. The mass deposited during Δt is ΔM_D and the mass de-stroyed is ΔM_E. However, for any given time unit, $\Delta M_D - \Delta M_E = 0.01M$. During the most recent time interval, Δt_1, the initial mass is $M - 0.01M$, the mass deposited is ΔM_D, and that destroyed is ΔM_E. Thus $\Delta M_E/(M - 0.01M) + \Delta M_D$ = fraction destroyed, and the fraction remaining is 1 minus fraction destroyed.

Repetition of this procedure for $\Delta t_2, \Delta t_3, ..., \Delta t_n$ yields a series for the fraction remaining of sediments deposited during any Δt_n:

$$\frac{M}{M + \Delta M_E} \times \frac{M - 0.01M}{(M + \Delta M_E) - 0.01M} \times \frac{M - 0.02M}{(M + \Delta M_E) - 0.02M} \cdots$$

$$\frac{M - (n - 1)\,0.01M}{(M + \Delta M_E) - (n - 1)\,0.01M},$$

where n is the designation of the particular Δt for which the fraction re-maining is desired. This series is easily solved; for example, in a model for which 5 times as much sediment has been deposited during geologic time as now exists, $M_D = 0.05M(100 \times 0.05M)$, M_E is $0.05M - 0.01M = 0.04M$, and the fraction remaining at Δt_{10} (10 time units before the present), or during the interval between 400 million and 360 million years ago, is

$$\frac{1.00}{1.04} \times \frac{0.99}{1.03} \times \frac{0.98}{1.02} \times \frac{0.97}{1.01} \times \frac{0.96}{1.00} \times \frac{0.95}{0.99} \times \frac{0.94}{0.98} \times \frac{0.93}{0.97} \times \frac{0.92}{0.96} \times \frac{0.91}{0.95}$$

$$\text{or} \quad \frac{0.94}{1.04} \times \frac{0.93}{1.03} \times \frac{0.92}{1.02} \times \frac{0.91}{1.01} = 0.66.$$

Figure 10.7 shows curves of percent remaining versus age for models in which the ratios of total mass deposited to mass accumulated are 2:1, 5:1, and 10:1. As for constant mass models, the curves for deposition of 2 or 10 times the existing mass give too much or too little old sediment remaining, and the curve for 5:1 gives a reasonable mass distribution.

DISCUSSION OF MODELS

The predicted age distribution of sedimentary rocks is almost independent of the model used if the total mass deposited is much larger than that preserved today. The effect of the presence of an initial mass of sediment is wiped out by the continuous processes of deposition and destruction. Figure 10.8 shows the distribution of mass as a function of age as it would have been at various times in the earth's history according to a constant

Figure 10.7 Percentage of sediments originally deposited that remain today, according to the linear accumulation model.

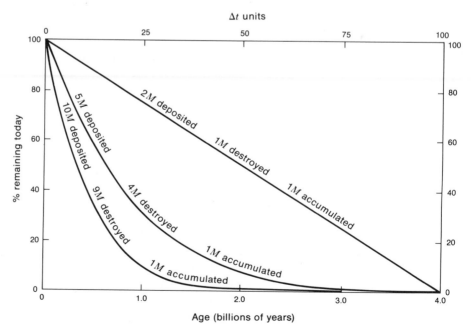

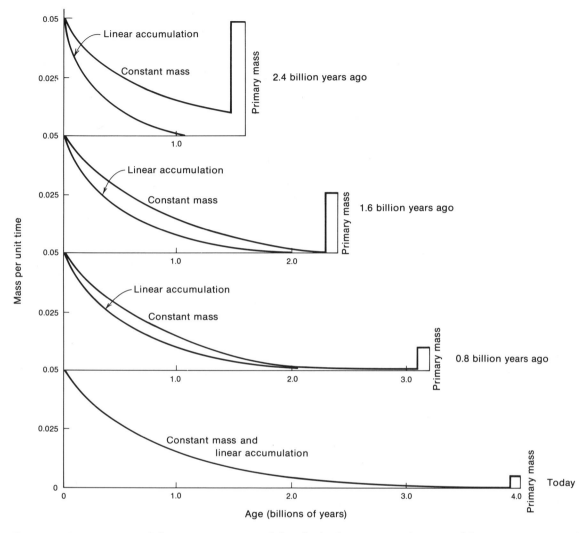

Figure 10.8 Comparison of the constant mass model with the linear accumulation model at various stages of their development toward the present. The curves for the two models, each drawn for the condition that the total sedimentary mass that has been deposited is 5 times that existing today, become indistinguishable at about 500 million years before the present.

mass model as opposed to a linear accumulation model. In both cases, the depositional rate chosen is sufficient to have deposited 5 times the mass of sedimentary rocks existing today. In the uppermost diagram, which is drawn to show the distribution of mass 2.4 billion years ago, the influence of the initial nearly instantaneous deposition of a sediment mass M is still strongly apparent, and more than 30 percent of the sedimentary rocks at that time would still be "primeval." The lower diagrams show successively how, with reworking, the constant mass and linear accumulation models converge, until today they are apparently indistinguishable.

It is an interesting and somewhat ironical conclusion that man apparently evolved just at the time of coincidence of the two models and too late to have much chance of finding evidence of a primeval sediment mass, if such existed. On the other hand, we have arrived in time to find at least fragments of very old sedimentary masses. Let us hope, for the sake of future geologists, that the present-day rate of destruction of the sedimentary record does not last too long—another 100 million years at the present rate would destroy half the currently existing deposits.

Figure 10.9 shows a comparison of the mass distribution with age based on the linear accumulation model (fraction remaining $\times M_D =$ mass existing per age unit), assuming a ratio of 5 units deposited for every unit accumulated, with an estimate of the probable type of distribution of sedimentary masses with age during the Precambrian. The presence of maxima and minima, fairly well established for the sedimentary rocks of the past 600 million years, suggests that there are time intervals during which a high percentage of the sediments deposited during the interval are destroyed, i.e., that one or more of the assumptions made do not hold. One kind of explanation for a minimum in the mass distribution curve would be deposition of a large proportion of eroded sediments in terrestrial environments, such as intermontane basins, which are subsequently stripped out during regional uplift. Under these conditions, the probability of destruction of the mass deposited during the time interval considered would be much greater than that of pre-existing rocks. Similarly, maxima might represent

Figure 10.9 Estimated actual distribution of sedimentary rock mass compared with the distribution predicted by the linear accumulation model, using a ratio of total deposition to present-day preservation of mass of 5:1.

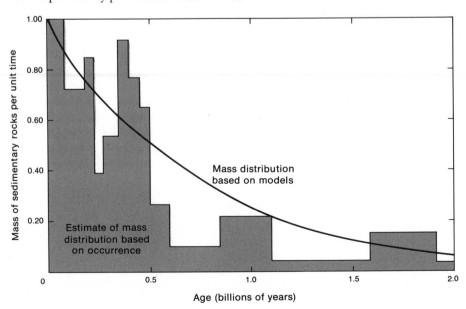

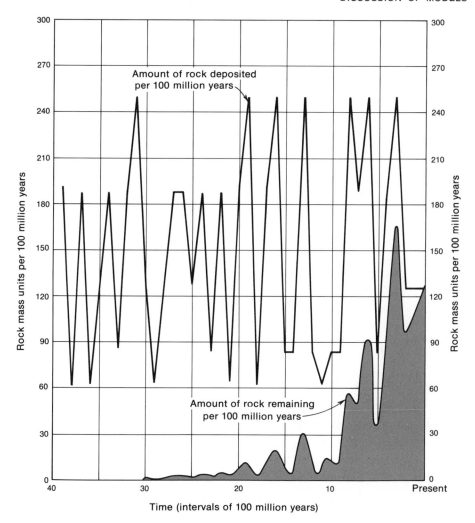

Figure 10.10 Relation of the ages of residual sedimentary rock mass, determined from a constant mass model in which the mass at any time was assumed constant at 1000 units; depositional rates of 250, 187.5, 125, 83.3, and 62.5 rock mass units / 100 million years were varied randomly through time. The rock mass deposited is about 6 times that remaining.

the reverse condition, when a large proportion of the sediments of a given time interval are being dumped into deep basins of one kind or another and hence have a smaller probability of being destroyed than the older sedimentary mass during subsequent erosion.

Maxima and minima also could be related to high or low depositional rates, respectively, although it seems unlikely that the era boundaries would be times of low depositional rates. Figure 10.10 shows an attempt to simulate the actual distribution of sediments by varying the depositional rate through time. The model used is a constant mass model in which the mass deposited per unit time is randomly varied about a mean. Notice that times of high depositional rate correspond with maxima in the resulting mass distribution.

It is of interest that the maximum possibility of destruction seems to correspond with the classic era boundaries and that era boundaries, although they may represent high depositional rates, may also be times of high erosional rates and low accumulation (rapid turnover).

In summary, the distribution of the mass of sedimentary rocks as a function of age is consistent with either constant mass or linear accumulation models. On the other hand, the distribution, even though poorly known, is consistent with the deposition, through time, of 4–6 times as much sedimentary rock mass as now exists. It is obvious that the assumption of a constant depositional rate oversimplifies reality but that an average rate through time of about one fifth the present rate, or about 50×10^{14} g/year, is consistent with the major relations that must be satisfied: that the mass of rocks 600 million years old or younger is roughly of the same size as that of the rest of geologic time, that the mass existing per age unit becomes smaller and smaller with increasing age but that there are some rocks almost as old as the crust itself, and that the mass preserved per unit time diminishes only slightly for the short time span from Cambrian to the present.

SOME IMPLICATIONS OF THE MODELS

Deposition Versus Destruction

We dare not go too far in applying the erosional-depositional models to reality, but they do seem to offer a point of departure for discussion of their applicability. One consequence of both models is that a given mass of rock tends to be destroyed at a rate proportional to its remaining mass; the real variations in mass distribution from the smooth curves we have drawn are surely measures of the amount of range of earth conditions from time to time. Although, as pointed out, the models permit the explanation of a small preserved mass per unit time as either a time of low deposition or as a consequence of selective destruction, most geologic thinking would correlate the major periods of high or low preserved mass with selective destruction rather than with changes in depositional rates.

During the mid-Paleozoic, for example, we find a maximum of preserved mass per unit time, plus an unusually high percentage of carbonate rocks. As pointed out in Chapter 5, chemical deposition tends to be more nearly constant with time than deposition of clastics. Thus we could interpret the mid-Paleozoic (as have many geologists for other reasons) as a time of low-lying lands, little deposition of clastics, and perhaps even a reduced rate of deposition of carbonate rocks. All of these considerations point to a minimum total rate of deposition, yet we find a maximum in the mass per unit time preserved. The implication is that mid-Paleozoic rocks, and especially the limestones, were deposited in places where their chances of destruction have been relatively slight as compared to both older and younger rocks.

Metamorphism

Another aspect of the model studies emphasizes some balances between the kinds of rocks being eroded and the circulation of acid volatiles in the crustal system. In the constant mass model we created an initial sedimentary mass by neutralizing all the acid gases with primordial igneous rocks, and the ensuing discussion did not explicitly consider whether or not the neutralized volatiles stayed neutralized from the first days of mass formation.

It is worth considering the *destruction* of the sedimentary mass in more detail, because we pointed out that destruction could be either by erosion or *metamorphism*. By metamorphic destruction we mean change of chemical composition back to an igneous composition, so that the rock is no longer distinguishable *chemically* from an igneous rock. The following discussion should be followed with reference to Figure 10.11.

If an initial sedimentary mass is half-destroyed in unit time and as a consequence a new mass is created equal to the destroyed mass, what are the differences, if any, in the residual masses, if destruction is entirely by erosion or by both erosion and metamorphism (e.g., granitization)? We shall permit half of the initial mass A to be destroyed equally by metamorphism and by erosion. Then new mass B must have its source half from old mass A and half from weathering of igneous rocks. As shown in Figure 10.11, the acid volatiles necessary to weather the igneous rocks to sediments must come from the metamorphism of part of mass A, which releases the volatile acids that were neutralized originally to form the sediments. During time stage 2, the process is repeated, and the new sedimentary mass C is made up half of deposits derived by erosion of sediments and half of deposits derived by weathering of igneous rocks. The distribution of sediment mass at the end of two time units is illustrated in Figure 10.11. In part II of the diagram, all new sediment comes from erosion of old sediment. The process is self-explanatory, and it can be seen that there is no difference in the sedimentary rock mass distribution at the end of two time units, whether metamorphism has taken place or not.

One consequence of interest is that the models suggest that the total mass of sedimentary rock that has been *deposited* is an extreme upper limit for the amount of igneous rock that can be formed by metamorphism. If our two estimates are approximately correct—that the total mass of sedimentary rocks is about $32,000 \times 10^{20}$ g and that 5 times the existing mass has been deposited—the total deposits are of the order of $160,000 \times 10^{20}$ g, about two thirds the mass of the whole crust. If only one fourth of the total deposits have been destroyed by metamorphism and three fourths by re-erosion, then only one sixth of the crust is of secondary origin from sedimentary rocks. This tentative conclusion applies only to the time interval—perhaps the most recent 2.5 billion years of the earth's history—in which we can be fairly confident we are dealing with "normal" sedimentary processes.

These simple models, of course, should not be overinterpreted, but they do suggest situations that may have occurred during the earth's history that might have predictable consequences. For example, what would happen if metamorphism released acid volatiles to the surface environment but no igneous rocks were available to be changed to sediments? The acid of the volatiles would have to be expended on the existing sedimentary mass, and the result would be a smaller total sedimentary mass made up of minerals such as kaolinite and gibbsite plus more free silica as quartz or chert. This is tantamount to saying that the average sedimentary rock would have to have a composition requiring more acid for its formation; i.e., the Na content of the average shale would be lower. On the other hand, the total free chloride would remain the same, and no change in the chlorinity of the oceans would necessarily occur.

Mass Half-Age

A simple concept that comes from the distribution of sedimentary mass is that of the mass half-age, which owes its origin to the roughly exponential decrease of preserved sedimentary mass with increasing age. Roughly half the total sedimentary rock mass, according to our estimates, is younger than Precambrian and half is older; the mass half-age of the total mass is thus approximately 600 million years. From this concept it is easy to estimate that roughly three fourths of the total mass is younger than 1200 million years, seven eighths less than 1800 million years, and so on. We shall use the term extensively in the following discussion of differential cycling rates.

Differential Cycling Rates

It is apparent that little of the rock deposited in the early days of the earth remains today, and the possibility that the rock types that do remain owe their relative abundances to selective destruction becomes important. There are many factors that could contribute to selective destruction, such as position relative to continental borders, the sequence in which beds of various kinds are deposited in sedimentary basins, and the ease of weathering and erosion of a given rock type. Today we have a striking example of selective removal of a given rock type in areas of Karst (see Chapter 6), where limestones are being removed from the rest of the rocks by solution processes, leaving sinks and caverns behind. Broadly speaking, we can classify the rock types in order of their expected resistance to chemical

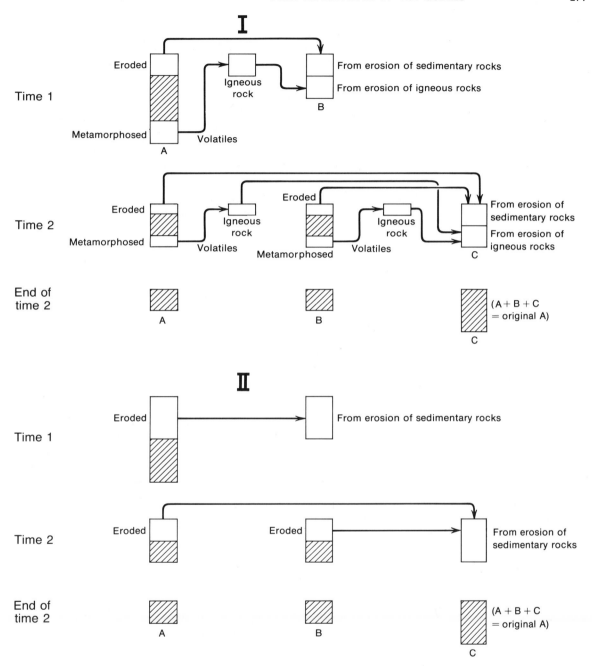

Figure 10.11 Demonstration that the distribution of sedimentary rock mass as a function of age is the same whether all of the deposited mass is derived from pre-existing sediments or some part is derived from weathering of igneous rocks.

weathering and transport. Evaporites are the most easily soluble, with their content of salt and gypsum; calcareous limestones are next; dolomitic limestones third; and shales and sandstones a poor fourth. Volcanogenic sediments are probably like shales because of their high clay mineral content. If we are correct that chemical processes have been more intense during the past than they are today, then relative ease of solution becomes an important factor in the differential solution of these types by erosive processes. Therefore the individual components of the sedimentary mass may have their individual cycling rates, and although the total sedimentary mass has a mass half-age of 600 million years, evaporites may have a much shorter one.

From Ronov's 1964 estimates of the distribution of various rock types as a function of time, the mass half-ages of major sediment types can be found. The mass half-age of evaporites is about 200 million years, that of carbonate rocks about 300 million, and that of shales and sandstones about 600 million. The number for shales also is in accord with our calculations in Chapter 9 which show that the mass of shales in the post-Precambrian is about the same as that in the Precambrian. It would appear, then, that the mass half-age of the total sedimentary rock mass is dominated by shales, sandstones, and volcanogenic sediments, which cycle through time more slowly than evaporites and limestones.

Figure 10.12 shows the result of a calculation for the relative percentages of evaporites, carbonate rocks, and shales and sandstones, based on the assumption that the total evaporite mass deposited is 15 times the existing mass, that that for carbonate rocks is 10 times, and that that for shales and sandstones is 5 times. For comparison we have redrawn the part of Ronov's diagram (Figure 10.3) that pertains chiefly to these rocks. It is evident, if we remove the kinds of perturbations that result from short-time variations in geologic processes, that the predicted distributions are in general agreement with the observed ones.

Selective Destruction Versus Primary Features

It is necessary now to try to present the current situation with respect to the distribution of rock types with time, as well as to look at the chemical trends of individual rock types. To what extent do these distributions reflect initial depositional conditions? To what extent do the ratios of types of a given age represent the ratios that existed at the time of deposition?

It is a consequence of the differential cycling model that the composition of the sediment mass deposited at any given time would always be the same if the model chosen is the constant mass model. The initial mass would be formed quickly, and the proportions of rock types would be immutable—the total mass at any time, regardless of the age distribution of types, would contain the same proportions. As soon as the initial mass began to be eroded and new sediments deposited, the proportions of rock types in the new deposits would reflect the differential rates of cycling of

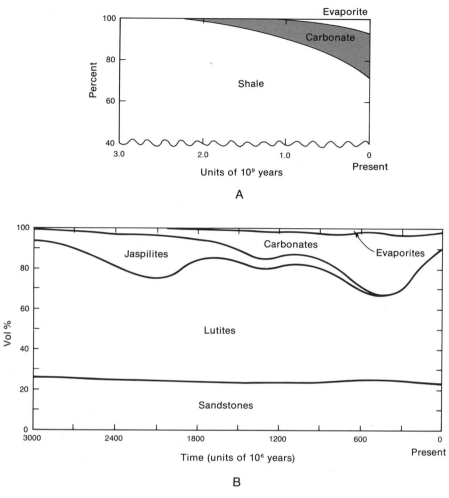

Figure 10.12 In A, the percentages of rock types as a function of age are shown as calculated from a model that assumes shales have been cycled 5 times in the last 3 billion years, limestones have been cycled 10 times, and evaporites 15 times. The overall percentages of evaporites and limestones were assumed at 3 and 10 percent, respectively. In B, Ronov's estimated measured percentages of several rock types are shown for comparison with the model.

the components, so that the deposits would have a different ratio of rock types from those of the original mass. The deposits would be enriched, relative to the original mass, in evaporites and carbonate rocks and impoverished in shales and sandstones. Because of the assumption of equal probability of exposure to erosion of all equal masses, the same source materials of the total mass would always be available to erosion, so that the result of deposition of a constant ratio of rock types, derived from a different ratio of source rock types, results in changes in the proportions of rock types as a function of age.

The present is probably not a good test of this concept; because of the immensity of geologic time, we can probably average the Cenozoic and Mesozoic deposits to derive the proportions of rock types that would be expected as the constant depositional ratios through time.

Such a model leads to a concept of chemical uniformitarianism; that is, the sediments deposited have always been the same in their proportion of types, and the differences we see in sedimentary rocks as a function of age are postdepositional in their origin, engendered by the continuous differential selection of contemporary materials from the total sedimentary mass.

There are a number of secular trends of individual rock types that are in harmony with this general concept. Mesozoic and Cenozoic rocks are diverse in their mineralogies; they contain higher proportions of carbonate rocks and of evaporites than the average for all sedimentary rocks. Mesozoic and Cenozoic lutites contain many minerals that disappear or diminish with time, such as aragonite, magnesian calcites, sepiolite, kaolinite, and abundant mixed-layer clays. The differences between Paleozoic lutites and the younger ones are consistent with a gentle water leach—loss of Ca, CO_2, Mg, Na, and Si (all relative to Al)—and a gain of K only. The number of minerals is decreased, and the number of incompatible phases also is decreased. It looks very much as if the Mg, once perhaps in Paleozoic lutites, has been transferred irregularly through a few hundred million years to the contemporaneous limestones and has made them dolomitic. The change in the relation of true age of lutites to K/Ar age, cited in Chapter 9, is further evidence for postdepositional changes extending over several hundred million years. Also, the decrease in the O^{18}/O^{16} ratios of carbonate sediments and in cherts from Cambrian to Recent has been interpreted as the result of extensive postdepositional exchange with fresh waters, another suggestion of important postdepositional modification.

We can now look at the other side of the coin. Let us interpret some of the same changes as primary deposition and add some examples of deposits that surely represent different initial conditions. The outstanding classic example is the cherty iron deposits. As shown in Figure 10.3, the jaspilites, as Ronov terms them, occur only in the Precambrian. Even though their relative abundance may have been enhanced since their deposition by differential removal of other rocks, they definitely composed at that time a significant part of the sedimentary materials deposited.

The occurrence of the jaspilites may be related to another aspect of the early Precambrian scene that is emphasized by Ronov, a high proportion of volcanogenic sediments and of graywackes. There is no particular reason these two rock types would have relatively slow cycling rates, which would be necessary to increase their percentages at the expense of lutites, for example, and thus their high early percentages are undoubtedly primary. The record of the early Precambrian is so fragmental that most of what has been said about cycling and rock types may break down entirely when we get to ages greater than 3 billion years. At any rate, the early high percentages of mafic lavas and graywackes may indicate that the average igneous source rock for sediments 3 billion years ago was considerably higher in Ca, Fe, and Mg than the one we chose for today, and the sedimentary mass would be expected to be enriched in carbonate rocks, chert, and iron compounds, corresponding to the alteration of volcanogenic rocks discussed in Chapter 9. Again, however, we would expect to have most of the carbonate rocks, which might have been an important fraction of the early deposits, to have been cycled forward in time, leaving the volcanogenic sediments behind.

We must remember that the large differences between the simple cycling models and the actual distribution of rocks, as shown in Figures 10.9 and 10.12, represent in part primary sedimentary mass differences. As pointed out, the high percentages of limestones in the middle Paleozoic probably mean a slowing at that time of the cycling rate of the clastic materials and thus an original depositional difference in rock-type ratios of that time.

The "noise" within Eras, that is, the large Periodic fluctuations in ratios and compositions of sedimentary types, also must be worldwide primary sedimentary features. They are modified, to be sure, by postdepositional differential cycling, but the initial variation of conditions still controls what we see. It is in the Precambrian rocks that selective loss of rock types has had its most important effects. If our estimates of rates of sedimentation are roughly valid, about 90 percent of the Precambrian once deposited is gone.

SUMMARY

The distribution of mass of sedimentary rocks has been examined. The center of mass of the post-Precambrian rocks is approximately at an age of 250 million years; the center of mass of all sedimentary rocks lies between 500 and 600 million years. The distribution of mass for the post-Precambrian shows a minimum at the beginning of the Paleozoic, a maximum in mid-Paleozoic, a minimum during Permian, and a high today. Our knowledge of Precambrian mass relations is not sufficient to permit us to make estimates in detail of its age distribution, except to note that the mass remaining today per unit of age diminishes irregularly and roughly exponentially with increasing age.

Simple deposition-destruction models for the history of sedimentary rocks were developed. It was shown that there is little basis for choosing between a model in which a sedimentary mass equal to the existing mass was formed early in geologic time or one in which the sedimentary mass, as well as the oceanic mass, increased regularly with time. The present mass distribution of sediments, on either model, is estimated to be about one fifth of the total sedimentary mass deposited through time. If so, the average rate of deposition through time must be lower than that today, perhaps one fifth to one third as rapid.

The cycling of sediments through time should lead to differences in the proportions of rock types of a given age as observed today and also those proportions at the time of deposition. In a broad and general sense, we proposed the concept of chemical Uniformitarianism, that the sediments deposited at any given time in the earth's history are the same as those deposited at any other time. This concept predicts approximately the kinds of long-term, postdepositional, selective changes expected in the sedimentary rock mass as a function of the ages of the rocks and helps to distinguish primary from secondary sedimentary features.

REFERENCES

Gregor, C. B., 1968, The rate of denudation in Post-Algonkian time: *Koninkl. Ned. Akad. Wetenschap. Proc. 71*, 22.

Holmes, A., 1965, *Principles of Physical Geology:* Prentice-Hall, Inc., Englewood Cliffs, N.J.

Kay, M., 1955, Sediments and subsidence through time: In Crust of the earth, A. Poldervaart, ed., *Geol. Soc. Am. Spec. Paper, 62*, 665–684.

Ronov, A. B., 1959, On the post-Precambrian geochemical history of the atmosphere and hydrosphere: *Geochemistry, 5*, 493–506.

Ronov, A. B., 1964, Common tendencies in the chemical evolution of the earth's crust, ocean, and atmosphere: *Geochemistry, 8*, 715–743.

11 | The Oceans

The broad features of the oceans, such as their total area, total mass, average depth, and chemical composition, have already been covered. Now we want to discuss the present ocean in terms of variations in its composition from place to place, the rate at which it mixes, the rate at which its water and dissolved solids are being renewed, and current views about its nature as a chemical system. Then we shall try to reconstruct its history.

CONSTANCY OF COMPOSITION

The investigations of Marcet (1819), Forchhammer (1865), and Dittmar (1884) established the empirical concept that the concentration *ratios* of the major constituents of sea water do not vary measurably laterally or vertically in the present oceans except in regions of major river runoff or in semienclosed basins. Most of the analytical work done on sea water since the late 1800's has confirmed this proposition for the major constituents Cl, Na, Mg, SO_4, K, Ca, Br, and Sr, and these constituents are termed conservative.

The total dissolved solids, usually expressed in parts per thousand (parts per mil,%$_0$), have an average value of about 35, ranging downward to about 30 near the land where major streams discharge and upward to about 40 in shallow waters of arid areas, where evaporation is faster than mixing. The vertical variations in a typical oceanic profile are 1 to 2%$_0$, with the deep cold waters being the less saline.

Minor constituents are more variable than the major ones, especially those such as phosphate, nitrate, and silica that are nutrients for plankton, and reflect the metabolic activities of the organisms that are abundant only in the well-lighted, upper few tens or hundreds of meters. Also, pH and dissolved gases, such as CO_2 and oxygen, show significant vertical variation. Figure 11.1 is a typical low-latitude vertical profile of the ocean. Temperature falls rapidly downward from the surface, reaching near-freezing values characteristic of most of the water mass at about 1000 m. pH also drops, from surface values slightly above 8 to deep-water values around 7.6 to 7.8. Oxygen diminishes downward to a minimum value at about the depth where water temperature becomes constant and then increases toward saturation values in the deeps. Phosphate, nitrate, and silica rise from near-zero concentrations in warm surface waters, where they are strenuously competed for by organisms, and become nearly constant as the temperature gradient disappears. There are minor differences in the absolute concentrations of nitrate and phosphate in the various oceans, but the gradients are alike. The average concentration of silica varies by a factor of about 3 from Atlantic to Pacific waters.

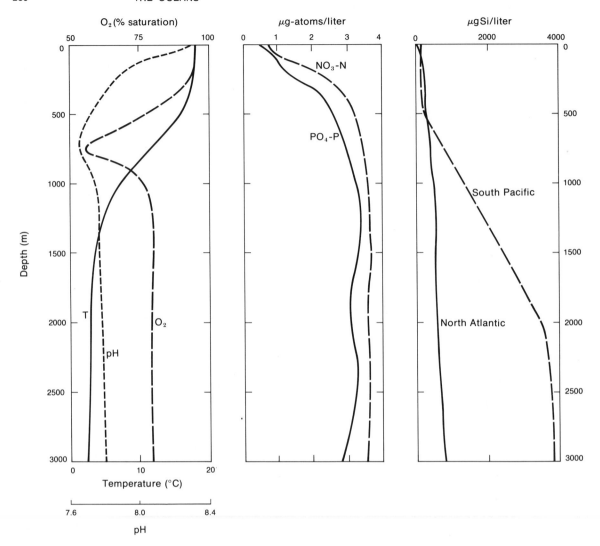

Figure 11.1 Generalized vertical profiles of some selected properties of sea water.

The gross circulation pattern of ocean water is from poles to equator and back; sea water is cooled in high latitudes at the surface, sinks, and slides downward along the basin floor. In the low latitudes it is heated at the surface and is displaced by the deeper waters. This overall convective pattern takes place through broad currents such as the Gulf Stream, which moves a great mass of warm surface water from the Gulf of Mexico across the Atlantic to Scandinavia, and the Kuroshio current in the Pacific, which flows along the eastern shore of Japan. Similar "rivers of the sea" are found in the southern oceans.

The cold deep waters are rich in nutrients and dissolved CO_2. Where they rise to the surface, biological productivity is high. The cold waters, with their high CO_2 and low pH, are undersaturated with calcium carbonate; as they are warmed, they lose CO_2 and become slightly supersaturated. Completely inorganic precipitation of calcium carbonate is kinetically difficult. If it were not for the carbonate-secreting organisms, a high degree of supersaturation might be achieved, but organisms seem to be capable of removing calcium at a rate sufficient to keep the waters close to saturation.

The oldest water yet found in the oceans has a carbon-14 age of a few thousand years, suggesting that "stirring" of the oceans by convection takes place in a thousands-of-years time scale. Communication between the Atlantic, Pacific, and Indian oceans today is good enough to keep them fairly uniform, although restrictions in circulation in the past must be considered as a possibility that might have caused development of several separate chemical systems.

On the average, all the water in the oceans is renewed every 40,000 years; that is, total present stream flow would equal oceanic volume in that span of time. If an experiment were to be performed to simulate the continent-ocean system throughout geologic time, the water in the container representing the oceans would have to be stirred a million times or so, and a volume of water would have to be evaporated from the container equal to 100,000 times the initial volume. The oceans, from the perspective of geologic time, have been an active chemical system.

CHEMICAL VIEWS OF THE OCEANS

The idea that the salts in the oceans, like the scale that accumulates in a constantly refilled boiling kettle, come from the accumulation of the tiny increments brought in by streams and are left behind when ocean water evaporates goes back to early historic times. The concept that the oceans are just an accumulation of the stream burden persisted for a long time. The mass of the oceans is so great it hardly seems finite, and as put so well by Russell and Jonge (1928), "there is enough salt in the oceans to yield fourteen and a half times the bulk of the entire continent of Europe above high water mark."

As the accepted age of the earth increased, the calculated total amounts of materials brought to the oceans by streams, small as they seem in a yearly basis, began to overwhelm even the ocean basins, when more than 3 billion years of continental erosion had to be considered. The present rate of salt addition by streams, as shown in Chapter 4, would require less than 100 million years to account for the mass of ocean salt. When the times required to accumulate the individual chemical species in sea water are calculated, these storage times are found to be but a few million years. Silica would require only 20,000 years to double its present oceanic amount.

The concept of the ocean then began to change from that of an accumulator to that of a steady-state system in which the amount of material coming in is continuously equal, or nearly equal, to the amount going out. This system permits influx to vary with time, but it would be matched by a nearly simultaneous and equal variation of efflux. Conceivably, in such a system, the rates of addition and removal of materials, although equal, could change and result in sea water of a different composition. Nearly everyone today agrees that the ocean is at least an *approximate* steady-state system, although there are residual arguments about the time that might be required for a change of addition rate to be equaled by a corresponding change in subtraction rate.

In 1961, Sillén published a paper on the physical chemistry of sea water in which he pointed out that the chemical composition of the oceans has a striking resemblance to the composition of a theoretical solution that has been brought to chemical equilibrium with a number of the minerals that are found in the sea. He was careful to distinguish between the real ocean and his theoretical solution, but the article has caused a great stir, because it implies that not only is the oceanic system steady state but that it is further restrained; even if the rate of addition of feed material is changed, or the proportions of the minerals in the feed, the chemical composition of sea water will be invariant. Sillén did not take quite so strong a stand; he pointed out that in his model the *ratios* of the important chemical species would be fixed, but their absolute amounts could increase and decrease with the concentration of NaCl.

Equilibrium in a static system cannot be applied to the oceanic system, but it is possible that precipitation of solids takes place at a rate sufficient to prevent significant supersaturation of the solution. Oversaturation must be achieved, but the departure from saturation need not be great. The relations for calcium carbonate are exactly in accord with this possibility, its influx is equal to its efflux, within the limits of measurement, and sea water is slightly oversaturated with calcite at the surface and a little undersaturated at depth. The important unknown, then, is the manner in which other dissolved materials brought in by streams are removed from the sea and, if the minerals form and can be identified, whether they can hold their components close to saturation values.

Chemical Mass Balance

In any current model of the oceans, the dissolved species transported to the oceans by streams eventually must precipitate from the oceans or be cycled back to the continents through the atmosphere. The authors (Mackenzie and Garrels, 1966) attempted to make a mass balance for stream water in which we assigned the various dissolved constituents of streams to minerals found in marine deposits. Several of our proposed species are still not documented as forming in sea water, but the estimate of the distribution of elements into minerals is the most complete we have. The procedure was to calculate the amount of material that streams could bring to the oceans in 100 million years by projecting today's rate and then to do a normative type of calculation to get rid of the constituents one by one.

To balance all the dissolved species we considered it necessary to remove part of the Na, Mg, K, and SiO_2 by reaction with solid "degraded" aluminosilicates of the suspended load of streams to make clay minerals typical of those found in lutites. The minerals chosen as the most likely candidates for removal of materials consisted of 28 molar percent evaporite minerals (NaCl and $CaSO_4$), 26 percent carbonate rock minerals, and 41 percent clay minerals: kaolinite, montmorillonite, illite, and chlorite. Small amounts of pyrite and silica also were formed. The point was emphasized that a constant composition ocean requires "reverse weathering." If the CO_2 consumed by the weathering process is not returned to the atmosphere by depositional processes, there will be a decrease in the atmospheric supply of CO_2, and sea water, because of the accumulation of bicarbonate ion, will tend to become alkaline. Because the weathering of calcium carbonate is known to be balanced by an equal rate of calcium carbonate deposition in the oceans, any net drain on atmospheric CO_2 would have to come from failure of reversal of the reactions of silicate weathering. About 20 percent of the HCO_3^- in streams is derived from the weathering of silicates (Chapter 5) according to the general reaction

$$\text{Siliceous silicate} + CO_2 + H_2O = \text{less siliceous silicate} + \text{cations} + \text{dissolved}$$
$$\text{silica} + HCO_3^-.$$

Field tests of the reversible nature of silicate weathering today are being made either to demonstrate the validity of the reversal of the reaction above by finding an appropriate amount of silicate regeneration taking place in the oceans today or to show that the dissolved silica brought to the oceans is precipitated as the tests of siliceous organisms, hence in the molecular form, thus proving indirectly that silicate synthesis cannot take place. Attempts have been made to apply both tests; the results to date are inconclusive. No clear-cut examples of regenerative reactions of the magnitude required have been discovered, and a controversy rages over the rate at which siliceous organisms are accumulating.

Silicate Mineral Reactivity

Another approach to the question of the controls for the removal of the dissolved materials added to the oceans is laboratory investigation of the reactions between clay minerals and aqueous solutions containing cations and silica. If it can be demonstrated that clay minerals react rapidly, if the new phases formed are typical of marine deposits, and if the resultant solution has a cation and silica content like that of sea water, then the feasibility of the maintenance of a steady-state oceanic system with a narrowly controlled composition will have been shown.

Figure 11.2 shows the results of adding the aluminosilicates, kaolinite, montmorillonite, and glauconite to sea water that was either deficient in silica or spiked to 25 ppm of dissolved silica (about 4 times the average

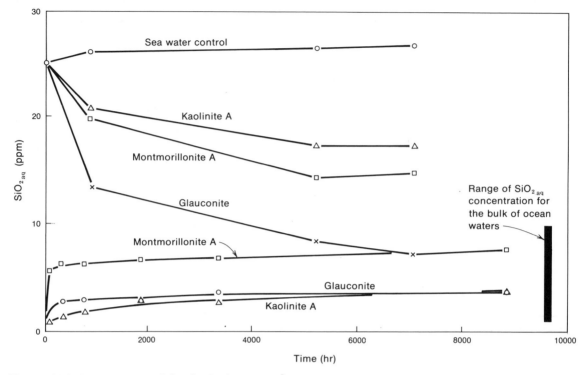

Figure 11.2 Concentration of dissolved silica as a function of time for sea water-clay mineral suspensions. The clays were added to sea water initially containing 0.03 and 25 ppm silica. In silica-deficient sea waters, the clays added silica, whereas for those solutions containing an excess of dissolved silica, silica was abstracted by the addition of clay minerals. The constant silica concentration values reached are similar to those found in ocean water (after Mackenzie et al., 1967).

concentration in the oceans). In both instances constant values of silica concentration were attained in about 6000 hr (8 months). For silica-deficient experiments, constant values of 4–6 ppm of silica were reached; for the spiked solutions the constant values were higher and more variable, ranging between 6 and 18 ppm. These concentrations are within or near the range of silica concentrations in ocean water. Presumably release of silica from the minerals into silica-deficient sea water was accompanied by formation of a second compound, less siliceous than the parent clay, whereas takeup of silica from the spiked solutions took place by adsorption of molecular silica or by the formation of a new phase more siliceous than the one that was reacted. The experiments show that the clay minerals of the detrital load of streams, when they enter the oceans, should be able to influence the dissolved silica content of the oceans. Unfortunately, the silica-uptake experiments did not demonstrate that a new silica phase is formed in accord with a "reverse weathering" reaction; if silica removal were by surface adsorption as molecular SiO_2, the effect on the HCO_3^- balance would be the same as that of SiO_2 removal as diatom tests.

Phase Equilibria

If sea water maintains near-equilibrium with the detrital silicates that have been falling through it for billions of years, its composition should correspond to that predicted by thermodynamic calculations of the composition of the aqueous solution in equilibrium with those phases. It is impossible, of course, for ocean water to have one equilibrium composition, because its temperature ranges from 0 to 35°C and its pressure from 1 to 1000 atm. Moreover, it cannot be in equilibrium simultaneously with all the minerals being delivered. Gibbsite and feldspar cannot coexist at equilibrium; i.e., if placed together in a solution they would react until one disappeared and the other was left, along with an intermediate phase (e.g., kaolinite) in the final solution.

Figures 11.3 and 11.4 show the phases that would be in equilibrium at 1 atmosphere total pressure with present-day sea water at 25 and 0°C, respectively. The stability fields of the phases are shown as functions of pH and dissolved SiO_2. The shaded areas denote the range of sea water compositions. Gibbsite, kaolinite, montmorillonite, K-feldspar, and illite all can be stable species within the commonly observed range of sea water composition. At a P_{CO_2} near that of today's atmosphere, calcite, too, is a stable phase. This relation does not prove that these minerals are the *cause* of sea water composition, but it shows, in accord with Sillén's article (1961), that if a mixture of these minerals were placed in a NaCl solution with the same Cl as sea water, maintained in equilibrium with the atmosphere, and allowed to come to equilibrium with the final solution, no matter which of the various compatible mineral assemblages remained, it would fall within the typical range of composition of sea water.

The chemical equilibrium relations predict that present-day sea water has a chemical composition compatible with that expected for a solution continuously reacting with the great variety of minerals being added to it. Of these minerals, it is nearly in equilibrium with the typically detrital clay minerals and out of equilibrium with most of the primary igneous rock minerals such as plagioclase feldspar, pyroxenes, and olivines. This situation may be transient and but an instant in a long-term chemical trend. On the other hand, it is compatible with continuous control of sea water composition by the synthesis and dissolution of a few silicate phases, resulting in sea water which never deviates far from saturation.

CHEMICAL HISTORY OF THE OCEANS

Evidence For and Against Constancy

Many, if not most, geologists today would contend that the chemical composition of ocean water has not varied markedly for an interval of time greater than a few million years, at least as far back as Cambrian time. The paleontological evidence since the Cambrian strongly favors, but does not prove, that during the past 600 million years ocean water has had a similar composition to that of today. Families, and even some species, of organisms have persisted from the early Paleozoic. The paleontological evidence must be used with caution, however, since some typically marine organisms, even on the species level, are able to adjust to large salinity

Figure 11.3 Activity diagram illustrating phase relations in an idealized nine-component ocean system at 25°C, 1 atm, and unit activity of H_2O. The activities of Na^+, K^+, Ca^{2+}, and Mg^{2+} are fixed at $10^{-0.50}$, $10^{-2.21}$, $10^{-2.64}$, and $10^{-1.79}$, respectively. The $\log f_{CO_2}$ scale denotes the fugacity of CO_2 for calcite saturation. The contours within the mixed-layer, illite-montmorillonite field show the percent of the illite component in the solid solution in equilibrium with sea water. The shaded area is the approximate range of sea water composition at this temperature (after Helgeson and Mackenzie, 1970).

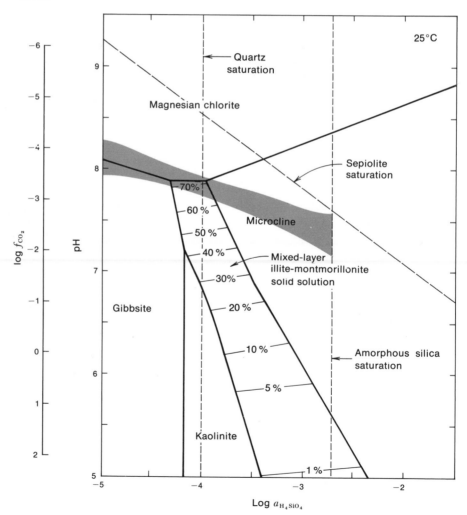

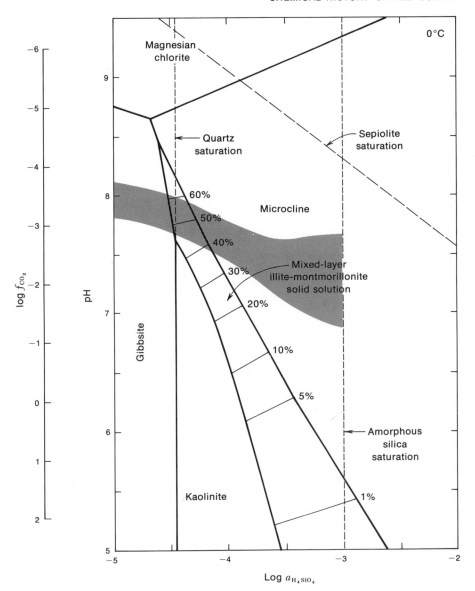

Figure 11.4 Activity diagram illustrating phase relations in an idealized nine-compo-nent ocean system at 0°C, 1 atm, and unit activity of H_2O. The activities of Na^+, K^+, Ca^{2+}, and Mg^{2+} are fixed at $10^{-0.50}$, $10^{-2.21}$, $10^{-2 \; 64}$, and $10^{-1.78}$, respectively. The $\log f_{CO_2}$ scale denotes the fugacity of CO_2 for calcite saturation. The contours within the mixed-layer, illite-montmorillonite field show the percent of the illite component in the solution in equilibrium with sea water. The shaded area is the approximate range of sea water composition at this temperature (after Helgeson and Mackenzie, 1970).

changes. For example, the shark *Carcharius* is found in the fresh water bodies of Lake Nicaragua and the Zambezi River; the sawfish is found in Lake Nicaragua. It has been suggested that the extinction of a number of genera, and even entire families and orders, of animals at the end of the

Permian was due to a reduction in the salinity of the upper circulating portion of the ocean and the development of a thin layer of very saline water at the bottom of the Permian seas (Fischer, 1966).

The geochemical evidence for constancy of ocean water composition also is equivocal. It has been shown that fossil brachiopods as old as Mississippian age show a chemical similarity to recent shells (Lowenstam, 1961). The O^{18}/O^{16} ratios and the $SrCO_3$ and $MgCO_3$ contents of fossil brachiopods are consistent with a model of sea water composition in which the O^{18}/O^{16} ratios, the Sr and Mg concentrations, and the Sr/Ca and Mg/Ca ratios of ocean water have remained nearly constant since at least Mississippian time.

Another piece of geochemical evidence supporting an ocean of nearly constant composition during geologic time is provided by the boron contents of illites in Precambrian rocks as compared with post-Precambrian sediments (Figure 11.5). It can be seen from Figure 11.5 that the boron content of Precambrian and Phanerozoic rocks is nearly the same, and, more significantly, there appears to be no systematic change of the boron concentration with geologic age. Because the amount of boron in illite is related to the boron concentration of the sea water in which the boron was first incorporated in the clay, the relative constancy of the boron contents of illites as a function of geologic age suggests that the boron concentration of ocean water has remained relatively constant over the past 2.5–3 billion years.

On the other hand, change in composition is implied by sulfur-isotope ratio studies. It was shown in Chapter 3, Figure 3.10, that there have been significant variations in the sulfur-isotope ratios of evaporites as a function of geologic age, although these variations fluctuate about a mean $\delta S^{34}/S^{32}$ of about 20 per mil, which is the average for present-day sea water. The minimum in $\delta S^{34}/S^{32}$ in the late Paleozoic requires a large transfer of light sulfur from shales to the ocean reservoir. Holser and Kaplan (1966) suggest that to account for the Permian $\delta S^{34}/S^{32}$ minimum of $+10$ per mil necessitates an enrichment in the sulfate in the Permian sea of about 45 percent; that is, the concentration of sulfate in the Permian sea was 3850 ppm instead of 2650. Notice, however, that this is only an increase in the total salinity of ocean water of about 1.2 parts per thousand, or 3 percent. There are some problems involved with the interpretation of the sulfur-isotope data, particularly since more than all the oxygen in the atmosphere would have to be removed to account for the excess oxidized sulfur derived from the weathering of pyrite in shales in the Permian. Nevertheless, the sulfur-isotope data are inconsistent with a constant sulfate content and raise the possibility of marked variation in other environmental parameters.

A Model for Ocean History

Earliest Oceans

Any reconstruction of earliest earth events is highly speculative, but it is possible to weave together some of the more likely interpretations of pri-

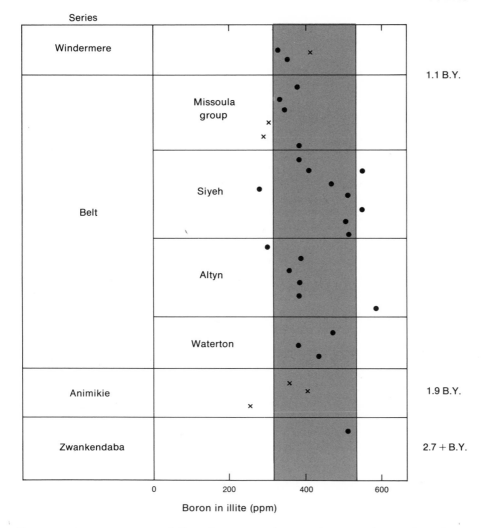

Figure 11.5 Boron contents of illites from Precambrian carbonate rocks (black circles) and shales (crosses). The shaded area encloses the approximate range of boron contents of Phanerozoic rocks (after Reynolds, 1965).

mordial and more recent conditions that lead to a coherent picture of oceanic evolution. Most investigators are agreed upon the loss of a primitive atmosphere (Chapter 1) and its replacement by one derived from loss of volatile substances from the earth's interior. If so, most of the loss should have occurred during the gross differentiation of the earth into core, mantle, and crust.

Higgins (1968) presents a mathematical model that predicts that nearly 80 percent of the crust, as we know it, was formed within 200 million years of the solidification of the mantle. If internal earth temperatures were high, the whole earth might well have approached or attained a molten state, with a thin but continuous siliceous crust. As the crust solidified and igneous minerals crystallized, volatile gases would be released to form an atmosphere. The atmosphere would contain H_2O, later to become the oceans: carbon gases, such as CO_2, CH_4, and CO; sulfur gases, probably mostly H_2S; and the halogen compounds HCl, HF, and HBr. Hydrogen and He would have been released from the earth's interior, but they have such low atomic weights that the earth's gravitational attraction could not hold them, and they would escape into space. Nitrogen may have been important and other gases would have been present in small amounts.

It is difficult even to guess at the sequence of events as the crust cooled from 600°C or so, at which temperature almost all these compounds, including H_2O, would have been in the atmosphere, to temperatures below boiling. Below 100°C, we can be fairly sure that all the H_2O would have condensed and that the acid gases would have reacted almost completely with the primordial crust, but there are at least two pathways by which these initial steps could have been accomplished.

One possibility is that the 600°C atmosphere would have had some 300 atm of H_2O pressure, to judge from the mass of H_2O now locked away in the oceans, 45 atm of CO_2 from the CO_2 in carbonate rocks and organic matter and 10 atm of HCl as estimated from the Cl in the oceans. When it cooled to the critical temperature of water, 373°C, H_2O would begin to condense. Most of the water would be liquid at a temperature of 200°C, where the equilibrium vapor pressure of water is only 15 atm. At this stage, most of the HCl would be dissolved in the ocean water, giving a concentration of about 1 mole/liter in solution; most of the CO_2 would be in the atmosphere, with about 0.5 mole/liter in solution. Total atmospheric pressure would be between 30 and 50 atm, with CO_2 pressure/H_2O pressure/HCl pressure in the approximate ratio of 30:15:1.

The reactivity of this ocean-atmosphere system would be awesome. Gaseous HCl is almost impossible to contain in modern experimental apparatus utilizing the most refractory materials known; an ocean at 200° C containing 1 mole/liter of dissolved HCl and ½ mole/liter of CO_2 would react vigorously with the newly formed crust. The crust would be strongly and continuously attacked, and the new ocean would dissolve out silica and cations, leaving at first only a leached out residue of alumina and silica.

Because we cannot model the system well enough to determine the degree of interaction between crust and atmosphere-ocean, we do not know whether a continuous reaction would ensue, during which the acid gases react almost completely with the crust at an elevated temperature, and then stay near equilibrium during cooling or whether cooling itself would consist of a long interval of continued reaction after less than 100°C temperatures had been reached. The version of the events we have been describing assumes that reaction is slow relative to cooling.

If cooling were so slow that the crust continuously "titrated" the acid volatiles, the possibility exists that at intermediate temperatures, such as about 400°C, most of the H_2O would be removed from the gas by hydration reactions with pyroxenes and olivines, and water might not condense until some unknown temperature was reached. If H_2O were locked up in mineral phases, the earth might have had, like Venus today, an atmosphere rich in CO_2 and no ocean. The Venus surface temperature is of the order of 350°C.

No matter which pathway was followed, it would not take a long time geologically for the acid gases to be used up after a nearly constant temperature below 100°C was reached. The presence of bacterial and possibly algal fossils in rocks more than 3 billion years old assures us that the crust had cooled to 100°C or lower at that time. Presumably 1 billion or so years had passed since the crust had formed, so we can be confident that the acid volatile gases had been neutralized. If their relative percentages were the same as they are today, the free chloride must have gone into evaporites or into the oceans, creating an ocean with a salinity comparable to that of today.

We must, however, take cognizance of a number of differences that may have affected the early period of the rock record. Figure 11.6 shows a calculated estimate of the proportions of gases in equilibrium with graphite as a function of temperature. Because it is likely that some of the carbon in the early earth was present as graphite, the calculated composition is a fair

Figure 11.6 Calculated estimate of the composition of gases emitted from a cooling earth. The gas phase is assumed to be in equilibrium with graphite and the hydrogen to oxygen ratio is maintained at 2:1. Total pressure is 1000 bars (after French, 1966).

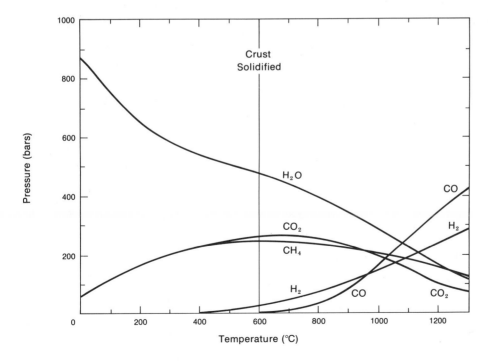

guess of the composition of gases emitted during cooling of the earth. Note that for temperatures below 1000° C, H_2O is the dominant component, and CO_2 and CH_4 are present in approximately equal amounts. As can be seen in the figure, oxygen would not have been a constituent of the atmosphere emitted from the cooling crust, but some would have been created from the photodissociation of water owing to absorption of ultraviolet light by the atmosphere:

$$2H_2O \overset{h\nu}{\rightarrow} 2H_{2(g)} + O_{2(g)},$$

where $h\nu$ represents the photon of ultraviolet light. Hydrogen would escape into space, and the oxygen would disappear by reaction with the reduced gases, by reactions such as

$$H_2S + O_2 = H_2 + SO_2.$$

The estimated rate of production of oxygen by photodissociation during geologic time is about 18×10^{10} moles/year (Dole, 1965). Thus it would take only about 400 million years to oxidize an amount of initial reduced sulfur gases to an amount of sulfate equal to that now found in sea water. The complexity of the primordial gaseous mixture makes it difficult to know if Dole's estimate of the rate of oxygen production would apply some 4 billion years ago, but oxygen production by photodissociation at least could have given the early reduced atmosphere a start toward present-day conditions.

Early Atmosphere and Life

Two of the major events in early earth history were the development of life and the evolution of photosynthetic organisms. Many hypotheses have been suggested to account for the first living forms; almost all of them accept an environment containing H_2O, CO_2, CO, H_2S, CH_4, and nitrogen, perhaps as N_2 or perhaps as HCN, or other compounds. Experiments have been performed to synthesize amino acids, the building blocks of proteins, from these raw materials under conditions reasonably attainable at the earth's surface. A variety of methods have produced amino acids. The first successful experiment was by S. L. Miller in 1953, who used a spark discharge through a typical gas mixture. The spark could be the lightning of the early days. Amino acids also have been synthesized by simple heating; the temperatures required are not impossible in near-surface environments.

One of the most interesting recent syntheses is by Abelson (1966), who polymerized cyanide compounds in alkaline solutions under the influence of high-intensity ultraviolet light to produce the amino acid serine. The reaction is of the type

$$5HCN + 7H_2O = C_3H_7O_3N + 4NH_3 + 2CO_2.$$
$$\text{Serine}$$

This method has some attractive aspects, particularly the one of ultraviolet light. One of the most difficult aspects of simulating the early earth environment is the estimation of the kinds of radiation to which incipient organisms would be exposed. Today, for example, oxygen absorbs a great deal of ultraviolet light from the sun by the formation of ozone in the upper atmosphere. In the absence of oxygen, the ultraviolet intensity would reach lethal levels. Thus the use of ultraviolet light to synthesize amino acids avoids a difficulty, although further organization of organic materials into living forms would require protection from direct radiation.

Records of organisms extend to the oldest surface rocks. Baarghorn and Schopf (1966) have described bacteria from 3.1-billion-year-old rocks of South Africa, where they occur preserved in chert. Engel et al. (1968) report the occurrence of spheroidal and cup-shaped alga-like bodies and other biological microstructures in 3.2-billion-year-old Onverwacht Series of South Africa. Life developed rapidly indeed, reaching the organism level within 1 billion years or so. Oxygen began to be an important component of the atmosphere with the rise of photosynthetic organisms. Estimates of the time of origin of these organisms range back as far as 3 billion years. With the appearance of photosynthetic organisms, the chief events of evolution had been accomplished. A photosynthetic organism represents a degree of evolution from which the development of man is but a small step compared to what had gone on before.

Only during the last 600 million years have we had a record of organisms with hard parts—shells and bones. Why, with a record of life extending 5 times as long, organisms discovered how to make shells so late in time is a continuing mystery. Perhaps, as Fischer (1965) suggests, early marine animals developed in "oxygen oases," that is near photosynthesizing marine plant communities. When the oxygen in the atmosphere became sufficiently great (late Precambrian), these small naked animal forms could leave their host plants and spread throughout the ocean. They then developed skeletons to protect themselves from radiation. On the other hand, our knowledge of soft-bodied forms of the Precambrian has increased a great deal recently. For a long time the only undisputed fossils of old Precambrian rocks were laminated and bulbous structures formed by blue-green algae. Alga-like bodies and microfossils of unknown biological affinities have now been found in the oldest rocks known. Perhaps the initial record of shelled forms, and even unicellular life, is lost to observation because of destruction of the rock record that contained it. Solution during differential cycling, metamorphism, and incorporation of vast areas of the sea floor into the crust-mantle system during sea floor spreading (Chapter 12) could completely obliterate parts of the sedimentary and biological record.

Oceans: Approximately 1.5–3.5 Billion Years Ago

We do have some evidence that can be interpreted as meaning that as recently as 2 billion years ago, oxygen was significantly lower than today. This evidence suggests that in the period between deposition of the first sediments of which we have a record (about 3.5 billion years ago) until perhaps 1.5 billion years ago, there were continuous chemical trends both in the sedimentary rocks formed and, more subtly, in oceanic composition. As pointed out so succinctly by Ronov in his excellent interpretation of the history of sea water (1968), from which we have drawn heavily, and by Engel and Engel (1964), the source rocks of sediments 3 billion years or more ago were more basaltic than later ones and the sediment-forming processes consisted of the alteration of these rocks by an oxygen-deficient atmosphere. We can write some chemical reactions that suggest the dominating mineral-ocean interactions at that time:

$$14H_2O + 6NaAlSi_3O_8 + 3FeCO_3 + 15MgSiO_3 + 3CaCl_2 = 22SiO_2$$
$$\text{Albite} \qquad \text{Siderite} \quad \text{Enstatite} \qquad\qquad \text{Silica}$$

$$+6NaCl + 3Mg_5Al_2Si_3O_{10}(OH)_8 + 3CaCO_3 + Fe_3^{2+}Si_2O_5(OH)_4,$$
$$\text{Chlorite} \qquad\qquad \text{Calcite} \qquad \text{Greenalite}$$

$$5MgSiO_3 + CaAl_2Si_2O_8 + CO_2 = Mg_5Al_2Si_3O_{10}(OH)_8 + CaCO_3 + 4SiO_2,$$
$$\text{Enstatite} \quad \text{Anorthite} \qquad\qquad \text{Chlorite} \qquad\qquad \text{Silica}$$

and

$$Fe_3^{2+}Si_2O_5(OH)_4 + 3CO_2 = 3FeCO_3 + 2SiO_2 + 2H_2O.$$
$$\text{Greenalite} \qquad\qquad \text{Siderite} \quad \text{Silica}$$

The chief difference between these reactions and those involved in the production of modern sedimentary minerals is the emphasis on an important role for ferrous iron. In today's oceans dissolved iron is extremely low in its concentration because of the insolubility of oxidized iron oxide. During the stage when the atmosphere was oxygen-deficient, anaerobic conditions in streams and in the oceans were probably widespread. A reduction in atmospheric oxygen today from 21 percent of the atmospheric gases to 1 or 2 percent would have drastic effects in increasing the percentage of the marine depositional environments that are oxygen-free. Depletion of oxygen permits ferrous iron to be an important constituent in the mineral species resulting from the alteration of volcanogenic rocks. The three reactions written above include mineral phases that are more important in sedimentary rocks 2 billion years old or more than they are today. They are the iron carbonate siderite and the iron silicate greenalite. Siderite and greenalite, in close association with chert and pyrite, are characteristic of many of the iron formations of the middle Precambrian. The chert was initially amorphous silica; the calculated CO_2 pressure for equilibrium at 25°C among siderite, greenalite, and amorphous silica is about $10^{-2.5}$ atm—roughly 10 times what it is today.

The chemical system that seems to fit the older Precambrian, then, is one in which basaltic and andesitic source rocks were more important in contributing to sediment formation than they are today; anaerobic depositional conditions were more common, and the internal CO_2 pressure of the depositional systems, and hence perhaps that of the atmosphere, was 10 times that of today. If so, oceanic pH would have been a little lower, near neutral, and dissolved calcium somewhat greater than today. The accumulation of silica in the water would have reached saturation with amorphous silica (ca. 120 ppm) in many places.

The early oceans, then, can be thought of as the solution resulting from an acid leach of basaltic and andesitic rocks. The rocks were leached in situ, for the most part, by submarine processes, because the neutralization of the volatile acid gases was not restricted to land areas, as it is today. As time went on, the waters of this ocean were modified by the continuous cycling of continental debris through time.

A crude simulation of the earliest reactions can be imagined by emptying the Pacific basin, throwing in great masses of broken-up basaltic material, then filling it with HCl, letting the acid be neutralized, and finally carbonating the solution by bubbling CO_2 through it. Oxygen would not be permitted into the system. The HCl would leach the rocks, resulting in the release and precipitation of silica and producing a chloride ocean containing Na, K, Ca, Mg, Al, Fe, and reduced sulfur species in the proportions present in the rocks. As complete neutralization was approached, Al could begin to precipitate as hydroxides and then combine with precipitated silica to form cation-deficient aluminosilicates. The aluminosilicates, as the end of the neutralization process was reached, would combine with more silica and with cations to form minerals such as chlorite, and ferrous iron would combine with silica and sulfur to make greenalite and pyrite. In the final solution, Cl would be balanced by Na and Ca in roughly equal quantities, with subordinate K and Mg; Al would be quantitatively removed, and Si would be at saturation with amorphous silica. If this solution were then carbonated, Ca would be removed as $CaCO_3$, and the Cl balance would be maintained by abstraction of more Na from the primary rock. The "sediments" produced would contain chiefly silica, ferrous iron silicates, chloritic minerals, calcium carbonate, calcium-magnesium carbonates, and minor pyrite.

If the HCl added were in excess of the CO_2, the resultant "ocean" would have a high content of $CaCl_2$, but the pH would still be near neutrality. If the CO_2 added were in excess of the Cl, Ca would be precipitated as the carbonate until it reached a level approximately that in the oceans of today—a few hundred parts per million.

If this newly created "ocean" were left alone for a few hundred million years, its waters would evaporate onto the continents, and streams would transport their loads to the ocean. The sediments created would be uplifted and incorporated into the continents. Gradually the influence of the continental debris would be felt and the pH might shift a little. Iron would be oxidized out of the ferrous silicates to make iron oxides, but the water composition would not vary a great deal.

The primary minerals of igneous rocks are all mildly "basic" compounds. When they react in excess with acids such as HCl and CO_2, they produce neutral or mildly alkaline solutions plus a set of altered alumino-silicate and carbonate reaction products. It is improbable that ocean water can have changed through time from a solution approximately in equilibrium with these reaction products—clay minerals and carbonates.

"Modern" Oceans

About 1.5–2 billion years ago sedimentary rocks took on "modern" characteristics. Sandstones, shales, and carbonate rocks, with chemical and mineralogical compositions differing little from their Paleozoic counterparts, are common in the middle and late Precambrian record. A few evaporite deposits of Proterozoic age show that sulfate was present and that oxygen production by upper atmospheric radiation and by photosynthesis had been enough to oxidize the reduced sulfur gases. Iron deposits, in which chemically precipitated hematite is abundant, also attest to available free oxygen, whatever its atmospheric percentage. These, and other comparable criteria, although modified by differential cycling, lead us to believe that continuous cycling of sediments like those of today has been in effect for about 2 billion years and that they have controlled oceanic composition.

To discuss the chemistry of the oceans during this time interval, it is advantageous to return to the present and to work backward to the Proterozoic. The stream load coming into the oceans today is dominated by the materials from Mesozoic and Cenozoic rocks, which occupy a greater area of the continents than the Paleozoic rocks. The Precambrian is exposed over 25 percent of the land; of this, only 20 or 30 percent is sedimentary rocks. The detritus of the rocks being eroded is removed mechanically, as well as being chemically weathered and leached. Bicarbonate, Ca, SiO_2, SO_4, Mg, Na, and Cl are the major species of the dissolved load, and their concentrations are in the order listed. These are the elements that must be removed from the average Mesozoic-Cenozoic shale to change it to the average Paleozoic shale (see Figure 9.6). The most recently uplifted rocks are being destroyed by mechanical erosion and selectively leached by chemical erosion to form a new sedimentary mass very similar to their original compositions. In the course of time the remaining Mesozoic and Cenozoic rocks should take on the characteristics of today's Paleozoic rocks, with the formation of a new sedimentary mass with a chemical and mineralogical composition like the present Mesozoic and Cenozoic rocks. The rocks of this new Era, to judge from the chemistry of the dissolved load of present-day streams, would be enriched in their proportions of limestone and evaporites over the original Mesozoic-Cenozoic deposits. If this picture is accurate in its broad long-term relationships, no matter how much they are modified by shorter-term Period-to-Period relationships, the kinds of detrital and dissolved materials being brought to the oceans have not changed greatly since the Proterozoic, and a mean sea water composition approximately in equilibrium with calcite, K-feldspar, illite-montmorillonite solid solution, and chlorite is implied.

We assumed a mass half-age (Chapter 10) for carbonate rocks of about

300 million years. Accordingly, the total carbonate rock mass deposited in the last 2 billion years should be about 6 times the existing mass, or about $6 \times 3360 \times 10^{20}$ g $= 20,160 \times 10^{20}$ g. The present rate of $CaCO_3$ deposition is 12×10^{14} g/year. In 2 billion years the total $CaCO_3$ deposited would be $12 \times 10^{14} \times 2 \times 10^9 = 24,000 \times 10^{20}$ g. The agreement in the order of magnitude of the calculations, which are based on different procedures for estimating total deposition, is at least mild supporting evidence in favor of a rate of limestone deposition through time averaging out at about today's rate and also is in agreement with the conclusion in Chapter 5 that chemical denudation rates should vary less with time than mechanical denudation rates.

In summary, the proportions of species in the present solution load of streams and the mineralogical makeup of the detrital load of streams today may be representative of those of late Precambrian and of Phanerozoic time. If so, sea water with ion ratios close to those of today may be the "mean sea water" of about half of the geologic past.

PERIODIC COMPOSITIONAL EXCURSIONS

It is premature to try to evaluate the fluctuations in sea water composition that would have resulted from time to time as coincidences of various sets of physical and biological conditions changed the composition of the material going into or coming out of the oceans. If there were times when almost no erosion of the land was occurring, sea water might have had a chance to react more completely with detrital primary rock materials, as opposed to being continuously exposed to the altered products of weathering. At best we can cite a few limits that have been established. In general, they are negative evidence, and the argument is based on the absence of a mineral that would form if sea water composition were changed in a given way. Our knowledge of such limitations is scant; little experimentation on changing sea water composition (except for evaporating it) has been done.

The silica content of sea water cannot increase above 25 ppm at pH 8 without precipitating a magnesium silicate such as sepiolite [Mg_2^- $Si_3O_6(OH)_4$] (Wollast et al., 1968). Sepiolite is found as a mineral in oceanic sediments, but its occurrences are sparse. It is common in the deposits of saline lakes but relatively unimportant in sedimentary rocks.

If the pH of sea water were increased to about 10, brucite [$Mg(OH)_2$] would precipitate. Brucite, like sepiolite, occurs in sedimentary rocks but never has the extensive distribution that would result from precipitation throughout the whole mass of sea water.

It can be calculated that instantaneous addition and dissolution of about 7 percent of the evaporite deposits into the oceans would saturate them with respect to gypsum. On the other hand, the total salinity would be increased only a few parts per thousand. There is no record of gypsum in deposits interpreted as open-ocean deposits; all the evaporites have formed in restricted portions of the sea. Gypsum seems to be a mineral whose concentration in the open oceans always has been less than the saturation value because it is "drawn off" by local high evaporation. Furthermore, the absence of evidence in the rocks for open-ocean gypsum requires the continuous storage of most of the sulfate in evaporites, and hence Uniformitarianism.

These are a few of the kinds of restrictions that can be suggested for limits of sea water chemical variation; eventually many more will be investigated by experimentation.

SUMMARY

The historical background of interpretation of the origin of ocean water chemical composition has been presented briefly. The concept of the oceans as a chemical system has changed from one in which the oceans have been thought of as a continuous accumulator of the salts brought down by rivers to a view in which the oceans are chiefly a mechanism of transfer of material from the continents to the sea floor. Today it is generally agreed that the ocean is a steady-state system, with influx of dissolved species approximately continuously equal to their efflux. Current controversy concerns the degree to which sea water composition can vary through time as a result of changing influx or efflux rate or composition.

Information is fairly good about the material being added to the oceans, both in terms of the minerals in the detrital load and the concentrations of species in the dissolved load of streams. Similar information, except for carbonate minerals, is not available for the substances that are lost. Although estimates of the minerals formed and their relative amounts have been made, the estimates have not yet been upheld or disproved by observation.

An interpretation of the chemical history of the oceans was presented. The history was divided into three stages. First, there was an early stage in which the crust was cooling and reacting with reducing acid volatile gases to produce the oceans and an initial sedimentary mass. This stage lasted until about 3 billion years ago. The second stage is a transition from the first to essentially modern conditions and is estimated to have ended 1.5–2 billion years ago. Since that time we proposed that there has been little change in ocean water composition.

Emphasis was placed, in discussion of the first stage, on the alteration of mafic volcanics to produce a sedimentary mass high in altered volcanogenic rocks, carbonate rocks, chert, and silicates and carbonates of ferrous iron and an ocean with a composition not yet predictable in detail but presumed to be nearly neutral and fairly salty.

For the second stage, emphasis was placed on the addition of oxygen to the originally reduced oceanic system, with accompanying oxidation of reduced sulfur compounds to put SO_4 in the oceans and the formation of ferric, instead of exclusively ferrous, compounds. The second stage was accompanied by the change from a sedimentary mass created largely by submarine processes to one dominated by continental erosion.

Finally, the last 1.5–2 billion years were assessed as an interval in which deposits qualitatively like those of today have been continuously formed and the long-term chemical trends seen in rocks explained as the result of selective erosional cycling of the constituents of the sedimentary rock mass. Ocean composition was judged to have averaged, through time, about that of today. Some of the possible constraints on variation of composition with time were mentioned.

REFERENCES

Abelson, P. H., 1966, Chemical events on the primitive earth: *Proc. Natl. Acad. Sci.*, *55*, 1365–1372.

Baarghorn, E., and Schopf, W. J., 1966, Microorganisms three billion years old from the Precambrian of South Africa: *Science, 152*(3723), 758–763.

Dittmar, W., 1884, Report on the Scientific Results of the Exploring Voyage of H.M.S. *Challenger*, Physics and Chemistry, vol. 1; H. M. Stationery Office, London.

Dole, M., 1965, The natural history of oxygen: *J. Gen. Physiol.*, *49*, 5–27.

Engel, A. E. J., and Engel, C. G., 1964, Continental accretion and evolution of North America: in *Advancing Frontiers in Geology and Geophysics*, A. P. Subramaniam and S. Balakrishna, eds.: Osmania Univ. Press, Hyderabad, India, 17–37.

Engel, A. E. J., Nagy, B., Nagy, L. A., Engel, C. G., Kremp, G. O. W., and Drew, C. M., 1968, Alga-like forms in Onverwacht Series South Africa: Oldest recognized lifelike forms on earth: *Science, 161*(3845), 1005–1008.

Fischer, A. G., 1965, Fossils, early life, and atmospheric history: *Proc. Natl. Acad. Sci.*, *53*, 1205–1215.

Fischer, A. G., 1966, Brackish oceans as a cause of the Permo-Triassic marine faunal crisis: in *Problems in Paleoclimatology*, R. Fairbridge, ed.: Interscience Publishers, Inc., New York, 566–577.

Forchhammer, G., 1865, On the composition of sea water in different parts of the ocean: *Phil. Trans., 155*, 203–262.

French, B., 1966, Some geological implications of equilibrium between graphite and a C-H-O gas phase at high temperatures and pressures: *J. Geophys. Res.*, *4*, 233–253.

Helgeson, H. C., and Mackenzie, F. T., 1970, Silicate-sea water equilibrium in the ocean system: *Deep Sea Res.*, (in press).

Higgins, G. H., 1968, Rate of evolution of the crust of the earth: *Program 1968 Annual Meetings:* Geological Society of America, Boulder, Colorado, 135–136.

Holser, W. T., and Kaplan, I. R., 1966, Isotope geochemistry of sedimentary sulfates: *Chem. Geol., 1*, 93–135.

Lowenstam, H. A., 1961, Mineralogy, O^{18}/O^{16} ratios and strontium and magnesium contents of recent and fossil brachiopods and their bearing on the history of the oceans: *J. Geol., 69*, 241–260.

Mackenzie, F. T., and Garrels, R. M., 1966, Chemical mass balance between rivers and oceans, *Am. J. Sci., 264*, 507–525.

Mackenzie, F. T., Garrels, R. M., Bricker, O. P., and Bickley, F., 1967, Silica in sea water: Control by silica minerals: *Science, 155*, 1404–1405.

Marcet, A. M., 1819, On the specific gravity and temperature of sea water in different parts of the ocean and in particular seas: with some account of their saline contents: *Phil. Trans., 109*, 161–208.

Miller, S., 1953, A production of amino acids under possible primitive earth conditions: *Science, 117*, 528–529.

Reynolds, R. C. Jr., 1965, The concentration of boron in Precambrian seas: *Geochim. Cosmochim. Acta, 29*, 1–16.

Ronov, A. B., 1968, *Probable Changes in the Composition of Sea Water During the Course of Geological Time:* Invited Lecture, VII International Sedimentological Congress, Edinburgh, Scotland.

Russell, F. S., and Jonge, C. M., 1928, *The Seas:* Frederick Warne and Co., Ltd., London.

Sillén, L. G., 1961, The physical chemistry of sea water: in Oceanography, M. Sears, ed., *Am. Assoc. Adv. Sci. Publ., 67*, 549–581.

Wollast, R., Mackenzie, F. T., and Bricker, O. P., 1968, Experimental precipitation of sepiolite at earth surface conditions, *Am. Mineralogist, 53*, 1945–1962.

12 Growth
of the
Continents

Some of the liveliest topics of research and discussion today are those dealing with continental drift, sea floor spreading, thermal convection in the mantle, and mountain building. These concepts are often intertwined in models employed to explain major surface features of the earth and continental evolution. However, a one-to-one correlation may not exist between these processes; for example, sea floor spreading could conceivably occur without continental drift, mountain building without convection, and so forth. The time scale is also important. Continents may have drifted in the past but may not be doing so today. Thermal convection was certainly an important process in the early history of the earth, although today it may be of only minor significance in the mantle. Models will be proposed later in this chapter to account for the major surface features of today. We wish to emphasize, at the risk of some repetition, perhaps, that the syntheses presented are not necessarily correct but are of importance from the standpoint of seeing the type of thinking that enters into studies of the nature of "why mountain ranges?" Such synthesis depends on a vast amount of data, much of which is inconclusive or fragmentary at present, and therefore the very nature of the problem does not permit a final, and perhaps not even a satisfying, answer.

In this chapter we shall look in greater detail at the geologic history of the crust, particularly at the great belts of deformed rocks (geosynclines) and their relations to continental growth and movement.

GEOSYNCLINES AND MOUNTAIN RANGES

In 1857, in a presidential address to the American Association for the Advancement of Science, James Hall made one of the most significant early contributions to the field of geology. He pointed out that the Appalachian Mountains in the eastern United States form a linear belt of folded Paleozoic rocks trending approximately NE-SW that are many times thicker than flat-lying contemporaneous strata in the interior lowlands to the northwest. A generalized picture of this relationship is shown in Figure 12.1. Notice that carbonate rocks and quartzites are typical rocks found in the interior lowlands, whereas lutites and "dirty" sandstones (e.g., graywackes) are characteristic of the thick prism of sediments contained in the Appalachians. The rocks of the interior lowlands contain structures and fossils indicating shallow water deposition. The depth of water in which the sediments of the thick prism were deposited is less well known. Certainly some of the sediments accumulated in shallow waters; the rate of sedimentation approximately equaled the rate of subsidence of the linear trough in which the sediments were deposited. However, certain structures and textures and lack of in situ shallow water fossils in associated graywacke-lu-

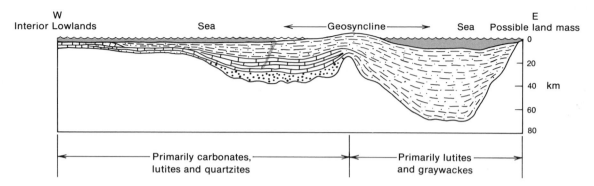

Figure 12.1 Diagrammatic cross section of the Appalachian geosyncline at the close of Ordovician time. Notice thickening of contemporaneous strata from the interior lowlands toward the east (section adapted from Kay, 1951).

tite strata in the prism suggest that these sediments may have been deposited by deep-water currents and turbidity currents.

The vertical and horizontal movements of the crust that can be read from detailed study of the rocks of a geosyncline and the correlative rocks of the adjacent platform are highly complicated. On the platform or *craton*, such as the area to the west in Figure 12.1, layers of a given rock type tend to be thin, even bedded, and to have a large lateral extent relative to their thickness. A layer of the craton may represent as much time as a whole group of layers in the trough, with 10 or 20 times the thickness of the platform layer. Furthermore, there are many erosional surfaces between the nearly horizontal layers of the platform, indicating oscillatory vertical movements which permitted shallow seas to creep in and drain out on a time scale of millions or tens of millions of years. The crustal movements were not entirely uniform; on a state-wide scale, basins and broad arches can be deciphered. The state of Michigan, during Paleozoic time, became a sediment-filled basin, accumulating as much as 7000 m of sediments in the center of the downwarp. The gross structure of Wisconsin, on the other hand, is that of a broad arch. The present-day disposition of sedimentary rocks in these states reflects the irregular, dominantly up-and-down movements of the crust. In Michigan the total actual oscillation in the basin center, both up and down, is several times the net downward movement of 4 km, yet at no time was the depth of the sea or the height of the land more than a few hundred meters.

Movements in the geosynclinal trough were similar in their oscillatory character to those on the platform, but with greater amplitude and fewer exposures to erosion. As a result, the rocks of the deeper parts of the trough tend to give a nearly uninterrupted record of the time interval they represent. In the Appalachian geosyncline, a wave of crustal deformation moved spasmodically southward along the trough; the trough itself reached its maximum depth progressively from northeast to southwest. As shown in Figure 12.1, the trough had two major subdivisions. The section is necessarily drawn to represent a particular moment in the geologic history of the trough; sections drawn at other times could perhaps show a single trough, a mountain range projecting high above the sea within the trough, or some other configuration resulting from active rising or sinking within the so-called mobile belt.

The eastern margin of the Appalachian trough has been a vexing problem ever since Hall recognized the nature of the geosyncline. Because some of the sedimentary rocks in the eastern part of the trough have their source areas to the east, a land mass was required where the Atlantic Ocean now exists. Proponents of "permanent" continents postulated a land called "Appalachia" and have had to sink it to depths precluding recognition of its remnants today. The continental-drifters would use the European continent as the sediment source. Both views currently have serious difficulties: the supporters of Appalachia are forced to an explanation that is uncheckable because of lack of evidence; the drifters are hard put to find source areas in western Europe that could provide the right kinds of sediments in the right places.

Hall also pointed out that major mountain ranges the world over are genetically related to linear belts of thick sedimentary rocks; that is, mountain ranges arise through the deformation of these thick prisms of rock. In the time since Hall, the geosynclinal concept has become an integral part of our hypotheses concerning mountain building.

Development of Mountains

An interpretation adapted from Griggs (1939) of the processes leading to the development of mountains is shown in Figure 12.2. Convection in the upper mantle tends to drag down a portion of the crust, which responds plastically to create a depression that accumulates sediments and a geosyncline is formed. The light crustal material is dragged downward until a "root" of granitic crust and sediments protrudes far into the mantle. As the sedimentary rocks move downward, they enter higher temperature and pressure regimes, and under such conditions they may flow or break, producing various sizes of folds and faults. At depth, the rocks recrystallize to metamorphic rocks and are intruded by magmas that later cool to form igneous rocks. These magmas are derived in part from the mantle and in part from the melting of the sedimentary rocks under the high-temperature conditions. When the currents abate, the large mass of light crustal rock rises slowly as isostatic equilibrium is re-established. As the mass rises, the sedimentary rocks are further contorted and tend to slide downhill under the influence of gravity. Contemporaneous and subsequent erosion of this mass produces a mountain range.

When a sequence of sedimentary layers thickens by accumulation and sinking, the lower part gradually heats as it is depressed into the crust. Geothermal gradients are of the order of 20–30°C/km; thus burial to 10 km or more produces rock temperatures of several hundred degrees, ac-

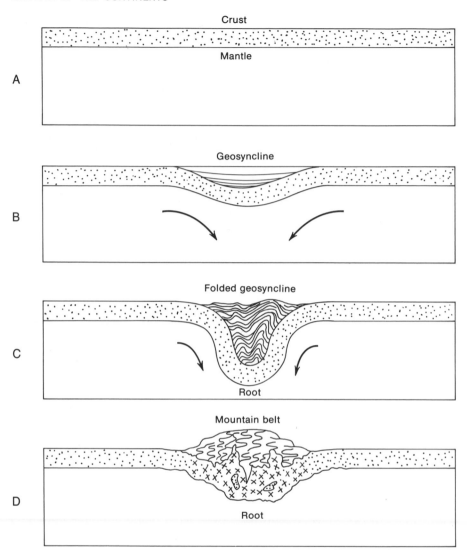

Figure 12.2 Development of mountains. A, Plastic crust overlying a fluid mantle. B, Period of slow convection in upper mantle and development of a geosyncline. C, Period of fastest convection in the mantle, major downwarp of crust, folding of geosynclinal region, and formation of a mountain "root." D, Convection ended. Buoyant rise and erosion of mass of folded, thickened, and intruded crust produces a mountain range. The X-pattern represents granitic rock formed by the cooling of magma derived from melting of the deeply buried sediments and from the more mobile components in the upper mantle. Vertical scale exaggerated (after Griggs, 1939).

companied by pressures of more than 1000 atm. As burial progresses, water trapped in the sediments, with its dissolved materials, tends to be squeezed out of the pores and migrates generally upward through the section. At great depth partial melting may occur, forming a magma which may then intrude the overlying rocks before it solidifies, or may, in some instances, reach the surface. Such partial fusion should produce a liquid richer in silica than the average of typical sediments and leave a silica-poor residue. Processes of this type, which are unobservable directly, undoubtedly are responsible for the heterogeneity of the lower crust as deduced from gravity and earthquake-wave studies.

Magma formation and intrusion are characteristic of the late stages of geosynclines; that is, they occur when the bottom of the trough reaches its maximum depth.

The "rebound" of geosynclines that follows their maximum filling commonly takes place as a broad regional uplift. Movement upward is sporadic and may cover a time span of tens of millions of years. In the case of the Appalachian geosyncline at about the latitude of Washington, D.C., the trough reached its maximum depth during the late Pennsylvanian and early Permian, and the sediments were metamorphosed, crumpled, and intruded by molten rock. Broad uplift took place during Triassic time, accompanied by tensional faulting, which resulted in accumulation of terrestrial sediments in local basins or depressions. By Cretaceous time, erosion had nearly leveled the area, and the sea crept across it as it submerged uniformly. During Tertiary time the area rose as a broad arch, with the axis of greatest uplift roughly coincident with the former deepest part of the trough. Today the Cretaceous erosion surface (or its reconstruction), representing approximately sea level at that time, is several kilometers beneath the sea at the outer edge of the continental shelf and several kilometers above it in Western Virginia.

Metamorphism

The typical chemical and mineralogical changes that accompany burial, heating, folding, and partial fusion are called diagenesis in the early stages and metamorphism in the later, more drastic stages. No one knows exactly where diagenesis ends and metamorphism begins; very broadly, diagenesis includes those changes that accompany simple burial without strong deformation or changes in the texture and structure of the sedimentary rocks as a whole, whereas metamorphism begins when the whole aspect of the rock is altered. Thus diagenesis deals with the squeezing out of pore waters, the cementing or decementing of original grains, and the development of minor quantities of new minerals from old. In metamorphism, lateral pressures may fold the rocks, all the minerals may be recrystallized, often with an orientation dictated by the stresses applied, and pore space is reduced to almost zero. In some cases, partial fusion may occur; material may migrate from one place to another, drastically changing the initial bulk composition.

The details of the chemistry of diagenetic changes have been discussed in previous chapters. If burial is sufficiently deep and the temperature reaches several hundred degrees centigrade, the original minerals of the sediment may be converted from various clay minerals to the typical minerals of igneous rocks: feldspars, micas, quartz, amphiboles, or pyroxenes. The original sedimentary chemical composition is fairly well preserved, and a chemical analysis is indicative of origin. The development of the new minerals usually takes place in an environment in which the pressure is not the same in all directions. When the minerals recrystallize, some, such as mica, which has a sheet structure, tend to orient to form plates perpendicular to the maximum pressure direction. There is also a tendency for the minerals to segregate into individual layers or bands.

The assemblage of minerals produced when a rock undergoes metamorphism depends on the overall chemical composition of the metamorphic rock and the pressure and temperature of metamorphism. Metamorphic rocks formed under similar pressure-temperature conditions, regardless of chemical composition, constitute metamorphic facies. Eskola originally proposed the facies concept; its validity rests on the observation that mineral assemblages in metamorphic rocks obey the laws of chemical equilibrium. Eskola recognized five metamorphic facies in 1939. In recent years the number of facies has been extended and subfacies within several facies have been recognized. Rocks formed at low temperature and moderate pressure are characterized by the minerals chlorite, muscovite, and biotite, whereas those metamorphosed at moderate temperatures and pressures are characterized by aluminosilicates such as sillimanite and kyanite. Very high temperature and extreme pressure produce a garnet-pyroxene-bearing rock. The important point is that metamorphic mineral assemblages can be used to determine the pressure and temperature conditions of metamorphism. These assemblages can be mapped and belts and aureoles of similar metamorphic conditions located.

When shales are metamorphosed, there is a sequence of changes related to the processes of recrystallization and mineral orientation that permits classification and assignment of rock names indicative of the stage reached. Only the three major types, slate, schist, and gneiss, will be considered here.

When a shale has been completely recrystallized and there is sufficient mineral orientation to cause the rock to cleave, yielding planar surfaces perpendicular to the stress that caused them, it is called a slate if the grains are still too small to be visible to the unaided eye. The cleavage planes may be remarkably extensive laterally and may take an angle to the original bedding, which is commonly still visible.

With increasing metamorphism, grain size increases, original bedding is destroyed, and there is a tendency for segregation of the minerals into layers. Individual grains can be resolved by the eye; the rock splits into irregular sheets. Such a rock is a schist and is said to be "foliated."

In a gneiss, recrystallization is extreme; the mineral grains are commonly a few millimeters across and sometimes much larger. The rock now has a banded appearance rather than a laminated one.

Metamorphism of limestones and sandstones usually does not produce the characteristic cleavages exhibited by shales and schists. The chief minerals—calcite and quartz, respectively—have structures that do not permit development of plates or rods under stress. The result of recrystallization of a limestone is to produce a coarse-textured marble in which the grains are roughly equidimensional and securely locked together by interpenetrating boundaries. Quartz-rich sandstones respond texturally in much the same way to yield quartzite.

One of the unfortunate consequences of metamorphism, from the viewpoint of those trying to reconstruct the earth's history, is that fossil remains are usually destroyed. In some cases, the original rock may be so extensively altered by partial fusion that it looks like an igneous rock. Some large granitic batholiths are believed to have formed by intense metamorphism of sedimentary rocks. Minerals containing radioactive elements are recrystallized, so that radioactive age dating may discover the time of metamorphism of the rock.

CRUST AND MANTLE MOVEMENT

Convection

In the Griggs' hypothesis of mountain building, heat-driven convection plays a major role. Other mechanisms to create the forces necessary to generate mountains have been proposed, such as a shrinking or expanding earth and continental drift in which the edges of continents act as great plows in the substratum, the edges being pushed up into mountain ranges as the continental plate moves through the mantle (the forces necessary in this case are thought to be tidal and centrifugal). However, the convection mechanism seems to have gained increasing favor over the last decade, perhaps because it helps to explain other phenomena affecting the earth's crust.

Much of current controversy concerning major earth movements depends on acceptance or rejection of thermal convection in the mantle. Some of the most powerful evidence for subcrustal currents of large magnitude and similar velocities comes from a compilation of the rates of differential movement of major sections of the crust. Table 12.1, extended from one by Holmes (1965, p. 1032), lists the displacement rates and the structural relations from which they have been deduced. The general concordance of the results, despite questions concerning the validity of some of the individual rates cited, certainly is suggestive of the presence of subcrustal movements at a finite and important geologic rate.

Just as the sea floor was once thought to be a featureless plain, until many accurate measurements of depth were made, our concept of the mantle continues to change from an initial picture of a radially uniform mass toward one with greater and greater local variability. Such variability is also compatible with thermal convection. On the other hand, there is still

Table 12.1

Some Rates of Crustal Movement

Phenomenon	Rate (cm/year)
Continental drift (averaged over 200×10^6 years)	2–4
Uplift of northwest shore of Pacific by movement of Pacific floor beneath crust	3
Development of Hawaiian Islands from north to south over 100 million years (3000 km)	3
Movement of Atlantic Islands westward from Mid-Atlantic Ridge	3
Displacement of southwest side of San Andreas fault to northwest relative to northeast side (present rate)	5
Great Alpine fault of New Zealand (present rate)	5
Magna Fossa fault, Japan (160 km in $4–8 \times 10^6$ years)	2.5
Algerian faults (180 km in $6–12 \times 10^6$ years)	2–4
Rafting of Baja California, to northwest (260 km in 4×10^6 years)	6
Spreading of Mid-Atlantic Ridge	2–3

enough good evidence of the presence of symmetric earth shells with different properties to provide strong arguments for the opponents of convection.

Intermediate and Deep-Focus Earthquakes

Some additional evidence for mantle convection comes from deep-seated earthquakes and gravity anomalies. Nearly all of the intermediate and deep-focus earthquakes have their epicenters in the circum-Pacific area and tend to occur along island arcs, such as the Japanese and Marianas Arcs in the Pacific and in the Andean Mountain chain of South America. When the focal depths of these circum-Pacific earthquakes are plotted on a depth section that trends at right angles to the trend of the arc or the Andean range, it is found that the foci occur along a line that dips inward toward the continent, that is, away from the Pacific Ocean basin (Figure 12.3). It is possible that convection currents are dragging the oceanic crust underneath the continental masses and that the alignment of earthquake foci with depth represents a major shear along which oceanic crust and sediments are slowly being pushed beneath the land mass. Numerous active volcanoes surround the Pacific basin and may be a surface expression of the activity at depth in the circum-Pacific region, reflecting heating and mobilization of the deep-seated crustal materials and emanations from deeper regions that rise along fractures.

Gravity Anomalies

An added support for the conclusion that the crust is being dragged down beneath island arcs is afforded by gravity measurements across such areas

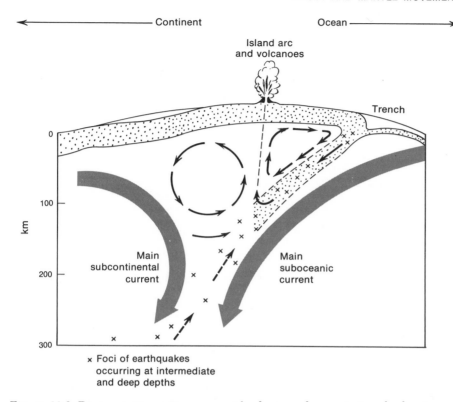

Figure 12.3 Diagrammatic section across island arc and oceanic trench showing a possible convection current pattern at depth that would account for the location of intermediate and deep earthquakes. The stippled area shows the crust and the area where oceanic crust and sediments are slowly being pushed beneath the land mass. The dashed arrows represent mobile materials, such as H_2O and CO_2, derived from within the mantle and eventually, in part, released to the earth's surface through volcanism. Some of the materials dragged to depths by convection are heated to the melting point and are intruded back into the earth's crust or escape through volcanoes (after Arthur Holmes, *Principles of Physical Geology*, Second Edition. The Ronald Press Company, New York. Copyright 1965).

(Figure 12.4). Sharp negative gravity anomalies occur in the East and West Indies and along island arcs in the circum-Pacific region. These anomalies may indicate that a large mass of light rocks occurs at depth; this light rock may represent crustal material that has been dragged to great depth by convection currents. Some scientists believe that the oceanic deeps bordering the oceanward sides of island arcs represent modern geosynclines and that the great mass of light crustal material (the root of Figure 12.2) beneath may become a mountain range of the future, when the convection current that holds this mass down abates.

However, until the nature of the mantle is better documented, the origin of the forces that move large crustal segments cannot be definitely ascribed to convection cells, especially in terms of the rather simple idealized models now suggested. There is a great deal of subcrustal activity, but in-

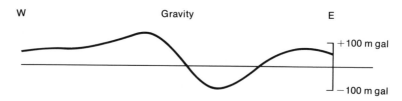

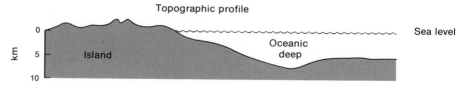

Figure 12.4 Generalized gravity and topographic profile across an island arc and oceanic deep.

terpretation of the mechanisms with any real assurance must wait for a while.

Continental Drift

The concept of continental drift is intimately related to today's ideas concerning motion in the mantle and mountain building. Most of us are familiar with the demonstration that the configuration of the coastline of the Americas can be fitted against the coastlines of Europe and Africa by moving the Americas eastward. Figure 12.5 shows a recent attempt at fitting the continents together; the results are remarkably good. The amount of overlap or underlap is small. The obvious deduction is that these continents were once joined together and at some time in the past drifted apart. Nearly 100 years ago, Antonio Snider made this deduction; in the early 1900's, Alfred Wegener revived and developed the idea and is usually credited with its dissemination. Wegener suggested that the continents were moved about on a dense, relatively low-viscosity substratum by centrifugal and tidal forces. He further stated that differences in centrifugal force at different latitudes would create differential stresses resulting in the continents drifting toward the equator, or in other words, "fleeing the poles" (polflucht). The force created by tidal attraction of the sun and moon would tend to drag the continents westward. Collision of continents would result in the development of mountain ranges as the continents were piled up one against the other. For example, he felt that the great Alpine-Himalayan mountain chain was formed by collision of Eurasia with Africa and India. However, it has been pointed out that both of the forces thought by Wegener to be responsible for creating the stresses to move continents are too weak to break crustal rocks and move them through a substratum of the strength of the mantle.

Besides the apparent "fit" of the continents, evidence for drifting has

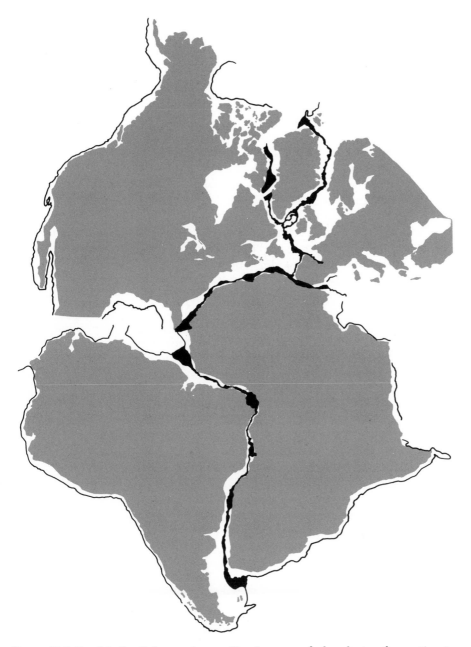

Figure 12.5 Possible fit of the continents. The fit was made by placing the continents together along the 500-fathom subsea contour. Gray shading shows underlap; black shading represents overlap (after Bullard et al., 1965).

been inferred from paleomagnetic data. These data provide information as to the ancient latitudes and orientations of continents; ancient longitudes are indeterminable. A large number of measurements of the magnetism of rocks of post-Precambrian age have been made, and the data have been plotted in various ways. Figure 12.6 shows one interpretive plot of the data, in which the north geomagnetic pole position, as obtained by averaging measurements of the magnetization direction of rocks of a particular age, is plotted as a function of geologic time and of the continents from which the paleomagnetic data were obtained. The pole positions are plotted relative to the earth's present geographic pole. If the earth's magnetic field has been singularly dipolar throughout the earth's history, and if the magnetism of rocks does not change appreciably with time, then Figure 12.6 shows that there has been some apparent movement of the earth's geomagnetic pole and that the path described by polar wandering through time is different for each continent. If only polar wandering had occurred, the loci of the geomagnetic pole positions should describe a single line instead of different paths for each continent converging on the present pole position. Thus it appears that the continents have also moved.

The number of observations of the magnetic orientation of rock minerals has increased greatly in the past 10 years. The accuracy of the data and their meaning for polar wandering and continental drift are subjects of intense discussion today. It is likely that there has been some continental movement, although the direction and magnitude are not fully known.

The original hypothetical protocontinent that broke apart to form the present continents is called Gondwana. Some advocates of drift propose two original continents: Laurasia and Gondwana. Innumerable evidence has been cited for breakup of Gondwana and subsequent drift. Glacial deposits of Permo-Carboniferous age are found on every continent in the southern hemisphere. If the continents are fit back together, the glacial deposits form a neat package situated near the southern boreal zone. Other paleoclimatological evidence has been used to prove drift, for example, the occurrence of coal, salt, desert sandstone deposits, and fossil animals and plants in climatic regions of today that are presumably incompatible with the environments inferred for their formation. When the continents are regrouped, the distribution patterns form a coherent picture.

Perhaps one of the most compelling recent arguments for drift has been the observation that age dates of rocks in West Africa are almost identical to dates obtained for rocks at opposite locations in South America. Figure 12.7 illustrates the situation after West Africa and South America are fitted together. The boundary between 2000- and 500-million-year-old rocks on the African continent is nearly coincidental with the same boundary on the coast of the South American continent, and the age provinces are aligned with each other. This evidence for drift seems compelling; however, the boundary in South America is not well defined. The apparent fit could simply reflect coincidental igneous-metamorphic events on continents separated by the Atlantic Ocean.

Figure 12.6 Apparent movement of the geomagnetic pole of the earth relative to the various land masses. Notice convergence of the geomagnetic pole positions toward the present pole. Letters refer to (P€), Precambrian; €, Cambrian; O, Ordovician; S, Silurian; D, Devonian; C, Carboniferous; P, Permian; Tr, Triassic; J, Jurassic; K, Cretaceous; LT, MT, and UT, Lower, Middle, and Upper Tertiary (after Deutsch, 1966).

The recognition that Wegener's forces were too weak to promote continental drift nearly dealt a death blow to the concept but three other mechanisms of drift have more recently been proposed. The first is that the earth has been expanding, and as a consequence, the continents have drifted apart as the circumference of the earth increased. The maximum

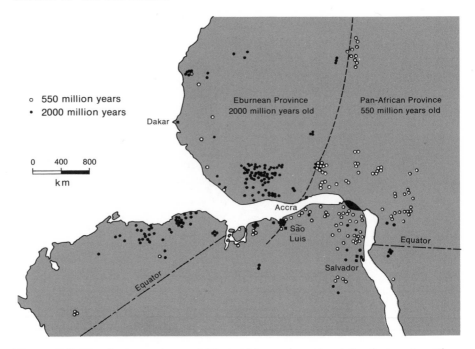

Figure 12.7 Fit of age provinces of West Africa and northeast South America. The boundary between 500- and 2000-million-year-old rocks in West Africa is nearly co-incidental with the same boundary on the coast of South America, suggesting that the two areas were once joined (after Hurley et al., 1967).

rate of expansion of the earth seems to be at least an order of magnitude less than is needed to account for the separation of Africa and South America. The second mechanism for drifting continents is thermal convection. If, for example, the Mid-Atlantic Ridge is the crustal expression of upwelling limbs of a convection cell in the mantle, the current would diverge to the east and west under the Atlantic basin and drag the Americas and Europe-Africa to the west and east, respectively. The pattern of apparent drift as evidenced in the fit of these continents then would be explained.

The convection mechanism has a major pitfall. The mantle does not appear to be radially homogeneous. A major inhomogeneity in the mantle occurs at 400–1000 km; this transition zone between the upper and lower mantle is most likely a region of rapid density change, rapid earthquake velocity change, phase change, and, according to Press (1968), chemical change. The boundary representing the contact between a lower iron-rich mantle and an upper magnesium-rich mantle would prevent mantle-wide convection. However, it would still be possible for convection to occur in the upper mantle, but the size of the convection cells would be reduced. Several discontinuities occur in the upper mantle; the existence and nature of these discontinuities is not fully known. If one or more of these discontinuities represent compositional breaks, convection in the upper mantle would be limited to smaller and smaller cell sizes, and the convection mechanism would be ineffective in drifting continents. However, if the mantle layer in which convection is occurring is heated internally, as well as from below, by energy released during radioactive disintegration of U, Th, and K concentrated in a particular region of the upper mantle, the size of the convection cell would increase, and a layer about 200–300 km thick could have convection cells of a horizontal dimension sufficient to account for large crustal movements such as drifting. At any rate, because of the apparent compositional change at the transition zone, it is likely that thermal convection, if it occurs in the mantle, is restricted to the upper 400 km.

A third mechanism which may promote drift is that of advection, or horizontal movement, of material in the mantle. A zone having relatively low earthquake-wave velocities, the low-velocity (LV) zone, has been recognized in the upper mantle beneath deep oceanic basins at depths of about 50–250 km. It has been suggested that high temperature gradients in the oceanic mantle, along with chemical and mineralogical inhomogeneities, are responsible for formation of the LV zone. The temperature distribution probably results in a decrease of density of material with depth within the upper 100 km of the mantle beneath the ocean basins. Thus, heavier material rests on lighter; the resulting gravitational instability could lead to mass movement within the upper mantle, particularly if horizontal temperature gradients exist between oceanic and continental masses, as seems to be the case. For example, mantle material, originating in the LV zone of the oceanic mantle, could move inward toward the present Mid-Atlantic Ridge, rise upward, and spread east and west. The resulting movement might result in continental drift and sea floor spreading.

The evidence for continental drift is growing but the mechanism of drift is uncertain. Thermal convection in the upper mantle is most favored today, but its presence depends on whether the upper mantle is homogeneous vertically down to 200–300 km and horizontally on a scale of several thousand kilometers. Also, for convection to occur, viscosity and strength of the upper mantle material should be low, and there must exist temperature differences in the upper mantle significant enough to promote instability and convective motion in the mantle. Even though thermal convection may occur in the upper mantle, the size of the convection cells may be too small to promote the amount of continental separation seen today.

Sea Floor Spreading

Hess (1962) and Dietz (1963) proposed the concept of sea floor spreading nearly simultaneously. Mantle material is assumed to move upward under the large oceanic rises and ridges because of thermal convection or advection (their models proposed convection) and spreads out under the oceanic crust. Basalt derived from differentiation of this mantle material intrudes the oceanic crust, and the sea floor spreads outward from the ridges as new basaltic rock is added. The sea floor spreading hypothesis has a large following today, primarily because of a number of observations that seem to corroborate the hypothesis. Magnetic profiles across some oceanic ridges show a series of anomalies that trend parallel to the ridge and are symmetrical about the ridge axis. The explanation of the pattern is that mantle material wells up from under the ridge axis, spreads laterally, and as it cools below its Curie temperature (the temperature above which a material cannot be magnetized) is magnetized in the direction of the geomagnetic field. The magnetic anomalies, deviations from the regional average strength of today's magnetic field, are thought to represent reversals in the earth's magnetic field. Belts of low magnetic strength are correlated with times of reversed field and those of high strength with times of normal field such as today. It has been shown that the earth's magnetic field has reversed itself several times in the past; based on paleomagnetic data and radiometric age dates, a time scale for reversals of the earth's magnetic field has been developed.

An excellent correlation between the magnetic reversal intervals and the geographic dimensions of the anomaly patterns near oceanic ridges exists and is consistent with a spreading rate of about one to a few centimeters per year. This rate is in accord with displacement rates of large blocks of the earth's crust as obtained from other structural relationships (Table 12.1).

Further evidence for sea floor spreading is provided by radiometric age dates of basaltic rock collected on a traverse across the Mid-Atlantic Ridge. Figure 12.8 shows the ages of basalts as a function of distance from the median, or rift, valley of the ridge at 45°N latitude. Notice that except for one sample, the ages of the basalts increase outward from the ridge. This trend is believed to be a consequence of sea floor spreading; the progressively older basalts represent material that upwelled and cooled near the median valley of the ridge and subsequently moved away from the ridge as newer basalt was added to the crust. The value of the spreading rate as obtained from the ages of the basalts and their distances from the median valley is 2.5 cm/year; this value also agrees with other estimates. Interestingly, the one sample that does not obey the trend is from an area that does not fit into the regional magnetic anomaly pattern and is believed to represent more recent volcanic activity than that responsible for the observed age-distance relationship. Also, it is possible that the progressively older ages away from the median valley do not represent movement of material away from the ridge but movement of basaltic volcanism toward the ridge. However, the apparent monotonic increase of basalt ages outward from the ridge in a region exhibiting a pattern of regular magnetic anomalies provides strong support for the sea floor spreading hypothesis.

Several other observations have been used by scientists to support the

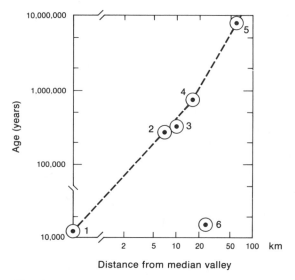

Figure 12.8 Age of basalts as a function of distance from the median valley of the Mid-Atlantic Ridge. Notice the monotonic increase in age away from the ridge except for sample 6 which is from a magnetically disturbed area (after Fleischer et al., 1968).

spreading hypothesis. Heat-flow values are higher over the crests of some ridges than away from them. This observation is consistent with the rise of hot basaltic magma under the ridge crests. Also, sediment thickness over ridge crests is less than on ridge flanks. One interpretation of this observation is that the process of spreading of the sea floor is intermittent and variable in rate. The present cycle of spreading, represented by the thin crestal sediments, is believed to have begun about 10 million years ago and followed an interval of relative quiescence during which the thick flank sediments were deposited. Ewing and Ewing (1967) have further suggested that the interval of quiescence was preceded by an earlier cycle, or cycles, of spreading that resulted in breakup of Gondwana and Laurasia and the movement of continents to their present locations. They also suggested that sea floor spreading was so extensive during this older cycle that most of the Paleozoic sediments were swept from the ocean floor.

Another observation seemingly confirming spreading is that fresh basalts less than 1 million years old are found nearly continuously along the crest of the East Pacific Rise in the eastern South Pacific Ocean. The width of this crestal band of basalt is similar to that of the most recent positive magnetic anomaly observed farther north on the crest of the rise. It is believed that these basalts were derived from the mantle and rose through fissures along the rise, as sea floor spreading occurred. The rise does not display a median valley as does the Mid-Atlantic Ridge; the lack of a valley is thought to be due to the extensive outpouring of basaltic lava in this region. If volcanism were less intense, a median valley could develop as a result of tensional forces due to sea floor spreading and an insufficient lava volume to fill in the resultant rift.

Some corroborative evidence for sea floor movement out from the East Pacific Rise is provided by the relative ages of Quaternary sediments west of the rise in a region immediately north of the equator. If sea floor spreading does occur, then sediments directly above the igneous basement of the ocean floor should decrease in age toward oceanic rises and ridges. The ages of Quaternary sediments west of the East Pacific Rise have been determined from assemblages of Radiolarian fossils. Younger Quaternary deposits are found near the rise and older sediments farther away, a distribution consistent with the spreading hypothesis. This observation seems to substantiate spreading outward from the rise, though some other data obtained from a region on and near the rise south of the equator are not consistent with spreading. Potassium-argon age dates of basalts dredged from the crest of the rise and from the flanks of three seamounts at varying distances from the rise do not exhibit a regular increase in age away from the rise, as was the case for basalts obtained on and near the crest of the Mid-Atlantic Ridge. The three seamounts are located east of the rise at distances of 500, 1200, and 1600 km and are apparently of the same age. Hess (1962) suggested that most seamounts (although not necessarily all) originate by basaltic volcanism near the crests of rises and, as the sea floor spreads, slowly move away from the rise crests. Obviously the age pattern obtained near the East Pacific Rise does not substantiate this suggestion. However, it must not be assumed that all basaltic volcanism on the sea floor is necessarily related directly to sea floor spreading. Intense, independent volcanic activity certainly can occur on the sea floor and not be coincident with intrusion of new basalt along rises and rifts and concomitant spreading.

Another argument against spreading is that the sedimentary layers below the ocean floor appear to be relatively undisturbed. If sea floor spreading does take place, the layers should be contorted, but major horizons within the deep ocean sedimentary column appear continuous and undeformed. If spreading is discontinuous and has only recently been more active after a long interval of quiescence extending back perhaps to the Cretaceous, then the relatively undisturbed layering could be explained. It is thought that the oceanic crust, in some regions near the margins of continents, is being slowly thrust under the continents. Some seismic reflection profiles in these areas reveal extensive crumpling of the sediments. Perhaps sea floor sediments ride passively on the underlying crust until they reach areas of down-buckling.

Spreading, like continental drift and convection, is an exciting hypothesis today. Many observations confirm it; some do not. Geologists are always reluctant to accept hypotheses that require earth movements to produce basins of oceanic dimensions. More data are needed from the ocean bottom; a drilling project (Deep-Sea Drilling Project) is now underway to obtain a set of cores of the deep ocean bottom. It is likely that our ideas concerning major motions in the earth's crust will evolve rapidly in the next decade.

MODERN AND ANCIENT GEOSYNCLINES

One difficulty with the idea that the oceanic deeps represent modern geosynclines is that these areas are regions of deep-water sediment accumulation. The sedimentation rate in these deeps is very slow and the trenches seem to contain only a thin layer of sediment. In actual fact, some analytical solutions of gravity data across oceanic deeps indicate a thin crust directly below the trench; in this interpretation, the crust beneath the trench is being pulled apart instead of down-buckled.

In contrast, the geosynclines of the past contain thick deposits of sedimentary rocks that accumulated rapidly and, to a large extent, in shallow waters. In recent years, Dietz (1963), in part to resolve this apparent enigma, has looked to the margins of present-day continents for a modern analogue of ancient geosynclines (prior to Dietz, the Gulf Coast Basin was referred to by some scientists as an example of a modern geosyncline). Figure 12.9 illustrates the essentials of the "actualistic" [1] model of geosynclines.

Convection takes place in the upper mantle and upwelling occurs beneath the midoceanic ridge. New materials are added to the oceanic crust as the sea floor spreads, and the continental plate is coupled to the mantle so that it moves along at approximately the same speed as that of sea floor spreading. Sediments accumulate on the continental margin; the sediments deposited on the shelf in the form of a wedge make up the continental terrace, whereas those deposited in deeper water construct the continental rise prism. These areas of accumulation of sediments slowly subside owing to the mass of the overlying sedimentary rocks.

For reasons not completely known at this time (perhaps the forward edge of the slowly moving continental block meets a convection current moving downward in the opposite direction), the continental plate becomes uncoupled from the underlying mantle convection cell, and the mantle shears beneath the continents. The continental rise is compressed, resulting in folding and faulting of the prism of sedimentary rocks as the rise is thrust against the continent and added to it. The terrace sediments are also deformed, but to a lesser degree than the rise sediments. Later, an oceanic trench forms and a shear develops between the continent and ocean basin along which oceanic sediments are thrust beneath the continent. Deep burial of the rise sediments results in their being heated to temperatures high enough to melt them. Magma derived from remobilized sediments and from the mantle intrudes the rocks at higher levels, and large, irregular igneous bodies are formed. When the current slackens, this

---

[1] A model in which a modern situation is used as a guide to an ancient one.

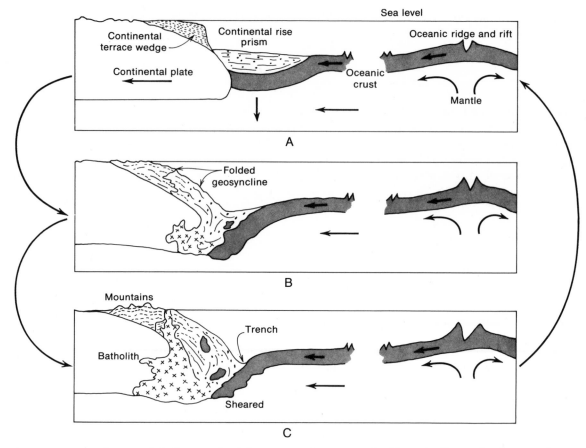

Figure 12.9 Actualistic model of geosynclinal formation and mountain building. A, Geosynclinal accumulation, convection in upper mantle, sea floor spreading, and continental drift. B, Deformation of geosyncline, mantle and continental plate uncoupled, no drift, mantle and oceanic crust shear beneath continent. C, Development of oceanic trench, formation of large granitic batholiths, and major mountain building (modified after Dietz, 1963).

mass of deformed terrace and rise sediments slowly rises, forming mountains and shedding debris to the oceans; at this time another cycle may begin.

Figure 12.10 shows the crustal structure of the Sierra Nevada in the western United States and a series of cross sections at various times in the development of the crust in this region. The regions termed miogeosynclinal and eugeosynclinal on this figure could be, respectively, the shallow water shelf area of accumulation of continental terrace sediments and the deep-water region of accumulation of rise sediments. This twofold division of a geosyncline into *mio* and *eu* portions is observed in most ancient geosynclines and Dietz equated these regions with the modern continental terrace and rise.

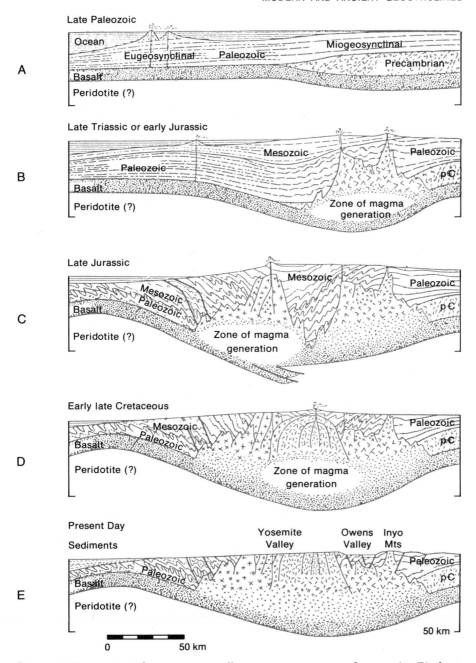

Figure 12.10 A series of cross sections illustrating a sequence of events (A–D) that led to the present crustal structure (E) of the Sierra Nevada in the western United States (Bateman and Eaton, 1967).

The actualistic model of geosynclinal development has become even more intriguing in recent years because of the concept of the "new global tectonics" (cf. Isacks et al., 1968; Le Pichon, 1968; Vine, 1969). The upper 100 km of the earth are visualized as a series of cold, rigid blocks (the lithosphere) which move about on a hot, perhaps partially molten layer (the asthenosphere) that has no strength and is approximately coincident with the low-velocity zone in the upper mantle. Underlying this second layer is the remainder of the mantle, the mesosphere, which is decoupled from movements involving the upper two layers. The lithospheric blocks are delimited either by midocean ridges or by trenches. The ridge crests are extensional features along which new oceanic crust is formed, and the trenches represent regions in which the lithosphere is slowly being dragged down or thrust beneath the trench systems. The ridges are loci of earthquake foci occurring at shallow depths and reflecting movement along vertical fracture planes termed transform faults, whereas deep-focus earthquakes down to a maximum depth of 700 km coincide with the trench systems. Thus these two types of lithosphere block boundaries reflect in their seismicity the different processes operating in the upper mantle and reflected in the crust—creation of new oceanic crust at ridges and resorption at trenches. Figure 12.11 is a pictorial summary of the six major lithosphere blocks and their directions of movement.

The present distribution of continental rises is also shown in Figure 12.11. Notice that the major continental rises coincide with margins of continents along which trench systems are not found. Thus it appears that the major rise prisms of today are within lithospheric blocks and are not undergoing deformation; in terms of the actualistic model these rises are still coupled to movements in the substratum. It is likely that former continental rise prisms have been completely swallowed up along trench systems and are in part resorbed into the mantle and in part added to the continental plates. Such a fate may await today's continental rises.

The important point to be emphasized here is that ancient geosynclines apparently are not unique; there are models that can be obtained from our knowledge of modern areas of thick deposits of sediments and deduction of possible processes in the earth's crust and upper mantle that are compatible with the sedimentary relations we observe in ancient geosynclines and with the tectonic history of geosynclines.

CONTINENTAL ACCRETION

The distribution of geosynclines and attendant belts of crustal deformation have been studied as functions of geologic time and spatial relationships to continents. The approximate age of formation of these belts has been obtained by radioactive dating of minerals in the granitic rocks that are found within these deformational provinces. Thus the age of mountain building for each belt is the time of cooling of minerals within the granitic

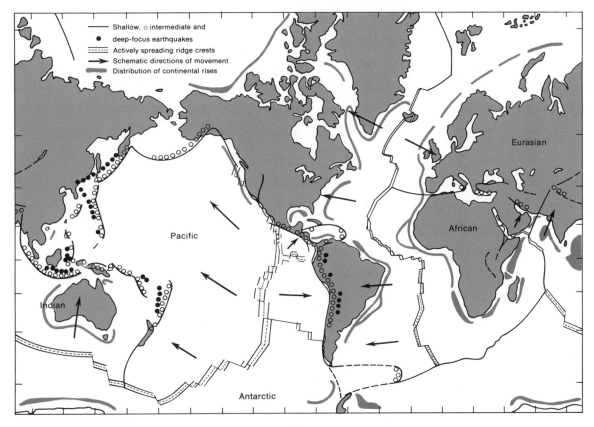

Figure 12.11 Major crustal plates of the earth and their direction of movement. The shaded areas near continental margins are the continental rises (after Vine, 1969).

igneous bodies that formed during crustal deformation. Figure 12.12 shows the patterns and ages of geologic provinces in North America. In general, these provinces exhibit a concentric relationship on the North American continent, with the oldest province near the center of the continent and the youngest near the margin, although in some areas an older belt occurs within a younger one. The provinces are defined on the basis of the ages of the major granite-forming episodes that accompanied mountain building, but may also represent later periods of metamorphism.

The radial pattern of geologic provinces in North America is cited by some scientists as evidence for continental accretion. The original continent, some 2.5–3.0 billion years ago, is thought to have been smaller in area than today. This protocontinent shed debris to the ocean around its edges and geosynclines developed. Deformation of these geosynclines, accompanied by granitic intrusion and mountain building, added new materials to the continental margins. These events took place episodically, resulting in the distribution of ages we now see.

This is the type of pattern that would be predicted from the actualistic

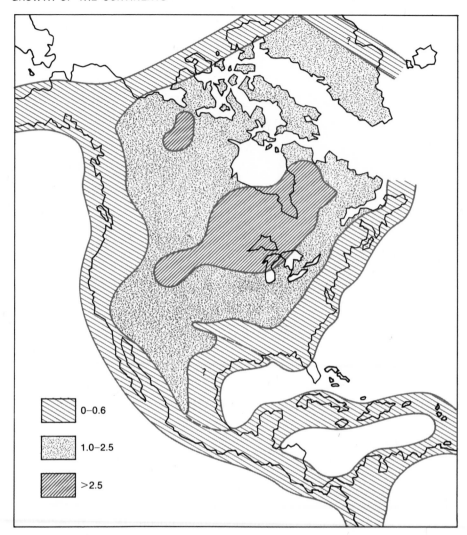

Figure 12.12 Distribution and ages of geologic provinces in North America. The provinces show the times when major granite formation and mountain building occurred. Ages are in units of 10^9 years (after Engel, 1963).

concept of geosynclinal development and mountain building. A repeat of the events developed in Figure 12.9 could produce this structural pattern; the development, collapse, and addition to the continents of successively younger and younger continental terrace and rise sediments and intrusive granites may result in the "accretion" of the continent. However, much the same pattern of the distribution of ages of geologic provinces can be obtained by assuming that the present total continental area does not differ greatly from that of the past and that continental accretion involves chiefly thickening of originally thin continental plates and only a minimum increase in area.

SUMMARY

Our ideas concerning the evolution of continents and ocean basins are advancing at a rapid rate. The concept of the new global tectonics has shattered the viewpoint, engendered by the Moho, that the continents and mantle are two separate systems acting independently of each other. There is an important degree of chemical and physical coupling between the two systems.

To formulate models of the history of continental growth at this stage of rapid advance of knowledge is hazardous. However, some major points have emerged in this book that need further consideration.

Evidence is accumulating that suggests that major degassing of the earth was an early event, resulting in formation of a sialic crust, a sedimentary rock mass, and the oceans. Crustal materials are being added today along the active rift zones of midocean ridges, but it is difficult to say how much of this material is new in the sense of having never been a part of the crust before and how much is recycled by resorption of crust into the mantle along trench systems. It is still a completely open question as to where the original continental plates were and whether at some time they formed a thin, nearly complete layer. Present sea floor spreading is such a rapid process that blocks of the lithosphere could have moved around the earth several times during geologic time. Today's continents may now occupy the sites of ancient ocean basins. This view would account for the approximately exponential decrease of the preserved sedimentary rock mass with age because the older the rock, the more time it has had to be destroyed by the processes of erosion and metamorphism that accompany mountain building and continental growth. The apparent episodic nature of the latter events may account for the maxima and minima observed in the sedimentary rock mass distribution. In this interpretation the maxima would represent times of greatest preservation of sediments owing to deep burial in geosynclines or to widespread, tectonically stable conditions in which seas inundated a large area of the continents. The minima are times of geosynclinal rebound in which seas leave the continents and sediments are trapped temporarily in terrestrial environments, later to be eroded and deposited at continental margins. It is also possible that the minima reflect times of active sea floor spreading in which sediments are swept under the continents and resorbed into the mantle. The sedimentary mass preserved is a small part of that deposited; it is even possible that events once recorded in sedimentary rocks are no longer observable because these rocks have moved beneath continental blocks and have been redistributed within the mantle and lower crust. Perhaps the sediments of former ocean basins in which shelled organisms first evolved are part of this missing record. On the other hand, sea floor spreading may be short-lived and occur infrequently, so that today's events may be representative of few other epochs in the earth's history.

Continents show a successive decrease in age of surface rocks toward their margins, indicating continuous or pulsating thickening of the crust outward from an initial hub, which may or may not be at the center of the present continent. The processes by which this growth occurs are rather well documented in the vertical sense; great sinuous downwarps, or geosynclines, accumulate several kilometers of sedimentary deposits and are eventually folded and intruded by igneous rocks of intermediate to high silica content. The depressed mass then floats upward, creating mountains and shedding debris. A schematic representation of a series of events to account for the surface age distribution is shown in Figure 12.13. In this interpretation, the ocean basins would widen and deepen with time; the continental crust would be changed from an initial widespread thin layer into thick chunks alternating with basalt-floored ocean basins. The drive for these processes could be accounted for by the tendency for differentiation of the mantle under the influence of radioactive heating; a continuation, through cyclical convection, of the movement of one set of elements toward the surface and the concentration of the residual set at depth.

The sequence of events leading to the distribution of continents and ocean basins is certainly more complex than shown in Figure 12–13, and depends on the geometry of spreading on the earth. An upwelling beneath an original continental plate may split it asunder, forming an ocean basin where the continent was originally located. It is conceivable that two continents could collide at a trench system and form mountain ranges. The Alps and Himalayas may have resulted from collision of Africa and India with Eurasia. The geometry of spreading today has only recently been worked out; that of the past is essentially unknown.

Much of this book can be summarized by modeling the crust-ocean system as a large factory. Figure 12.14 is such a model. The cyclic nature of the system is obvious. The factory can be divided into two major plants with a connecting conveyor. In the first plant, primary igneous rocks are produced by the cooling of magma generated during the long-term differentiation of the earth. The source of energy for magma generation is the

Figure 12.13 A hypothesis of sequence of major events in the history of the earth's crust. In this view, the areas of the continents may actually shrink while their masses are increasing. The thin crustal plate of A is split apart by two rising convection currents, B. Above the upwelling currents a midocean ridge develops and new crust is added. As illustrated, trenches form along two leading edges of the continental plates, whereas sediments are deposited as continental terraces and rises along the trailing edges of the plates. Where the convection currents move downward, a geosyncline is formed and in C is folded. Abatement of the currents, D, permits the geosynclinal mass to isostatically rise and shed debris to marine basins to form new geosynclines, E. In F and G the above processes are repeated, except trench systems, separated somewhat in time, develop seaward of the continents, resulting in resorption of lithosphere into the mantle and folding of the geosynclinal masses.

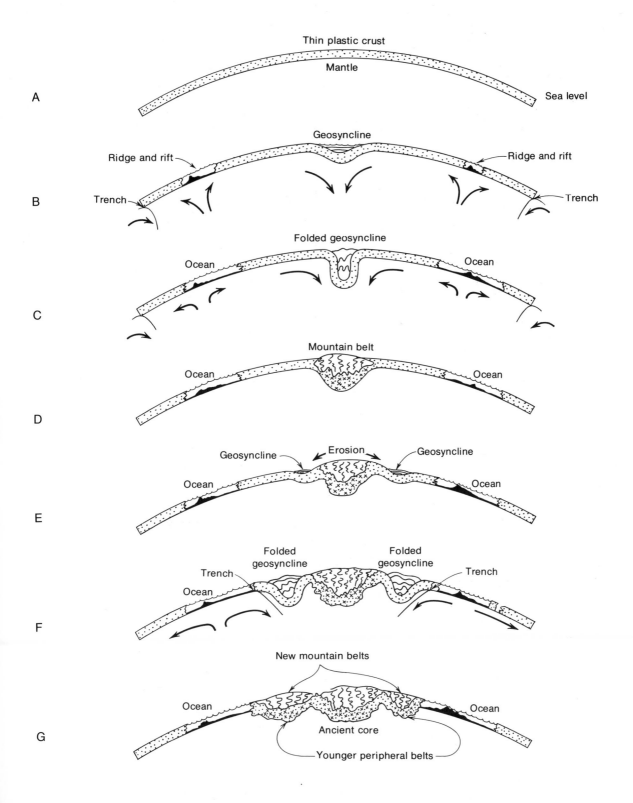

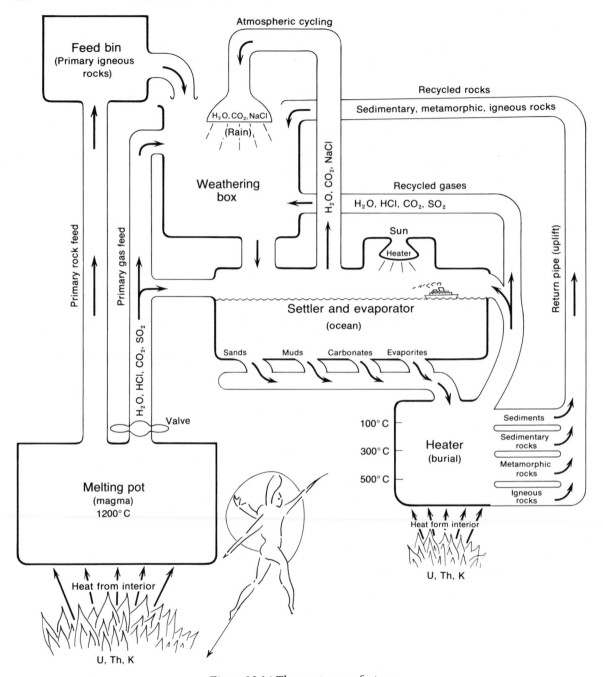

Figure 12.14 The crust-ocean factory.

radioactive disintegration of the elements U, Th, and K in the earth's interior. By-products are acid gases and water that eventually find their way to the earth's surface to react with the igneous rocks and produce weathered products. The reaction of primary igneous rocks with acid gases and H_2O through geologic time has produced the present mass of sedimentary rocks and the oceans.

In the second factory, sediments are cooked by burial; the source of heat energy again is from the decay of radioactive elements. Along the conveyor, primary igneous rocks and sedimentary, metamorphic, and secondary igneous rocks are uplifted to the earth's surface and weathered. The solid debris and dissolved substances produced by chemical weathering are swept into the oceans by streams, where in the huge settler and evaporator—the ocean—they are deposited and accumulate to form thick wedges and prisms of sedimentary rocks. Some materials are recycled through the atmosphere to fall back on the continents. The sedimentary rocks slowly sink under their own weight and are carried to even greater depths by convection currents where they are contorted and fractured. At depth, some of the sediments are little altered, others are lithified with little change in chemical composition, some are recrystallized to metamorphic rocks, while others are melted to form magma that is intruded into the crust to cool as igneous rock. Gases, including H_2O, generated during metamorphism, escape to the earth's surface, where they become agents of chemical weathering. The deformed masses of sedimentary, igneous, and metamorphic rocks slowly rise to form mountain ranges and to provide debris to the oceans once more. The crust-ocean factory has been operative for 4 billion years and there is no end in sight.

REFERENCES

Bateman, P. C., and Eaton, J. P., 1967, Sierra Nevada batholith: *Science, 158*, 1407–1417.

Bullard, E. C., Everett, J. E., and Smith, A. G., 1965, The fit of the continents around the Atlantic: *Phil. Trans. Roy. Soc. London, 258*, no. 1088, 41–51.

Deutsch, E. R., 1966, The rock magnetic evidence for continental drift: in Continental drift, G. D. Garland, ed., *Roy. Soc. London Spec. Publ.* (University of Toronto Press), 9, 28–52.

Dietz, R. S., 1963, Collapsing continental rises: an actualistic concept of geosynclines and mountain building: *J. Geol., 71*, 314–333.

Engel, A. E. J., 1963, Geologic evolution of North America: *Science, 140*, 143–152.

Eskola, P., 1939, *Die Entstehung der Gesteine*, T. F. W. Barth, C. W. Correns, and P. Eskola, eds.: Springer, Berlin, 344.

Ewing, J., and Ewing, M., 1967, Sediment distribution on the mid-ocean ridges with respect to spreading of the sea floor: *Science, 156*, 1590–1592.

Fleischer, R. L, Viertl, J. R. M., Price, P. B., and Aumento, F., 1968, Mid-Atlantic Ridge: Age and spreading rates: *Science, 161*, 1339–1342.

Griggs, D., 1939, A theory of mountain building: *Am. J. Sci., 237,* 611–650.

Hess, H. H., 1962, History of ocean basins: in Petrologic studies: a volume to honor A. F. Buddington, *Geol. Soc. Am.,* 599–620.

Holmes, A., 1965, *Principles of Physical Geology:* The Ronald Press Company, New York.

Hurley, P. M., de Almeida, F. F. M., Melcher, G. C., Cordani, U. G., Rand, J. R., Kawashita, K., Vandoros, P., Pinson, Jr., W. H., and Fairbairn, H. W., 1967, Test of continental drift by comparison of radiometric ages: *Science, 157,* 495–500.

Isacks, B., Oliver, J., and Sykes, L. R., 1968, Seismology and the new global tectonics: *J. Geophys. Res., 73,* 5855–5899.

Kay, M., 1951, North American geosynclines: *Geol. Soc. Am. Mem., 48,* 48–60.

Le Pichon, X., 1968, Sea-floor spreading and continental drift: *J. Geophys. Res., 73,* 3661–3697.

Press, F., 1968, Density distribution in earth: *Science, 160,* 1218–1220.

Vine, F. J., 1969, Sea-Floor spreading—New Evidence: *J. Geol. Ed., 17,* 6–16.

Appendix A
Minerals

This appendix is designed to provide a brief discussion and review of some of the important mineral species for readers with a background in geology or the minimum required amount of basic information for others.

The naturally occurring chemical compounds of the earth are minerals. Some important minerals and their characteristics are given in Table A.1. The behavior of rocks, which are aggregates of these natural compounds, depends on the chemical, physical, and geometric properties of the individual minerals, as well as on the way in which the mineral aggregates are held together.

Only a dozen elements make up 99 percent of the chemical composition of the earth as a whole—iron, oxygen, silicon, manganese, nickel, sulfur, calcium, aluminum, chromium, sodium, potassium, and titanium. The same dozen in different relative abundance make up a comparable percentage of the crust. They are organized into five major types of compounds: elements, oxides, silicates, aluminosilicates, and sulfides. When we focus our attention on the surface environment, we need to add hydrogen and carbon to our list of elements and carbonates and sulfates to the types of compounds.

In general, minerals are natural examples of what the chemists call solid *phases;* that is, they are solid substances that are homogeneous throughout and separated from other substances by a well-defined boundary. Minerals usually occur in grains ranging from submicroscopic size up to giants several meters across. However, in most rocks the maximum size is of the order of a few centimeters or less.

The grains are usually *crystalline,* which is to say that they are made up internally of repetitive groups of atoms in a fixed geometrical array. In some instances, where the minerals have had an opportunity to grow freely and reflect their internal structure in the development of planar faces, the grains are referred to as *crystals.* Also, a grain may be homogeneous throughout by any method of measurement we can devise but not consist of regularly repeated atomic groups; such minerals are called *amorphous.*

Both composition and structure are required to define a mineral; for example, the single compound SiO_2 occurs as the minerals opal, alpha-quartz, beta-quartz, cristobalite, tridymite, coesite, and stishovite, depending on the way in which the silicon and oxygen atoms are arranged within the grains. Figure A.1 shows the structures of some of these *polymorphs* of SiO_2. The word *polymorphs* (often *dimorphs* if there are only two known structures) is used in general to characterize minerals of the same chemical composition but of different crystalline structures. Studies of polymorphs have been of great significance in investigating the earth, because structural changes without chemical change result from changes of the temperature and pressure of the environment, and knowledge of structure may give a clue to the *P-T* conditions of formation.

Table A.1

Some Important Minerals

Mineral	Composition	Crystal system
1. Iron	Fe	Cubic
2. Nickel	Ni	Cubic
3. Gold	Au	Cubic
4. Silver	Ag	Cubic
5. Platinum	Pt	Cubic
6. Copper	Cu	Cubic
7. Quartz	SiO_2	Hexagonal
8. Hematite	Fe_2O_3	Hexagonal
9. Spinel	$Me^{2+}Me_2^{3+}O_4$	Cubic
"The" spinel	$MgAl_2O_4$	Cubic
Hercynite	$FeAl_2O_4$	Cubic
Magnetite	$FeFe_2O_4$	Cubic
Trevorite	$NiFe_2O_4$	Cubic
10. Ilmenite	$FeTiO_3$	Hexagonal
11. Corundum	Al_2O_3	Hexagonal
12. Gibbsite	$Al_2O_3 \cdot 3H_2O$	Monoclinic
13. Olivine	$Me_2^{2+}SiO_4$	Orthorhombic
Forsterite	Mg_2SiO_4	Orthorhombic
Fayalite	Fe_2SiO_4	Orthorhombic
14. Pyroxene	$Me^{2+}SiO_3$	
Enstatite	$MgSiO_3$	Orthorhombic
Wollastonite °	$CaSiO_3$	Triclinic
Diopside	$CaMg(SiO_3)_2$	Monoclinic
Hedenbergite	$CaFe(SiO_3)_2$	Monoclinic

° pyroxenoid

Formula weight	Density (g/cm^3)	Remarks
Elements		
55.85	7.9	Density of solid iron about 12 under conditions of earth's core
58.71	8.9	Probably makes up about 5–10% of core
197.0	19.3	Alloys commonly with Ag
107.9	10.5	Too dense for core at core pressures
195.1	21.5	Densest of all minerals
63.54	9.0	Too dense for core at core pressures
Oxides		
60.09	2.65	Most important oxide mineral in crust; chemically and physically durable at low T
159.70	5.3	Insoluble under most surface conditions
142.3	3.58	The close-packed cubic structure and high density, even at surface conditions, make the
173.8	4.26	spinels candidates for important mantle minerals
231.6	5.21	
234.4	5.37	
151.8	4.79	
102.0	3.99	Chiefly in metamorphic rocks
156.0	2.44	Product of extreme weathering of aluminosilicates
Silicates		
140.7	3.21	Olivines exhibit all compositions from pure forsterite to pure fayalite; other divalent cations also substitute; e.g., Ni^{2+}, Mn^{2+}
203.8	4.39	
100.4	3.12	Important rock-formers, both in low-silica rocks of crust and in mantle
116.1	2.91	
216.6	3.28	
248.1	3.55	

Table A.1 *continued*

Mineral	Composition	Crystal system
15. Amphibole		
Tremolite	$Ca_2Mg_5Si_8O_{22}(OH)_2$	Monoclinic
Iron tremolite	$Ca_2Fe_5Si_8O_{22}(OH)_2$	Monoclinic
Glaucophane	$Na_2Mg_3Al_2Si_8O_{22}(OH)_2$	Monoclinic
16. Feldspar		
Orthoclase (K-spar)	$KAlSi_3O_8$	Monoclinic
Plagioclase		
Albite (Na-spar)	$NaAlSi_3O_8$	Triclinic
Anorthite	$CaAl_2Si_2O_8$	Triclinic
17. Zeolite		
Analcime	$NaAlSi_2O_6 \cdot H_2O$	Cubic
Laumontite	$CaAl_2Si_4O_{12} \cdot 4H_2O$	Monoclinic
Phillipsite	$KAlSi_2O_6 \cdot 2H_2O$	Monoclinic
18. Chlorite	$Mg_5Al_2Si_3O_{10}(OH)_8$	Monoclinic
19. Montmorillonite	$Na_{0.33}Al_{2.33}Si_{3.67}O_{10}(OH)_2$	Monoclinic
	$Ca_{0.17}Al_{2.33}Si_{3.67}O_{10}(OH)_2$	Moncolinic
20. Muscovite (mica)	$KAl_3Si_3O_{10}(OH)_2$	Monoclinic
21. Kaolinite	$Al_2Si_2O_5(OH)_4$	Monoclinic
22. Garnet	$Me_3^{2+}Me_2^{3+}Si_3O_{12}$	Cubic
Almandine	$Fe_3Al_2Si_3O_{12}$	Cubic
Grossular	$Ca_3Al_2Si_3O_{12}$	Cubic
Pyrope	$Mg_3Al_2Si_3O_{12}$	Cubic
Spessartine	$Mn_3Al_2Si_3O_{12}$	Cubic
23. Calcite	$CaCO_3$	Hexagonal
24. Aragonite	$CaCO_3$	Orthorhombic
25. Dolomite	$CaMg(CO_3)_2$	Hexagonal
26. Magnesite	$MgCO_3$	Hexagonal
27. Gypsum	$CaSO_4 \cdot 2H_2O$	Monoclinic
28. Anhydrite	$CaSO_4$	Orthorhombic
29. Pyrite	FeS_2	Cubic
30. Halite	$NaCl$	Cubic

Formula weight	Density (g/cm^3)	Remarks
812.5	2.98	Note presence of $Si_4O_{11}^{-6}$ groups and OH^-; especially important in sheared crustal rocks; composition highly variable; Al commonly replaces some of Si
970.1	3.40	
783.6	2.91	

Aluminosilicates
Framework silicates

278.4	2.55	Most important rock-forming minerals of crust; plagioclase exhibits all compositions from albite to anorthite
262.2	2.62	
278.2	2.76	

220.2	2.26	Zeolites comprise very large group of hydrated aluminosilicates; cations generally almost completely replaceable
470.4	2.2–2.3	
254.3	2.2	

Layer silicates

555.7	2.6–2.8	Most important minerals of sedimentary rocks; those given here are representative of a much larger group; generally not stable at high T and P
257.5	2–3	
256.6	Variable	
398.3	2.8	
258.1	2.65	

443.7	4.25	Probably important upper-mantle and lower-crust minerals
396.4	3.53	
349.1	3.51	
441.0	4.18	

Carbonates

100.1	2.71	Chief minerals of limestones; aragonite is high P form but is common at low P and T
100.1	2.93	
184.4	2.87	
84.3	3.01	

Sulfates

172.1	2.32	Major products, with NaCl, from evaporation of sea water
136.2	2.96	

Sulfides

120.1	5.02	Much of sulfur of sedimentary rocks occurs in pyrite

Halides

58.45	2.16	Major mineral in basins of evaporation

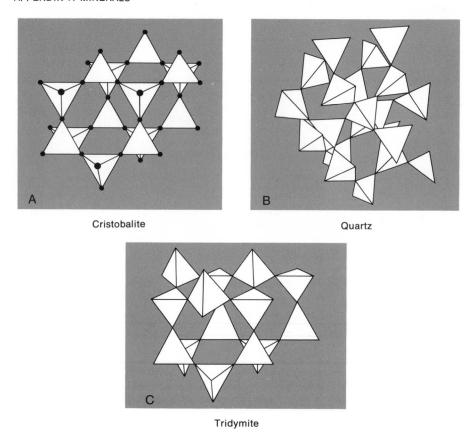

Figure A.1 Structures of cristobalite, quartz, and tridymite; three of the polymorphs of silica. The individual tetrahedra are made up of a small central silicon atom (not shown) surrounded by four oxygen atoms. In A, the oxygen atoms at the corners of the tetrahedra are shown.

MINERAL TYPES (SEE ALSO TABLE A.1)

Elements

Many minerals are simply crystalline grains of the element. Among them are iron, nickel, silicon, sulfur, copper, silver, gold, arsenic, bismuth, and platinum. "Native" iron, of great importance in the deeper portions of the earth, has only a few occurrences in rocks exposed at the surface, as at Ivigtut, Greenland. Native silicon has never been observed but may be an important alloying element in the earth's core.

Oxides

Some of the most important oxide minerals are quartz (SiO_2), hematite (Fe_2O_3), corundum (Al_2O_3), ilmenite ($FeTiO_3$), and spinel (which includes a wide variety of compositions in a characteristic structure). The spinel structure is shown in Figure A.2. The large oxygens are in cubic array in such an arrangement that the interstices can be occupied by cations of two distinct size groups. Small cations such as Al^{3+} or Fe^{3+} fit into one position and somewhat larger ones such as Fe^{2+} or Mg^{2+} fit into the other. Representative chemical formulas are $MgAl_2O_4$ (or spinel proper), $Fe^{2+}Fe_2^{3+}O_4$ (magnetite), and $MgFe_2^{3+}O_4$ (magnesioferrite). Neither of the two positions need be filled with a single cation. Spinels probably are important minerals in the mantle, and their compact structure and high density, combined with their versatility in accepting a variety of cations, make them natural candidates as materials for the mantle. In the crust they are not a major constituent of rocks, probably because there is enough SiO_2 there to permit formation of silicates rather than oxides.

Most important of the oxide minerals in the crust is quartz. The low-temperature modification (alpha-quartz) is by far the most abundant in the crust and in fact makes up about 10–15 weight percent of crustal rocks. Quartz is probably scarce or absent in the lower continental crust and in the oceanic crust, which nevertheless contain about 40–50 weight percent of SiO_2 combined with other elements. Nearly pure quartz rocks are accumulated by depositions from wind, waves, and streams; quartz is rather unreactive at earth-surface temperatures. Hematite is rare in the earth as a whole, but is concentrated locally in the crust as nearly pure masses that are exploited commercially.

Figure A.2 Structure of spinel.

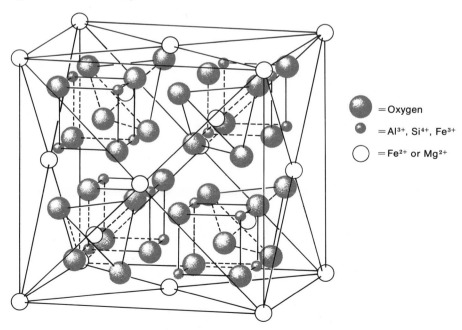

= Oxygen

= Al^{3+}, Si^{4+}, Fe^{3+}

= Fe^{2+} or Mg^{2+}

Silicates

The important silicate minerals belong to three groups: the olivines, the pyroxenes, and the amphiboles. In the olivines, individual SiO_4^{-4} tetrahedra are linked by various cations, and the general formula is $Me_2^{2+}SiO_4$. The pyroxenes are made up of chains of SiO_3^{2-} groups, whereas the amphiboles have a double-chain structure of repeating $Si_4O_{11}^{-6}$ groups and characteristically have OH in their structure (Figure A.3). A great variety of cations are found balancing the negative charge of the silicate groups, and compositional variation is further complicated by substitution for Si, most importantly by aluminum. The ratio Si/O increases in the sequence olivine, pyroxene, amphibole.

　　Most investigators think that the mantle is largely composed of olivine and pyroxene material. In the crust these minerals occur abundantly in low-silica rocks, such as the lavas found in the ocean basins. The olivines form a continuous solid solution series from Mg_2SiO_4 to Fe_2SiO_4. The amphiboles are most characteristic of relatively shallow rocks that have been squeezed and distorted in the presence of H_2O. Various investigators are beginning to hypothesize that amphiboles may be important species in the upper mantle.

Aluminosilicates

The aluminosilicates are not clearly separable from the silicates, but it is convenient to distinguish the important feldspar and zeolite groups, which are highly aluminous, from the olivines, pyroxenes, and amphiboles, in which aluminum is commonly present but is not an essential structural element.

Figure A.3 Structural framework of pyroxenes (A) and amphiboles (B).

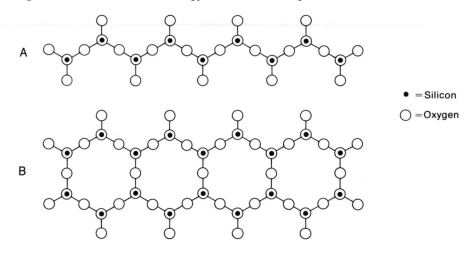

● = Silicon

○ = Oxygen

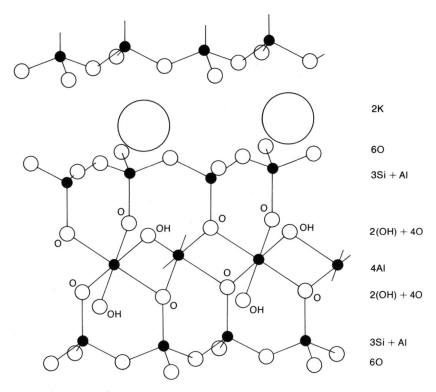

2K

6O

3Si + Al

2(OH) + 4O

4Al

2(OH) + 4O

3Si + Al

6O

Figure A.4 Structure of muscovite mica.

The feldspars are divided into the potassium feldspars, $KAlSi_3O_8$, and the plagioclase feldspars, which form a continuous solid solution from $NaAlSi_3O_8$ to $CaAl_2Si_2O_8$. Potassium feldspar may contain several percent sodium substituting for potassium, and potassium enters the plagioclase structure to an even more limited extent. The feldspars are the most important mineral group in the crust,, making up more than 50 percent of many rocks. They are probably unimportant in the mantle and core.

The zeolite group contains about 25 distinct species. Many resemble hydrated feldspars from the compositional viewpoint. A major characteristic of the zeolites is their structure, in which $(Al,Si)O_4$ tetrahedra are linked together in such a way as to leave spaces in the crystal structure. Consequently, whereas the cations of feldspars are tightly enclosed and cannot be removed without disrupting the structure, the zeolites can replace one cation by another without disturbing the basic structure. Zeolites are not very important as rock-forming minerals, but they are representative "upper-crust" or surface environment species. They are found abundantly in rocks of basaltic composition and form an important portion of some deep-sea sediments.

Layered Aluminosilicates

There is a large group of hydrated aluminosilicates with a sheet structure. Mica is a well-known example. The micas are made up of stacks of sheets built from Al-O and Si-O tetrahedra, with various cations between the layers. Many show exchange properties like the zeolites in that the interlayer cation can migrate in and out between the layers. Figure A.4 illustrates the structure of mica. Soils and sediments tend to be dominated by clay minerals, layer silicates characteristically with a grain size of the order of 1μ. The layer silicates are typical of the surface environment of the earth and in general are products of the weathering of other minerals. Some, such as mica and chlorite, exist as important rock-forming species in the upper crust, but at high temperatures and pressures, more compact structures are formed and the OH-groups in the layered aluminosilicate structure are expelled.

Carbonates

The carbonate minerals calcite ($CaCO_3$), its dimorph aragonite, and dolomite [$CaMg(CO_3)_2$] make up about 10 percent of the sedimentary rocks of the upper crust. Magnesite ($MgCO_3$) is relatively rare as a rock-forming mineral. In general, the carbonate minerals are products of surface processes, and $CaCO_3$, with or without some Mg, is precipitated by organisms to form shells and other types of skeletons. Carbonatites, rocks made up of calcium carbonate formed at many hundreds of degrees centigrade, are rare.

Sulfates

Only two sulfates will be considered here—anhydrite ($CaSO_4$) and the dihydrate, gypsum ($CaSO_4 \cdot 2H_2O$). Both occur abundantly where sea water has undergone extensive evaporation.

Sulfides

There are many sulfide minerals, and they have great economic importance, yielding much of the copper, lead, zinc, and many other metals of industrial importance. However, except for pyrite (FeS_2), they form an insignificant fraction of earth materials. Pyrite is formed abundantly in the surface environment in areas where bottom sediments are oxygen-deficient. It is a minor constituent of crustal rocks, but it or pyrrhotite (FeS) may well be important in the lower mantle.

Appendix B

Mineral Chemistry

In this appendix some of the fundamental chemistry currently used in describing minerals and their relations to their environments will be presented. Special emphasis is placed on the descriptive variables most commonly employed, together with the graphical techniques used in representing them. The material presented should be useful in understanding reactions between minerals and waters.

CHEMICAL NOTATION FOR MINERALS

Mineral formulas are written in the usual manner for chemical compounds, i.e., $CaCO_3$, Fe_2O_3. Oxide formulas are in common use; the composition of the mineral albite is often written as $Na_2O \cdot Al_2O_3 \cdot 6SiO_2$ instead of the familiar form $NaAlSi_3O_8$. If structural information is available, the oxide formula is superseded; using $Al_2Si_2O_5(OH)_4$ instead of $Al_2O_3 \cdot 2SiO_2 \cdot 2H_2O$ for the mineral kaolinite specifies that the H and part of the O are present in the structure as hydroxyl ion. Note that the use of the oxide formula may change the formula weight; $Al_2O_3 \cdot 3H_2O$ is $2[Al(OH)_3]$. This point is particularly important when calculations are made involving mole ratios or energy contents per mole.

The chemical formula alone may not be adequate to describe a mineral species; subscripts are used as required. The notation SiO_2 does not distinguish among the polymorphic forms of silica, but $SiO_{2\alpha}$ can be used to designate alpha or low quartz. Subscripts are also used to designate state; i.e., H_2O_g, H_2O_l, and H_2O_{hex} refer to steam, liquid water, and hexagonal ice, respectively. Most of the subscripts are self-explanatory; a list of those in common use is as follows:

amorph	amorphous
gls	glass
cr	crystalline
g	gas
l	liquid
aq	dissolved

The degree of polymerization of a chemical species, if known, is shown either by increasing the numerical subscripts, as $Al_{20}(OH)_{60}$, or by using a coefficient and parentheses or brackets, as $20[Al(OH)_3]$.

Solid solution or ionic substitution is usually shown by fractional subscripts and parentheses $(Fe_{0.7}Mg_{0.3})SiO_3$; sometimes the subscripts are increased proportionally yielding whole numbers, as $Fe_7Mg_3Si_{10}O_{30}$, but in this text whole numbers will be reserved for minerals with a small range in composition. $CaMg(CO_3)_2$ thus indicates the mineral dolomite and further that it is a discrete phase; $(Ca_{0.5}Mg_{0.5})CO_3$ indicates a solid solution of $CaCO_3$ and $MgCO_3$ of intermediate composition.

Ion exchangers are a special kind of *solid solution* in which a relatively indestructible negatively or positively charged structural unit of a compound can rapidly *exchange* the cations or anions that balance the negative or positive charge of the structural framework. The charge on the framework originates from partial substitution within the framework of an element of a given charge for an element with a higher or lower charge. The charged framework is often written in brackets, with the net charge shown as a superscript. The formula $Na_{0.33}Al_{2.33}Si_{3.67}O_{10}(OH)_2$, or $Na_{0.33}[Al_{2.33}Si_{3.67}O_{10}(OH)_2]^{-0.33}$, represents the composition of one of the montmorillonite minerals in which there is an aluminosilicate group with a net charge of -0.33 and the exchangeable ion is Na^+. Additional structural information is sometimes given by writing the formula $Na_{0.33} \cdot [(Al_2)(Al_{0.33}Si_{3.67}O_{10})(OH)_2]^{-0.33}$, which shows that part of the Al may be considered as substituting for Si (in solid solution in the silica tetrahedra of the mineral structure). The specific additional information imparted is that there are two kinds of Al in the crystal structure.

GEOMETRIC REPRESENTATION OF MINERAL COMPOSITIONS

Currently there is increasing use of geometrical methods for portraying the compositions of minerals. These graphical devices are highly efficient in depicting in limited space a great deal of information on compositions and in addition are useful mnemonic devices. Also, they form the basis for illustration and prediction of chemical relations among minerals.

Bar Diagrams

The smallest number of chemical entities that can be used to describe the compositions of a group of minerals can be called the *components* of the group. In the simplest case, it is clear that SiO_2 is adequate to describe low quartz, high quartz, tridymite, cristobalite, coesite, and stishovite, the polymorphs of SiO_2. These minerals make up a one-component system. Compositional representation is not a problem.

Next consider the minerals corundum (Al_2O_3), diaspore ($Al_2O_3 \cdot H_2O$), and gibbsite ($Al_2O_3 \cdot 3H_2O$). They contain three elements—Al, H, and O —but can be described by only two components—Al_2O_3 and H_2O. A bar diagram can be used to show their compositions in terms of components. Both weight ratios and mole ratios are used to illustrate compositional relations. In Figure B.1 the upper line has been drawn with one end representing 100 percent Al_2O_3 and zero H_2O and the other end 100 percent H_2O and zero Al_2O_3, with a linear scale between. Corundum lies at one end of the line, diaspore is at 50 mole percent of each component, and gibbsite is 75 mole percent of H_2O and 25 mole percent Al_2O_3. The lower line shows the same system but with compositions represented in weight percent. Mole percent diagrams are used more frequently than weight percent diagrams in this book.

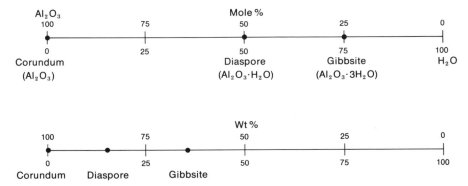

Figure B.1 Bar diagrams showing mineral compositions in the system Al_2O_3-H_2O. The upper line is plotted on a mole percent basis and the lower line on a weight percent basis.

Triangular and Orthogonal Diagrams

Although oxides are widely used as components, they are not necessarily always components, or there may be an alternative choice of components. For the minerals nahcolite ($NaHCO_3$), trona ($NaHCO_3 \cdot Na_2CO_3 \cdot 2H_2O$), and natron ($Na_2CO_3 \cdot 10H_2O$) the oxides Na_2O, CO_2, and H_2O are components, but $NaHCO_3$, Na_2CO_3, and H_2O are also components. For three component systems triangular or orthogonal diagrams are used. Figure B.2 shows alternative plotting of nahcolite, trona, and natron on a triangular diagram. Each side of the triangle represents a two-component system. Points representing compositions are plotted by obtaining the mole percent of each component, just as in the two-component systems.

Figure B.2 Triangular diagrams showing alternative plotting of the minerals nahcolite, trona, and natron, using two different sets of components.

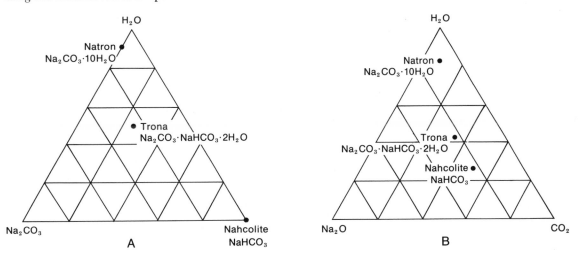

Orthogonal diagrams for three-component systems are plotted by using mole ratios (or weight ratios) of components as the axes, keeping one of the components as the denominator in both ratios. Figure B.3 shows the same sodium carbonate minerals as in Figure B.2 plotted on an orthogonal diagram.

Triangular diagrams are used almost exclusively in the United States; orthogonal diagrams are popular in the extensive Russian literature. Each has advantages and disadvantages that will appear later.

Systems of four components are represented either by equilateral pyramids or by orthogonal diagrams with three mutually perpendicular axes.

Figure B.4 shows some of the many minerals whose compositions can be described by the four components K_2O-$Al_2Si_2O_7$-SiO_2-H_2O. The minerals that plot on the edges or the faces of the pyramid are adequately portrayed, and their compositions can be read without a great deal of difficulty, but for "internal phases" whose compositions lie within the pyramid, the ratios of the components are not easy to determine. Because of this difficulty, three-dimensional diagrams showing four components are to be avoided if composition is the parameter of interest.

Many stratagems have been used to avoid three-dimensional representation for systems with more than three components. There inevitably is loss of information no matter what device is employed, but the parameters not shown often can be chosen to preserve the relations of particular interest. For the system shown in Figure B.4, the important compositional rela-

Figure B.3 Orthogonal diagrams showing the composition of the minerals nahcolite, trona, and natron, using mole ratios of oxides as axes.

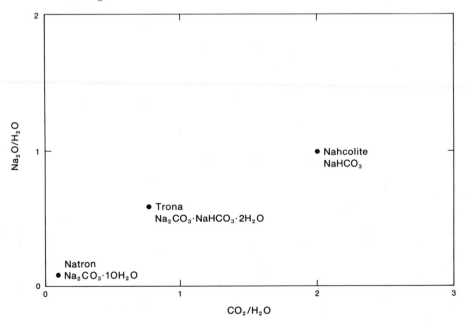

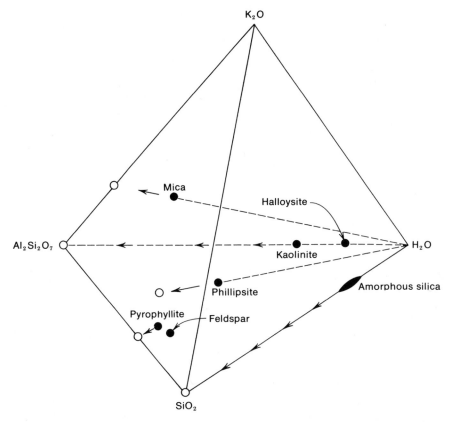

Figure B.4 Compositions of some of the minerals in the system K_2O-$Al_2Si_2O_7$-SiO_2-H_2O plotted on a triangular pyramid. Arrows show how compositions above the $Al_2Si_2O_7$-SiO_2-K_2O plane can be projected onto that plane by lines from the H_2O corner passing through the phase compositions. Note the difficulties in reading compositions.

tions among the minerals, for certain purposes, might be the mole ratios of K_2O, $Al_2Si_2O_7$, and SiO_2. If so, the compositions of all minerals that lie above the K_2O-$Al_2Si_2O_7$-SiO_2 basal triangle can be projected upon it by drawing a line from the H_2O apex through the mineral composition point to the basal triangle, as in Figure B.5. This geometric procedure is equivalent to writing the mineral formulas in terms of oxides, eliminating H_2O, calculating the mole fractions of K_2O, $Al_2Si_2O_7$, and SiO_2 on this water-free basis, and then plotting directly on the triangle. Such a representation assumes the presence of H_2O with all other phases.

Triangular diagrams will be used extensively in this book to show mineral composition relations and will be plotted on a mole basis unless otherwise indicated.

Figure B.6 is a triangular composition diagram in the system MgO-SiO_2-H_2O and is included here to emphasize the mnemonic aspects of composition diagrams. One can see on one diagram the compositional

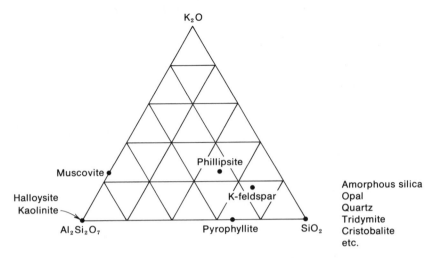

Figure B.5 Projection of mineral compositions in the system K_2O-$Al_2Si_2O_7$-SiO_2-H_2O to the K_2O-$Al_2Si_2O_7$-SiO_2 plane.

differences among a large group of minerals, and the mineral names and compositions can be remembered in terms of their patterns.

Compositional Variation in Minerals

The minerals considered so far have had fixed compositions and have been plotted as points on the composition diagrams. However, the basis of mineral classification today is chiefly structural rather than compositional. A

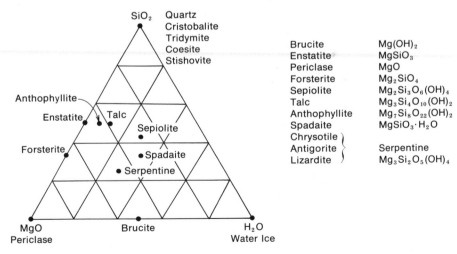

Figure B.6 Triangular composition diagram of some of the minerals in the system MgO-SiO_2-H_2O.

mineral may have a wide range of chemical compositions and still parade under the same name. A classic example is olivine, which ranges continuously in composition from Mg_2SiO_4 to Fe_2SiO_4 (plus small percentages of other elements). On the other hand, the crystal structure remains essentially unchanged throughout the range. The problem is a little like the old question of how far one can go in replacing the parts of a person's body before he becomes someone else. Sometimes olivine is treated as a mineral group, and various arbitrary ranges of composition are given individual mineral names.

Figure B.7 illustrates the problem of compositional variation and how it is handled. Let us suppose that several hundred mineral grains were collected and that each individual grain was shown to be crystalline and homogeneous throughout. Initial examination with a microscope showed that there were five distinctly different kinds of grains, with marked differences in crystal structure as indicated by indices of refraction and other

Figure B.7 Triangular diagram showing results of plotting compositions of 500 mineral grains. Note that the term *mineral* is essentially defined by the density of points. There is little question that the individual minerals can have any composition within the enclosed areas.

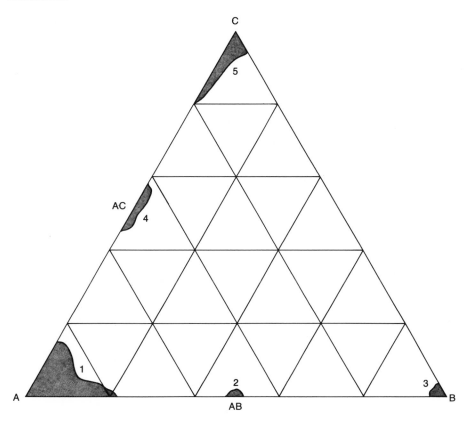

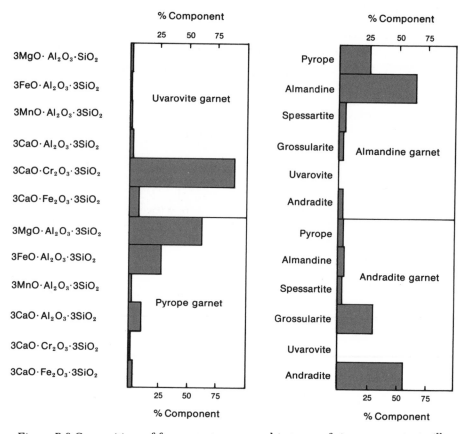

Figure B.8 Compositions of four garnets, expressed in terms of six components, to illustrate the tremendous range of chemical compositions possible within a single mineral structure (adapted from Deer et al., 1966).

optical properties. The grains within a group have similar but not identical optical properties. Chemical analysis of the individual grains showed that their compositions could be expressed, within the limits of the analysis, in terms of three components, A, B, and C. The dots on Figure B.7 are plots of the individual analyses.

There are five minerals present. The "pure" minerals have the compositions A, B, C, AC, and AB. AB and B show almost no compositional variation, whereas A exhibits limited *solid solution* of B+C. AC shows solid solution of A and C but not of B. Thus the concepts that go into the definition of a *mineral* are a particular internal structure (usually determined chiefly by X-ray methods) and a *continuous* range of composition. Thus minerals are distinguished from each other on the basis of a compositional gap and/or a structural break. There are many residual "gray areas" in the problem of classification, but plots of compositions supplemented by structural information usually give an operational base that bypasses many of the difficulties of nomenclature. The greatest difficulties arise when it is found that a given structure persists throughout compositional variations involving many components. The mineral (or mineral group?) known as montmorillonite is a fine example. At least seven or eight components are necessary to describe the compositional variations, yet the internal structure and grain morphology are only slightly changed. Another such case is the garnet group, in which a unique structure persists through extreme compositional variation. Figure B.8 shows some of the range of garnet composition.

MINERAL SOLUBILITY

In general, minerals are highly insoluble compounds. If they were not, the "Everlasting Hills" would disappear quickly; all streams and springs would be salty and undrinkable. On the other hand, mineral reactivities are sufficient in most cases to permit useful studies of solubility in the laboratory, even at room temperature. Most minerals do not dissolve directly and completely in an aqueous solution; instead they tend to react with the solvent to produce new minerals.

Congruent Solution

If a mineral reacts with an aqueous solution, and if the material dissolved from the mineral has the same composition as the mineral, it is said to dissolve *congruently*. The simplest and best known example probably is NaCl, the mineral halite. Halite can be added to water until about 6 moles of salt/liter of water have been dissolved. After this point is reached, halite no longer dissolves but remains as an excess solid maintaining its original composition, NaCl. The course of the solution process, at constant T and P, can be shown on a simple bar diagram (Figure B.9). When the solution composition becomes constant and is independent of the addition of further solid NaCl, the solution is said to be saturated, and the concentration of NaCl in the solution is the *solubility* of NaCl. To determine whether the solubility value as determined is the "true" or equilibrium solubility, it is usually necessary to approach it from "both sides," that is, from under- and oversaturation. After supposed saturation has been reached, a little pure

Figure B.9 Bar diagram of the system NaCl-H_2O showing the change of the solution composition as sodium chloride is added to H_2O until saturation is reached.

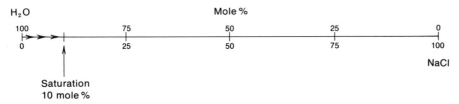

water can be removed from the solution by evaporation, and an increment of NaCl necessary to restore the saturation composition should precipitate. Conversely, a slight dilution of the solution by addition of water should result in the dissolution of more salt. Such an experiment is a classic test for the definition of equilibrium—a system in a state such that a small disturbance in the system is followed by a return to the original condition.

The concentration units that are commonly used for salt solutions, and which can be employed to describe saturation, are the molality, m, of the solution, the molarity, M, and the mole fraction, N.

Molality is moles/1000 g H_2O; molarity is moles/1000 ml solution, and mole fraction is moles per total moles of solvent plus solute. More formally,

$$m = \frac{\text{weight solute} \times 1000}{\text{formula weight solute} \times \text{weight water}}, \tag{1}$$

$$M = \frac{\text{weight solute} \times 1000}{\text{formula weight solute} \times \text{volume of solution}}, \tag{2}$$

and

$$N = \frac{\text{moles solute}}{\text{moles solute} + \text{moles solvent}}. \tag{3}$$

We can obtain from the *Handbook of Chemistry and Physics* a solubility of 35.7 g/100 g H_2O for a saturated NaCl solution at 0°C; the density of the solution is 1.20. The formula weight of NaCl is 58.45 and that of H_2O is 18.00. Substituting appropriately in equations (1), (2), and (3) yields

$$m_{\text{NaCl}} = \frac{35.7 \times 1000}{58.45 \times 100} = 6.1,$$

$$M_{\text{NaCl}} = \frac{35.7 \times 1000}{58.45 \times (\text{weight solution} / \text{density solution})}$$

$$= \frac{35.7 \times 1000}{58.45 \times [(35.7 + 100) / 1.20]} = 5.4,$$

and

$$N_{\text{NaCl}} = \frac{35.7 / 58.45}{(35.7 / 58.45) + (100 / 18.01)} = 0.10.$$

Molality and mole fraction are used almost exclusively in geochemical calculations; on the other hand, it is convenient to make up solutions in the laboratory by volume, and many data are reported as molarities. In dilute aqueous solutions, where the density of the solution approaches unity, m and M are very nearly the same.

For a system involving only NaCl and H_2O, there is a fixed equilibrium solubility at a given temperature and pressure, and the effects of T and P can be shown graphically. Figure B.10 summarizes temperature effects on

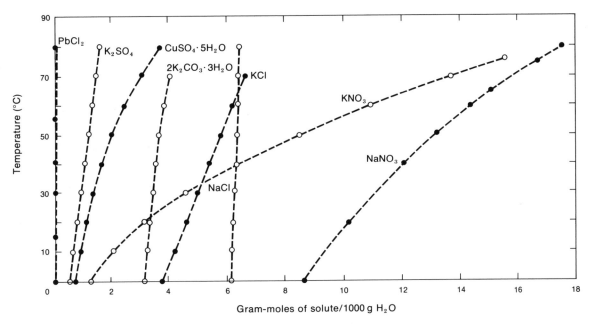

Figure B.10 Temperature dependence of the solubilities of some compounds that dissolve congruently in water.

the solubility of a series of compounds that dissolve congruently in water. Note that they show wide variations in their increases of solubility with increasing temperature.

Figure B.11, taken from Kennedy's work (1950), shows the congruent solubility of quartz as a function of both T and P. Note that the solubility, like that of many other substances, increases with increasing temperature and pressure.

Incongruent Solution

Most minerals, when placed in water, can be considered to react rather than to dissolve, and the meaning of an equilibrium solubility becomes difficult to define. For example, if the common clay mineral kaolinite [$Al_2Si_2O_5(OH)_4$] is put into water, aluminum and silicon dissolve into the solution. As time goes on, the solution reaches a constant composition that is independent of further addition of kaolinite, but the ratio of Si to Al in the solution is much higher than that in kaolinite. Consequently, a second solid phase must be formed, in this case an Al hydroxide. The reaction can be written

$$Al_2Si_2O_5(OH)_4 + H_2O = 2Al(OH)_3 + 2SiO_{2aq},$$

yet this reaction does not express a relation we may want to know; i.e., how much dissolved aluminum is in solution? And further, what is meant by the solubility of kaolinite when it dissolves to give another solid plus solution?

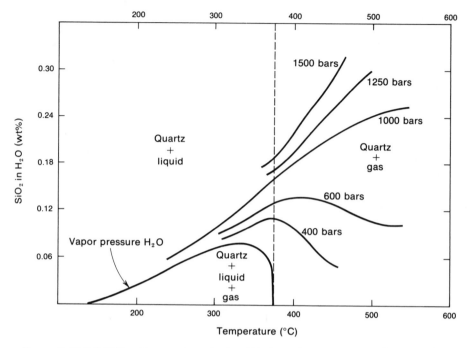

Figure B.11 Solubility of quartz in water as a function of temperature and pressure (after Kennedy, 1950).

We tend to think of "solubility" in terms of situations such as the one in which NaCl is added to water. In such a case, the composition of the solution can be correlated directly with that of the material that has been added. If a saturated NaCl solution is evaporated, the material dissolved can be recovered as solid NaCl. Natural solutions seldom behave so simply; it might be better to avoid using the word *solubility* and restrict our thinking to the composition of solutions in *equilibrium* with a given solid phase. In the kaolinite example, should we consider the "solubility" of kaolinite in terms of the concentration of dissolved SiO_2 in the solution or that of the dissolved Al_2O_3? We could choose either value arbitrarily, or alternatively; we could use both by asking the question "How many moles of kaolinite will be used up if kaolinite is added to water and permitted to come to equilibrium?" Consequently, henceforth the compositions of solutions in equilibrium with solids will be discussed, but solubility in the classic sense will rarely be employed.

REPRESENTATION OF EQUILIBRIA BETWEEN MINERALS AND AQUEOUS SOLUTIONS

The compositions of $NaCl-H_2O$ solutions have previously been shown on the bar diagram in Figure B.9. The $NaCl-KCl-H_2O$ system is a natural extension to three components. Figure B.12 summarizes the experimental results of adding KCl and NaCl in various ratios to water until a solid phase remains. In this system, at room temperature, the KCl and NaCl in equilibrium with the solutions are nearly pure end members of the two-component system NaCl-KCl. The diagram shows that NaCl alone is much more soluble in water than KCl (points 1,2). The line 1-3-2 gives the compositions of the solutions saturated with either NaCl or KCl, and at point 3, both NaCl and KCl are in equilibrium. The lines drawn from the composition of NaCl to the solution boundaries between points 1 and 3 are called tie lines and show that all solutions along the line 1-3 are in equilibrium with NaCl; similarly, the solution compositions along line 3-2 are in equilibrium with KCl, whereas at point 3 both minerals are in equilibrium with the solution.

The problem of defining *solubility* is well shown by the diagram. Between points 1 and 3, the solution is saturated with NaCl only. Na in solution diminishes, whereas Cl^- increases $(Na^+ + K^+ = Cl^-)$. The question

Figure B.12 Composition diagram of the system KCl-NaCl-H_2O. The line 1-3-2 is the composition of the saturated solution. The arrow starting at point A shows the changes in solution composition resulting from the evaporation of an initially undersaturated solution.

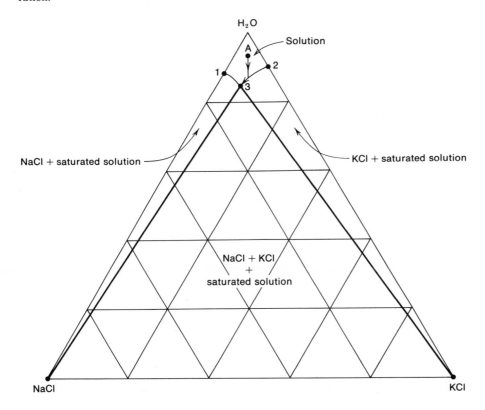

arises, should the "solubility" of NaCl be measured by Na or Cl? In this instance, we presumably would use the concentration of Na^+ as our measure.

Representations like Figure B.12 are sometimes known as "drying-up" diagrams. For example, if an NaCl-KCl solution originally had a composition at point A and then was evaporated, the NaCl/KCl ratio in solution would remain constant and the solution composition would then move along the arrowed line until it reached the saturation boundary for KCl. With continuous evaporation, KCl would precipitate and the solution composition would change along the line 2-3 until the solution also saturated with NaCl at point 3. At this point both phases would continue to crystallize until all the water was evaporated.

Figure B.13 shows the same system on an orthogonal plot that uses the mole ratios of $NaCl/H_2O$ and KCl/H_2O as axes. The various points are labeled exactly the same as in the preceding figure. In this case, the diagram is a little inconvenient because the compositions of the pure solids are at an indefinite distance along the axes.

A long chapter could be devoted to drying-up diagrams alone; in this book emphasis is placed on showing the compositions of solids and coexisting equilibrium solutions.

The great drawback in using typical triangular diagrams in representing solution equilibria is that they are based on mole or weight ratios, and

Figure B.13 Orthogonal diagram showing the composition of saturated solutions in the system NaCl-KCl-H_2O. The lines joining solution compositions to those of the pure compounds are drawn parallel to the axes because the pure solids plot at infinity in this representation.

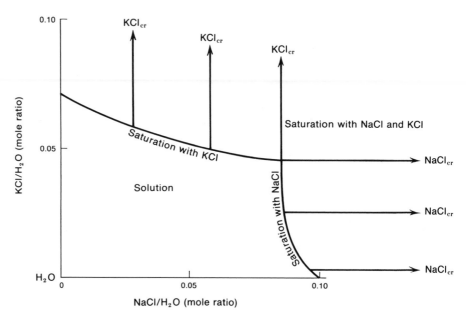

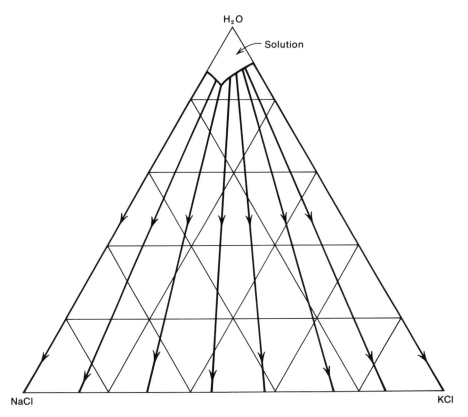

Figure B.14 Projection of the compositions of saturated solutions of the system NaCl-KCl-H₂O system onto the NaCl-KCl axis. Lines are drawn from the H₂O apex to the base of the triangle.

unless the solution carries a substantial concentration of dissolved material, the composition of the equilibrium solution is so crowded up against the H_2O corner that the diagram is not legible. This difficulty can be overcome in part by projecting the solution composition onto the base of the triangle to obtain the NaCl/KCl ratio in solution and then plotting a double bar diagram correlating NaCl/KCl in solution versus NaCl/KCl in the solids. Figure B.14 shows the projection of the NaCl/KCl system, and Figure B.15 is the resultant double-bar diagram. As usual in a projection of this kind, some information is not represented; in this case the actual concentrations in solution are unknown and only the ratios are retained.

Sometimes the H_2O corner of the triangle is magnified to show relationships, but this procedure is often confusing and, as will be seen later, vitiates many of the useful properties of the triangular diagrams.

Figure B.16 shows the equilibrium relations of kaolinite, gibbsite, and quartz in the three-component system Al_2O_3-SiO_2-H_2O—first true scale and then with the solubilities magnified by a factor of 10,000.

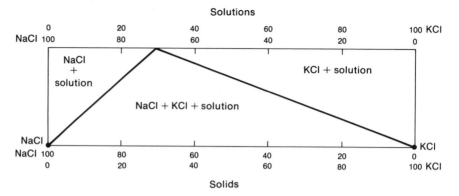

Figure B.15 Double bar diagram derived from the projections of Figure B.14. Note that only the ratios of NaCl to KCl in the saturated solutions are represented. Actual solubilities were lost in the projection process.

Solid Solutions

So far the minerals considered have had fixed compositions, but the same kinds of diagrams can be used to show relationships between aqueous solutions and mineral solid solutions.

In the NaCl-KCl system, the materials precipitated from aqueous solu-

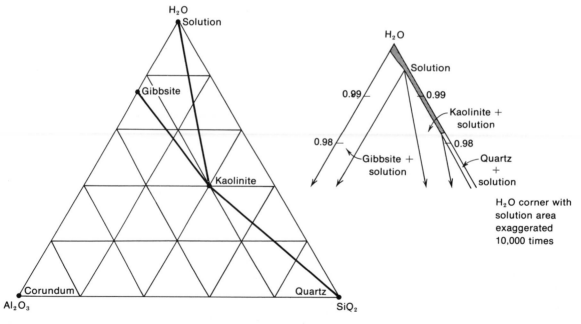

Figure B.16 Composition of solution in equilibrium with kaolinite, gibbsite, and quartz shown first true scale and then with the solution area exaggerated 10,000 times. Even this magnification plus exaggeration is not sufficient to permit satisfactory depiction of the equilibrium solution.

tion at room temperature are nearly pure end members of the two-component system, as shown in Figure B.12, but with increasing temperature, K^+ begins to enter the NaCl structure, and Na^+ goes into KCl. Figure B.17 is an idealized diagram illustrating how the compositions of solids change with changes in solution composition for a system of two chlorides. The lines connecting the compositions of the solids with those of the aqueous solutions can be used to determine the composition of the saturated solution and of the coexisting solid.

Even at elevated temperatures, the problem of low solubility makes representation of the actual solution composition difficult, and in most cases only the ratios of various dissolved constituents are shown.

Ion exchangers, such as the clay minerals, are rapidly responsive to their aqueous environment. In some waters, the exchange positions of montmorillonite-group minerals may be occupied by K^+, whereas in others the K^+ may be displaced by Na^+. For such an exchange, the Na form of the mineral can be considered as one component and the K form as another. Figure B.18 indicates that the double-bar diagram is well suited to show

Figure B.17 The system $NaCl$-KCl-H_2O at elevated temperature and pressure showing how compositions of solids change with compositions of coexisting solutions.

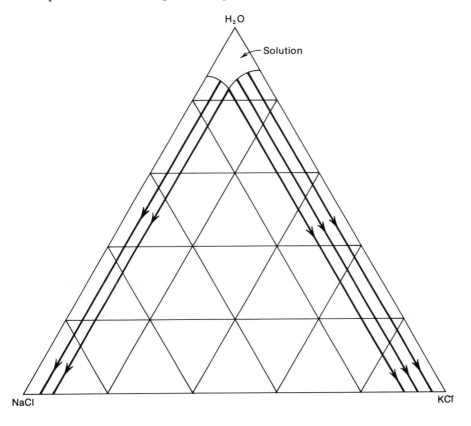

Relative mole % in solution

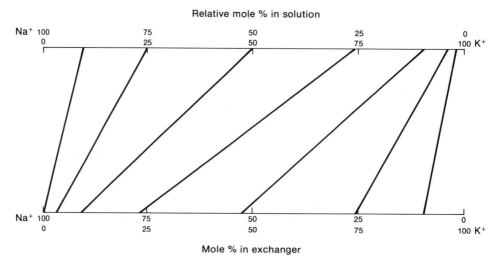

Mole % in exchanger

Figure B.18 Double-bar diagram showing the relative mole percent of Na^+ and K^+ in aqueous solutions in equilibrium with an ion exchanger showing complete exchangeability of Na^+ and K^+. The ratio of Na^+/K^+ in the exchanger to Na^+/K^+ in solution is constant at 10X.

the relations between Na^+-K^+ in the aqueous solution and Na^+-K^+ in the exchange positions of the montmorillonite. For ion exchangers which form complete solid solutions, however, a simple graph plotting ratios of ions in solution to those in the exchanger can be used, or, even better, a plot of the logarithms of the ratios. These relations are illustrated in Figures B.19 and B.20.

The Chemical Basis for Mineral Stability Diagrams

The chemistry of mineral stability relations is complex if put on a quantitative basis. Here it will be handled entirely qualitatively but in sufficient detail to show the descriptive variables commonly used. When two phases are in equilibrium at a given temperature and pressure, each is "satisfied" with respect to the other phase and the surrounding conditions. The components of the phases are so distributed that they have no net tendency to move from one phase into the other. For two coexisting feldspars, this means that if we consider K_2O as one of the components in the system, a very small content of K_2O in the crystal structure of the albite (Na-feldspar) has the same tendency to move into the orthoclase (K-feldspar) as the very large content of K_2O in the orthoclase has to move into albite. This tendency to move has been called the *chemical potential,* and chemical equilibrium can be redefined as a condition in which the chemical potentials of all components in the system are equal in all phases. Figure B.21 is a diagram of the two-component system $CaCO_3$-$MgCO_3$ at a particular temperature and pressure. Tie lines show that calcite can be in equilibrium with dolomite and that dolomite can be in equilibrium with magnesite but that magnesite cannot be in equilibrium with calcite. From the standpoint of chemical potential this means that if pure calcite and magnesite are mixed intimately together, under these conditions the chemical potential of $CaCO_3$ in the magnesite is originally zero, as is that of $MgCO_3$ in the calcite. Therefore, there would be a tendency for $MgCO_3$ to move into the calcite and vice versa until the chemical potentials are equalized. As this process takes place we can visualize an increase in the $MgCO_3$ content of the calcite and of the $CaCO_3$ content in the magnesite, but before the chemical potentials become equal, either the calcite or the magnesite acquires so much of the other component that it becomes unstable structurally and changes to the dolomite structure. If the experiment were started with an excess of magnesite over calcite the final products would be magnesite and dolomite; if calcite were in excess, the final products would be calcite and dolomite.

It is possible to use chemical potentials to describe equilibrium among

Figure B.19 Linear graphical representation of ratio of ions in solution to ratio in an ion exchanger. Note difficulty of showing very large or very small ratios.

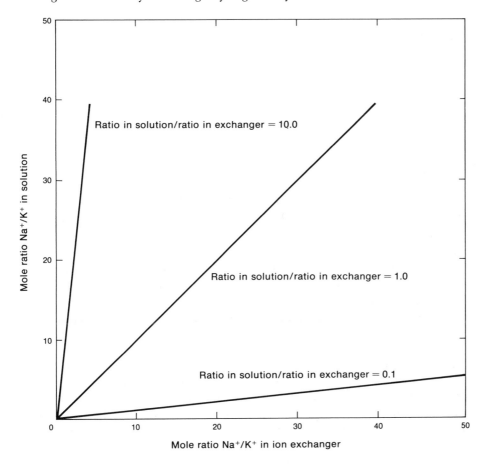

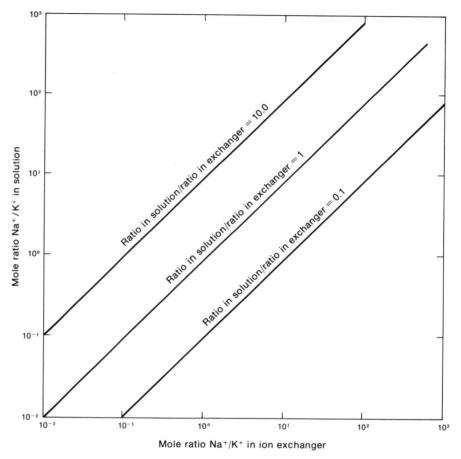

Figure B.20 Logarithmic plot of the same ratios shown in Figure B.19.

phases when magnesite and dolomite occur together at a given temperature and pressure. There must be a fixed ratio of the chemical potentials of $CaCO_3$ and $MgCO_3$. Similarly, coexistence of calcite and dolomite requires a different but also fixed ratio of the chemical potentials. Figure B.21 also shows a plot of the ratios of the chemical potentials of the components in this system versus the compositions of the phases.

The concept that each component in a mineral has a potential also requires that if the mineral is to persist at a given place the chemical potentials of its components in the surrounding environment must be equal to or greater than those in the mineral. If the chemical potential of $CaCO_3$ in an aqueous environment surrounding calcite is less than in calcite, $CaCO_3$ will tend to leave the calcite and go into the surrounding environment until the potentials are equal or the calcite disappears. Note that the chemical potential of a component in a mineral is an *intensive property;* that is, in the above example, it is independent of the amount of calcite present.

One of the convenient aspects of orthogonal diagrams for showing the compositions of minerals in three-component systems is that lines representing equilibrium ratios of chemical potentials of components of the phases are perpendicular to lines joining the compositions of the phases. This relation permits easy transformation from composition diagrams to chemical potential diagrams. Figure B.22A is an orthogonal diagram showing the compositions of some phases in the system CaO-MgO-CO_2-H_2O. The solid lines are the equilibrium relations or compatibilities of known phases; the dashed lines show the slopes of the lines representing equilibrium ratios of the chemical potentials. In Figure B.22B the chemical potentials have been plotted as axes and the stability fields of the phases shown as a function of them.

In Figure B.22A we see from the tie lines that calcite and dolomite can coexist at equilibrium with a solution saturated with CO_2, as can dolomite and magnesite, but magnesite and calcite cannot. Also, magnesite, dolomite, and hydromagnesite can coexist but not in the presence of a CO_2 gas phase.

The chemical potential diagram tells us that calcite, CO_2, and solution coexist at equilibrium at some fixed ratio of the chemical potentials of CaO and CO_2, which is independent of the chemical potential of MgO. On the other hand, equilibrium of dolomite and solution depends on both the chemical potential of CaO and that of MgO. Specifically, the relation is

$$\mu_{CaO} = k - \mu_{MgO}.$$

The physical picture evolved is simple; because dolomite contains MgO, CaO, and CO_2, all those components must be involved if equilibrium is attained between it and the solution.

Thus the concept of chemical potentials need not be restricted to the components of the minerals of the system. In an aqueous solution in equilibrium with potassium feldspar the chemical potential of K_2O in the solution must equal that of K_2O in the feldspar. The physical significance of

Figure B.21 The two-component system $CaCO_3$-$MgCO_3$ at approximately 800°C and 1 atm total pressure. Arrows are tie lines joining the compositions of coexisting phases.

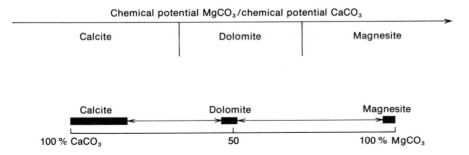

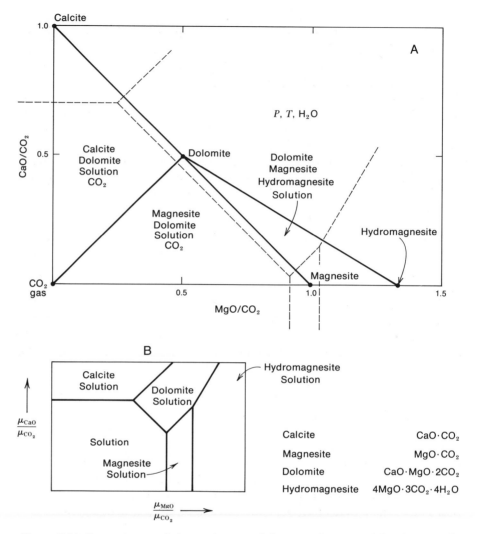

Figure B.22 Composition and chemical potential diagrams for some of the phases in the system CaO-MgO-CO$_2$-H$_2$O. In A the compositions are shown projected through H$_2$O, on the assumption that all phases are in equilibrium with nearly pure water. The dotted lines are drawn normal to the compositional tie lines and represent lines of equal chemical potential ratios. In B the dotted lines have been transposed to show the phase relations described in terms of the chemical potentials.

dissolved K$_2$O is somewhat difficult to understand; it can be translated into familiar terms as follows. We can write

$$K_2O_{aq} + H_2O_l = 2K^+_{aq} + 2OH^-_{aq},$$

and we see that dissolved K$_2$O can be expressed in terms of potassium ions and hydroxyl ions. Alternatively we could write

$$K_2O_{aq} + 2H^+_{aq} = 2K^+_{aq} + H_2O_l.$$

The equilibrium constants for these dissociations are

$$\frac{a_K^2 + a_{OH^-}^2}{a_{K_2O}\, a_{H_2O}} = K_1 \qquad \text{and} \qquad \frac{a_K^2 + a_{H_2O}}{a_{K_2O}\, a_{H^+}^2} = K_2.$$

We may write the expression for K_2 as

$$\frac{a_K^2 +}{a_H^2 +} = K_2 \frac{a_{K_2O}}{a_{H_2O}}.$$

Because the activity of water in most aqueous solutions in rocks is nearly unity, we can see that the activity of K_2O is approximately proportional to the ratio of $a_{K^+}^2 / a_{H^+}^2$.

Ion Activity Diagrams

Without going into the details of the chemistry, we see that the chemical potential of dissolved K_2O can be linked to the activities of ions in solution. Many diagrams are employed in which ion ratios are used to describe the stability fields of minerals. The relative sequences and slopes of the lines in the diagrams are identical to those of the chemical potential diagrams if the logarithms of the ion ratios are used as axes. Figure B.23 compares composition and ion ratio diagrams describing stability fields of some of the minerals in the system K_2O-Al_2O_3-SiO_2-H_2O. In this four-component system, as has been stated before, the chemical potential of H_2O

Figure B.23 In A, the compositions of some of the phases in the system K_2O-Al_2O_3-SiO_2-H_2O are shown at constant T and P, projecting hydrated phase compositions into the three-component system K_2O-Al_2O_3-SiO_2. In B, the chemical potential of K_2O has been transformed into $\log\left(a_{K^+}^2 / a_{H^+}^2\right)$, and the chemical potential of SiO_2 into $-\log a_{SiO_2}$. Use of activities permits obtaining values for the descriptive variables.

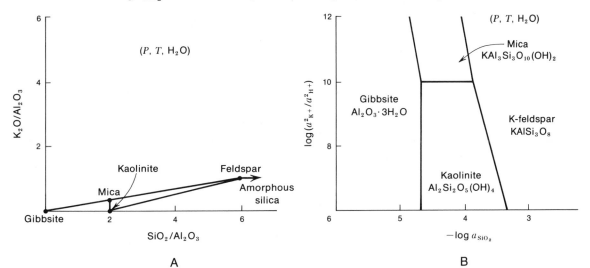

has been ignored. That is, it is assumed that the chemical potential of H_2O is a dependent variable, determined throughout the system by the presence of aqueous solution in equilibrium with the various solid phases.

This device of ignoring the chemical potentials of the components of a ubiquitous phase helps appreciably in representing complex systems. Many of the major minerals of the earth's crust are represented fairly well by the five-component system Na_2O-K_2O-Al_2O_3-SiO_2-H_2O. By considering only those conditions in which the minerals are in equilibrium with water, it is possible to reduce the number of variables required for representation at a fixed temperature and pressure to three. If one is willing, in addition, to represent only those mineral assemblages in which quartz (SiO_2) is in equilibrium with all other phases, the independent variables required for representation are reduced to two. On a chemical potential diagram, these would be the chemical potential ratios of K_2O to Al_2O_3 and Na_2O to Al_2O_3; on an ion activity ratio diagram one can use the logarithm of ratios, $a_{Na^+}^2/a_{H^+}^2$ and $a_{K^+}^2/a_{H^+}^2$. Figure B.24 shows stability relations for some of the min-

Figure B.24 Stability fields of some of the minerals in the system K_2O-Na_2O-Al_2O_3-SiO_2-H_2O at 25°C and 1 atm total pressure. It is assumed that quartz and water are ubiquitous.

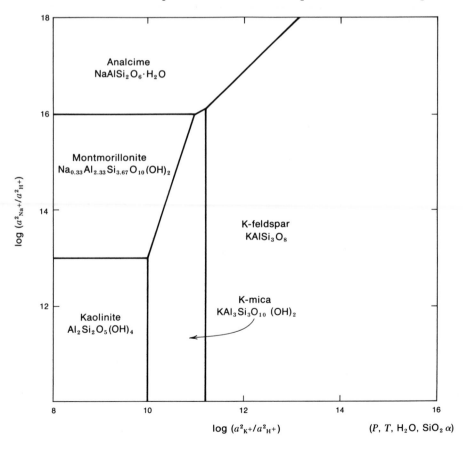

erals in the system in terms of these variables. Notice the stipulation that H_2O and quartz are assumed to be in equilibrium with every other mineral shown and further, that the representation is at a fixed temperature and pressure.

Representation in terms of ion ratios is particularly useful in investigating natural systems; ion-sensitive electrodes can be used to obtain values for ion activities in aqueous solution that approximate the true values. If the mineralogy of a given sediment is known, measurement of the ionic activities in the pore solutions, compared with those predicted from equilibrium diagrams constructed for the temperature and pressure of the sediment-water system, can be used to test for chemical equilibrium.

Mineral Reaction Diagrams

Mineral equilibrium diagrams can be used to show the complex relations among mineral phases as functions of temperature and pressure or as functions of the properties of a coexisting aqueous environment. We can also construct diagrams to portray the changes in solution compositions as minerals *react* toward equilibrium.

A classic example of a reacting system is illustrated by the evaporation of sea water. As evaporation takes place, more and more minerals precipitate, and the composition of the residual solution changes continuously. There are so many different chemical species involved that the system is difficult to represent graphically, and accurate theoretical calculation of the dissolved species at various stages of evaporation is not yet possible. Figure B.25 shows the changes in the concentrations of dissolved species, as well as the solid phases that are present as a function of the degree of concentration of the original sea water.

Another example of an important reacting system is that of soil formation. In this situation minerals are continuously exposed to a new supply of CO_2-rich oxygenated water. The CO_2 reacts to dissolve minerals incongruently, and the presence of oxygen further complicates the picture by changing the valence of various species, such as ferrous iron or manganous manganese, usually causing the formation of slightly soluble higher valence oxides or hydroxides.

We need many more actual measurements of the changes in compositions of solids and of the coexisting solutions in reacting systems to compare with calculated relations; they are difficult to obtain because the rates of change are usually slow. It is possible to calculate what should happen in simple, idealized systems, assuming that equilibrium is continuously maintained between solution and solid products as reaction takes place. This is more or less equivalent to treating the reactions like typical laboratory titrations, in which each point of the titration curve represents a temporary equilibrium.

As an example of a simple system, consider the hydrolysis of K-feldspar ($KAlSi_3O_8$) in initially pure water. Although the actual reaction in nature may take place between a large feldspar crystal and water, the system, for

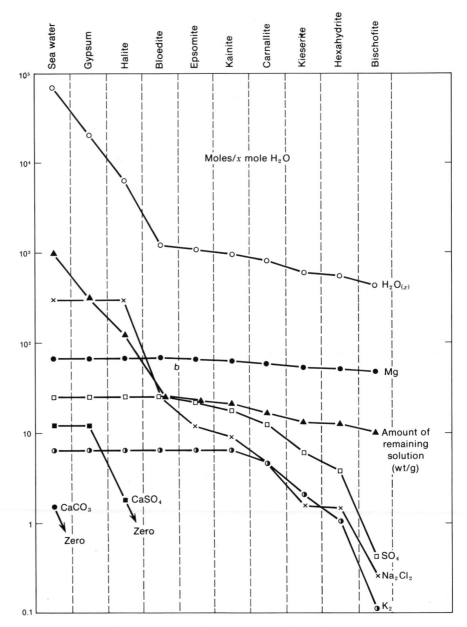

Figure B.25 Chemical composition of ocean water under isothermic evaporation at 25°C and at the start of precipitation of the individual mineral species (Degens, 1965; after Braitsch, 1960).

purposes of illustration, can be treated as if small increments of feldspar were added to a fixed volume of water and then permitted to achieve equilibrium after each addition until the solution finally equilibrates with feldspar itself. Qualitatively, the succession of events, which can be followed on Figure B.26, is as follows: The first increment dissolves completely to yield a solution that contains the same ratio of elements as feldspar. Eventually the concentration of aluminum builds up high enough to precipitate the aluminum hydroxide mineral gibbsite $[Al(OH)_3]$ (Point A). With further reaction progress, gibbsite continues to precipitate, and the potassium and silicon build up in solution (A→B). When silicon is sufficiently concen-

Figure B.26 Equilibrium diagram for some of the phases in the system K_2O-Al_2O_3-SiO_2-H_2O, with the conditions for phase stability described in terms of potassium ions, hydrogen ions, and dissolved silica. The course of reaction between feldspar and water is shown by arrows. Points are discussed in the text.

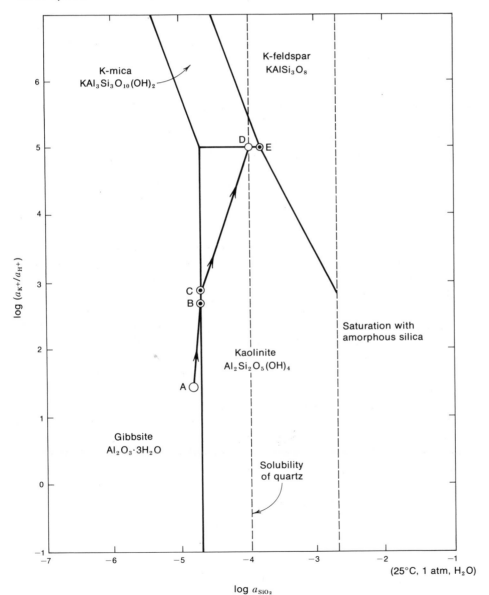

trated, the mineral kaolinite [$Al_2Si_2O_5(OH)_4$] begins to form at the expense of the gibbsite (B→C). Gibbsite finally disappears, and kaolinite continues to precipitate until the solution composition reaches equilibrium with mica (C→D). Mica and kaolinite continue to form until the solution equilibrates with K-feldspar (D→E). The assumption is made that quartz is inert under weathering conditions and that silica concentrations can rise above quartz solubility.

Another illustration of a reacting system of geologic importance is the succession of changes that take place when sediments are buried and heated. The sea water in the pore spaces of the sediments can be expected to react with the minerals. Figure B.27 shows the changes expected if K-feldspar is reacted with 1 liter of sea water at 300°C and 100 atm pressure. About 40 g of feldspar are converted into mica, chlorite, and quartz before equilibrium is reached.

Eh-pH Diagrams

Oxidation-reduction relations are so important, especially in explaining the environments of deposition of sediments, that they are given special treatment here. The variables used in describing systems containing elements in more than one oxidation state are usually Eh (oxidation potential) and pH.

One simple measure of the oxidation state of a chemical system is the equilibrium partial pressure of hydrogen. In an aqueous system, hydrogen pressure can be related to pH and oxidation potential by

$$\tfrac{1}{2}H_2 = H^+ + e^-.$$

Thus the chemical potential of hydrogen can be related to the combined potentials of hydrogen ions and electrons. The glass pH electrode can be used (in conjunction with a reference electrode) to measure an EMF that is proportional to the chemical potential of hydrogen ions, and a platinum electrode can be used to measure an EMF that is proportional to the chemical potential of electrons. In most instances, the EMF of the platinum electrode (Eh) is used directly, whereas the EMF of the glass electrode is transformed into $-\log a_{H^+} = pH$. Currently there is a tendency to transform the EMF of the platinum electrode into $-\log a_{e^-}$, or pe^-.

These variables can be illustrated by relations between the iron minerals magnetite and hematite in the presence of aqueous solution. Figure B.28A shows that magnetite will convert to hematite at a particular chemical potential or partial pressure of hydrogen. In Figure B.28B the mineral stabilities are described in terms of chemical potentials of hydrogen ions and electrons and alternatively in terms of pH and pe^-. In Figure B.28C the stabilities are described in terms of Eh and pH. The chemical reaction for the formation of hematite from magnetite can be written

$$2Fe_3O_4 + H_2O = 3Fe_2O_3 + H_{2g}$$

or

$$2Fe_3O_4 + H_2O = 3Fe_2O_3 + 2H^+ + 2e^-,$$

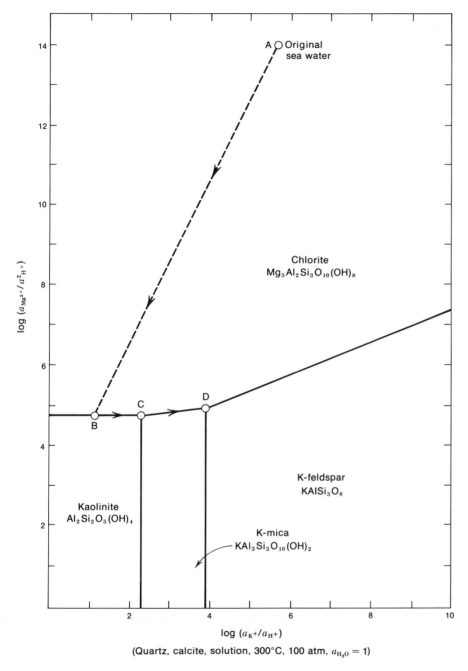

Figure B.27 Calculated results of reacting K-feldspar with sea water at 300°C and 100 atm pressure in the presence of quartz and calcite. The arrows show the progressive changes in solution composition as equilibrium is approached (adapted from Helgeson, 1967).

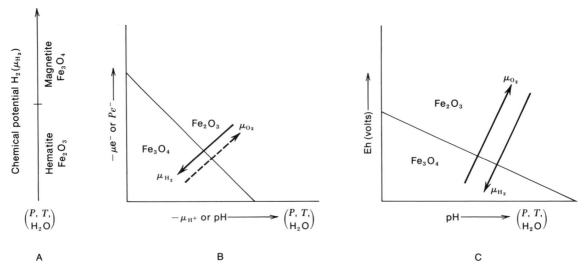

Figure B.28 Diagrams showing the relations among chemical potential of hydrogen, pH, Eh, and pe^- used as descriptive variables in mineral systems involving oxidation-reduction.

and shows the dependence of the equilibrium on a hydrogen pressure, or the alternative choice of H^+ and e^- as variables.

In the presence of water, we can write

$$H_2O = H_2 + \frac{1}{2}O_2$$

or

$$\frac{a_{H_2} a_{O_2}^{1/2}}{a_{H_2O}} = K.$$

Therefore,

$$a_{O_2} = K^2 \frac{a_{H_2O}^2}{a_{H_2}^2}.$$

If the activity of water is held constant, the oxygen partial pressure is reciprocally related to the square of the hydrogen pressure, as shown by the arrows on the figures.

SUMMARY

In this appendix, a chemical notation for minerals and for related aqueous solutions has been developed. The use of bar diagrams, triangular diagrams, and orthogonal diagrams for representing mineral compositions is illustrated. Mineral "solubility" in aqueous solutions is discussed, and the importance of incongruent solution of minerals is emphasized.

Typical mineral solubility relations in aqueous solutions are described in terms of pressure-temperature diagrams, ion activity diagrams, and Eh-pH diagrams, with a brief exposition of the chemical and mathematical basis for each. Also, plotting of the course of chemical reactions between minerals and coexisting aqueous solutions is illustrated. This appendix should provide background material for the understanding of many of the illustrations of chemical relations used widely today in the literature of the earth sciences.

REFERENCES

Deer, W. A., Howie, R. A., and Zussman, J., 1966, *An Introduction to the Rock-Forming Minerals:* John Wiley & Sons, Inc., New York.

Degens, E. T., 1965, *Geochemistry of Sediments:* Prentice-Hall, Inc., Englewood Cliffs, N.J.

Garrels, R. M., and Christ, C. L., 1965, *Solutions, Minerals, and Equilibria:* Harper & Row, New York.

Helgeson, H. C., 1967, Solution chemistry and metamorphism: in *Researches in Geochemistry,* vol. 2, P. H. Abelson, ed.: John Wiley & Sons, New York, 362–404.

Kennedy, G. C., 1950, A portion of the system silica-water: *Econ. Geol.,* 45, 629–653.

Krauskopf, K., 1967, *Introduction to Geochemistry:* McGraw-Hill, Inc., New York.

Mason, B., 1966, *Principles of Geochemistry,* 3rd ed.: Prentice-Hall, Inc., Englewood Cliffs, N.J.

Pourbaix, M., 1966, *Atlas of Electrochemical Equilibria in Aqueous Solutions:* Pergamon Press, Inc., New York.

Appendix C

Normative Mineral Calculations

The normative mineral composition of a rock can be calculated from its chemical analysis. The "norm" does not necessarily portray the actual complex composition of a rock but provides a feeling for the minerals present and permits comparison of rocks on a standard mineralogical basis. Differences and trends in normative mineralogy from rock type to rock type, along with chemical analyses, provide clues to the genesis and history of a rock. For example, the mineralogical changes accompanying the transformation of igneous rocks to sediments can be documented by normative mineral calculations. Furthermore, norms can be represented as bar diagrams to enable visualization of relationships.

The choice of normative minerals is not always a clear-cut matter, particularly with respect to sedimentary rocks. The choice is somewhat arbitrary and selection depends on the particular problem and degree of *a priori* knowledge of rock composition. The method of normative mineral calculations as applied to sedimentary rocks is discussed below. The norm of Clarke's average shale and a summary of the steps necessary to obtain the norm are given in Table C.1.

The first step is to recalculate the chemical analysis on a water-free basis in terms of the oxide components SiO_2, Al_2O_3, total iron as Fe_2O_3, MgO, CaO, Na_2O, K_2O, and CO_2. The weight percentages are then divided by the molecular weights of the respective components to obtain the moles per 100 g of rock of each component. The oxides are then distributed between normative minerals in the following way. All the CaO is calculated as $CaCO_3$ according to the reaction

$$CaO + CO_2 = CaCO_3, \tag{1}$$

provided there is enough CO_2 to satisfy the CaO in reaction (1). If not, the excess CaO is put into Ca-feldspar such that

$$CaO + 2SiO_2 + Al_2O_3 = CaAl_2Si_2O_8, \tag{2}$$

and the moles of SiO_2 and Al_2O_3 used subtracted from their original amounts.

If there is an excess of CO_2 after making $CaCO_3$, it is combined with enough MgO to make $MgCO_3$, according to the reaction

$$MgO + CO_2 = MgCO_3. \tag{3}$$

The remaining MgO, together with sufficient SiO_2 and Al_2O_3, are combined to make the typical sedimentary mineral chlorite:

$$5MgO + Al_2O_3 + 3SiO_2 = Mg_5Al_2Si_3O_{14}. \tag{4}$$

The SiO_2 and Al_2O_3 are reduced by the moles used in reaction (4). K_2O and Na_2O are combined with Al_2O_3 and SiO_2 to form K-feldspar and Na-feldspar, respectively:

$$\tfrac{1}{2}K_2O + 3SiO_2 + \tfrac{1}{2}Al_2O_3 = KAlSi_3O_8, \tag{5}$$

and

$$\tfrac{1}{2}Na_2O + 3SiO_2 + \tfrac{1}{2}Al_2O_3 = NaAlSi_3O_8. \tag{6}$$

The remaining Al_2O_3 is reacted with enough silica to form kaolinite:

$$Al_2O_3 + 2SiO_2 = Al_2Si_2O_7. \tag{7}$$

All the iron is assigned to hematite (Fe_2O_3) and excess silica to quartz. The moles of minerals formed are then converted into weight percentages by multiplying by the respective mineral molecular weights.

In that the actual mineralogy of sediments includes important percentages of the minerals montmorillonite and illite, an attempt can be made to include these minerals in the normative calculation. This procedure is tantamount to reacting some of the feldspars with kaolinite and requires some prior knowledge of the actual mineral percentages of a sediment. In some sedimentary rock analyses, S is given as FeS_2, and these constituents are included in the norm as pyrite.

Table C.1

Balance Sheet for Normative Mineral Calculation of Clarke's Average Shale
Constituent Balance (units of modes / 100 g rock)

	SiO_2	Al_2O_3	Fe_2O_3
	62.5%	16.6%	7.2%
Normative mineral reaction	1.0417	0.1628	0.0450
$0.0589CaO + 0.0589CO_2 = 0.0589CaCO_3$	1.0417	0.1628	0.0450
$0.0047MgO + 0.0047CO_2 = 0.0047MgCO_3$	1.0417	0.1628	0.0450
$0.0628MgO + 0.0126Al_2O_3 + 0.0377SiO_2 = 0.0126Mg_5Al_2Si_3O_{14}$	1.0040	0.1502	0.0450
$0.0226Na_2O + 0.0226Al_2O_3 + 0.1356SiO_2 = 0.0452NaAlSi_3O_8$	0.8684	0.1276	0.0450
$0.0372K_2O + 0.0372Al_2O_3 + 0.2232SiO_2 = 0.0744KAlSi_3O_8$	0.6452	0.0904	0.0450
$0.0904Al_2O_3 + 0.1808SiO_2 = 0.0904Al_2Si_2O_7$	0.4644	0.0000	0.0450
$0.4644SiO_2$	0.0000	0.0000	0.0450
$0.0450Fe_2O_3$	0.0000	0.0000	0.0000

The procedure for calculation of igneous rock norms is similar to that given above, except the following normative minerals are used:

K-feldspar	$KAl\,Si_3O_8$
Na-feldspar	$NaAlSi_3O_8$
Ca-feldspar	$CaAl_2Si_2O_8$
Wollastonite	$CaSiO_3$
Enstatite	$MgSiO_3$
Ferrosilite	$FeSiO_3$
Hematite	Fe_2O_3
Quartz	SiO_2

MgO	CaO	Na_2O	K_2O	CO_2	Normative mineral product		
2.7%	3.3%	1.4%	3.5%	2.8%	Moles/100 g rock		g/100 g rock
0.0675	0.0589	0.0226	0.0372	0.0636			
0.0675	0.0000	0.0226	0.0372	0.0047	0.0589	calcite	5.9
0.0628	0.0000	0.0226	0.0372	0.0000	0.0047	magnesite	0.4
0.0000	0.0000	0.0226	0.0372	0.0000	0.0126	chlorite	6.1
0.0000	0.0000	0.0000	0.0372	0.0000	0.0452	Na-feldspar	11.8
0.0000	0.0000	0.0000	0.0000	0.0000	0.0744	K-feldspar	20.7
0.0000	0.0000	0.0000	0.0000	0.0000	0.0904	kaolinite	20.0
0.0000	0.0000	0.0000	0.0000	0.0000	0.4644	quartz	27.9
0.0000	0.0000	0.0000	0.0000	0.0000	0.0450	hematite	7.2

Subject Index

Author Index

Evolution of Sedimentary Rocks

**by Robert M. Garrels
and Fred T. Mackenzie**

Evolution of Sedimentary Rocks approaches the study of physical and historical geology from an entirely new point of view. Treating the earth as a single huge geochemical factory, it considers many of the problems that have puzzled geologists for 200 years and works toward solutions through a synthesis of global data.

This book assumes of its readers only a modest background in chemistry and geology, at either the high school or college level. Upon that base the authors begin to build immediately a quantitative as well as qualitative interpretation of earth processes and earth history. Among the problems considered are the nature and amount of materials being deposited in the oceans, the nature and rate of continental denudation, the origin and cycling of sedimentary rocks, and the evidence for and causes of the evolution and migration of continents.

Untraditional as it is, *Evolution of Sedimentary Rocks* gives the serious introductory student a view of geoscience more challenging and more relevant than that provided by the descriptive approach so long accepted as the basis for an appreciation of our planet or for later specialization in geology. Because of its emphasis on transfer and cycling of water and materials on a global scale, it gives an excellent foundation for environmental studies.